INTERNATIONAL MEASUREMENT CONFEDERATION

5th TC7 Symposium on
Intelligent Measurement

(5th : 1986 : Jena, Germany)

Jena, German Democratic Republic
June 10—14, 1986.

PROCEEDINGS

IMEKO TC Event Series No. 10

Editors: T. Kemény and K. Havrilla

NOVA SCIENCE PUBLISHERS

COMMACK

NOVA SCIENCE PUBLISHERS, INC.
283 Commack Road
Suite 300
Commack, New York 11725

IMEKO TC Event Series No. 10

INTELLIGENT MEASUREMENT

Proceedings of the 5th International Symposium within the series
Measurement Theory and its Application to Practice
Organized by the IMEKO Technical Committee on Measurement
Theory (TC7) under the chairmanship of
Prof. Dr.-Ing. habil. Dietrich Hofman at the Friedrich Schiller
Universität Jena, German Democratic Republic, June 10–14,
1986.

Series editor: Dr. T Kemény
Editor: K. Havrilla

ISBN 0-941743-39-X

Cataloging in Publication Data Available Upon Request

IMEKO Technical Committee events series No. 10

IMEKO TC series include publications by the 17 Technical Committees of the International Measurement Confederation:

Higher Education (TC1)
Photon-Detectors (TC2)
Measurement of Force and Mass (TC3)
Measurement of Electrical Quantities (TC4)
Hardness Measurement (TC5)
Vocabulary Committee (TC6)
Measurement Theory (TC7)
Metrology (TC8)
Flow Measurement (TC9)
Technical Diagnostics (TC10)
Metrological Requirements for Developing Countries (TC11)
Temperature and Thermal Measurement (TC12)
Measurements in Biology and Medicine (TC13)
Measurement of Geometrical Quantities (TC14)
Experimental Mechanics (TC15)
Pressure Measurement (TC16)
Robotics (TC17)

Inquiries:
IMEKO Secretariat
H—1371 Budapest, P. O. Box 457 — Hungary

The International Measurement Confederation (IMEKO) is an international federation of 30 national member organizations individually concerned with the advancement of measurement technology. Its fundamental objectives are the promotion of international interchange of scientific and technical information in the field of measurement, and the enhancement of international co-operation among scientists and engineers from research and industry.

Latest editions:

IMEKO TC Series No. 7
Studies on Metrology – (TC8)
Publisher: OMIKK–Technoinform, 1428 Budapest, P. O. Box 12,
              Hungary (in two volumes)
650 pages, US $ 55.—

IMEKO TC Series No. 8
Proceedings of the 11th TC3 Symposium
on Mechanical Problems in Measuring
Force and Mass
held in Amsterdam (Netherlands), 1986
Publisher: Martinus Nijhoff, P. O. Box 163, 3300 AD Dordrecht,
              The Netherlands
333 pages, US $ 77.—

IMEKO TC Series No. 9
Proceedings of the 2nd TC12 Workshop
on Heat Flux Measurement
held in Budapest (Hungary), 1986
Publisher: OMIKK–Technoinform
325 pages, US $ 36.—

IMEKO TC Series No. 11
Proceedings of the 12th TC2 Symposium on Photon Detectors
held in Varna (Bulgaria), 1986
Publisher: OMIKK–Technoinform
615 pgaes, US $ 42.—

IMEKO TC Series No. 12
Proceedings of the 1st TC4 Symposium on Noise
in Electrical Measurements
held in Como (Italy), 1986
Publisher: OMIKK–Technoinform

IMEKO TC Series No. 13
Proceedings of the 3rd TC8 Symposium on Theoretical Metrology
held in Berlin (GDR), 1986
Publisher: OMIKK–Technoinform
410 pages, US $ 28.—

The above prices do not include postage.

IMEKO TC Series No. 14
Proceedings of the 1st Symposium on Laser Applications
in Precision Measurement
held in Budapest (Hungary), 1986
Publisher: NOVA Science Publishers, Inc.
              283 Commack Road, Suite 300
              Commack, New York 11725-3401
267 pages

IMEKO TC Series No. 15
Proceedings of the 4th TC13 Conferences on Advances
in Biomedical Measurement
held in Bratislava (Czechoslovakia), 1987
Publisher: Plenum Publishing Company Limited,
            88/90 Middlesex Street, London E1 7EZ, England

IMEKO TC Series No. 16
Proceedings of the 6th TC7 Symposium on Signal Processing
in Measurement
held in Budapest (Hungary), 1987
274 pages

IMEKO TC Series No. 17
Proceedings of the 13 th TC2 Symposium
on Photonic Measurements (Photon Detectors)
held in Braunschweig (FRG), 1987
Publisher: OMIKK–Technoinform
368 pages

IMEKO TC Series No. 18
Proceedings of the 1st TC15 Conference on the Measurement
of Static and Dynamic Parameters of Structures and Materials
held in Plzen (Czechoslovakia), 1987
404 pages

IMEKO TC Series No. 19
Proceedings of the 2nd TC4 Symposium on Industrial
of Electrical and Electronic Components and Equipment
held in Warsaw (Poland), 1987

IMEKO TC Series No. 20
Proceedings of the 4th TC10 Symposium
on Technical Diagnostics
held in Kupary (Yugoslavia), 1986

CONTENTS

2.   INTELLIGENT SENSORS, INTERFACES,
      MEASUREMENT MODULES AND INSTRUMENTS

4.   ON-LINE MEASUREMENTS AND QUALITY CONTROL

PREFACE

The 5th International IMEKO Symposium on "Intelligent Measurement" had the aim to present, discuss and summarize current developments and collected experiences in measurement theory and its application to practice.

The purpose of the symposium was to give more than 400 specialists in the field of measurement engineering an opportunity to meet at Jena and to become familiar with most recent developments in a rapidly expanding field.

With the increase of technical capabilities of hardware the solvable measurement problems are becoming more complicated. Academic methods of measurement information acquisition and processing are now applicable in plants. In particular, engineers are increasingly interested in multi-dimensional and multi-functional direct and indirect measurements.

Main reasons are the increasing flexibility of process and factory automation stepwise developing an unified production automation. Typical computer-aided technologies on that way are CAD, CAM, CAT, CAQ and CIM.

These new fundamental problems and far reaching applications are concerned with intelligent measurements IM. IM is any sophisticated computer-aided measurement information acquisition to enhance objectivity and productivity of information acquisition in research, development, production and consumption. IM from exotic horizons are coming down to earth.

The Plenary lectures given by noted specialists well reflect the growing variety of attractive IM systems. Session papers and Poster sessions present answers on special problems and raise new questions. So we were very motivated to have an interesting discussion.

The text of the Proceedings consists exclusively of reproductions from the original manuscripts. I have to thank the authors for their valuable work. Thanks are also due to our sponsors and co-sponsors for mutual encouraging co-operation. OMIKK - TECHNOINFORM deserves praise for the efforts to convert the manuscripts into book form.

It is our sincere hope that the readers of the Proceedings will not only benefit scientifically but also get a feeling about the warm IMEKO Jena atmosphere.

Prof.Dr.-Ing.habil. Dietrich Hofmann
Chairman IMEKO TC 7 Measurement Theory

1
MEASUREMENT THEORY AND INFORMATION PROCESSING

INTELLIGENT MEASUREMENT

FROM ART TO SCIENCE AND TECHNOLOGY

Dietrich Hofmann

Dept. of Technology for Scientific Instruments
Friedrich-Schiller-University of Jena
JENA, 6900, German Democratic Republic

MATTER, ENERGY AND INFORMATION

The dramatical increase of world population within few generations (Fig. 1) and substantial changes in consumer habits like cars, television, telephone, kitchen and sanitary technics are based on powerful streams of matter, energy and information.

Matter keeps man from a hostile environment and supplies industry with raw material. Energy keeps man in a thermostable condition and supplies industry with power. Information keeps man, production and social processes under controlled conditions.

Matter and energy are basic terms. They are well defined. They are transformable by EINSTEIN's law. Information is another basic term. It's clear definition and eventually transformation with mass or energy is still open [1].

INFORMATION AND MEASUREMENT

Revolutionary inventions intensified first the streams of artificial matter and energy production like iron and steel, synthetics, fertilizer, coal, gasoline, steam, electro and atom energy. Now we entered the information age. Information stands for news, report, data, account, notice, intelligence, signal, communication, message, measurement value, measurement result.

Subjective information is gained by observations. Observation is the experimental comparison of states and processes with the sensory apparatus of man using subjective internal virtual standards. The result of observations is recognition.

Objective information is gained by measurements. Measurement is the experimental comparison of a measurable quantity with a known external metrological, particular or

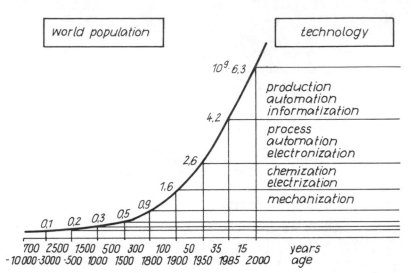

Fig. 1 Development of world population and technology

virtual standard. The result of measurements is knowledge [2].
Objective information is the general basis to control computerized production, transport, distribution, administration,
management, consumption and education processes in any high-
technology and information-oriented society. Objective information acquisition and information engineering getting
now a revolutionary development by the quick expanding
possibilities of intelligent measurements.

Intelligent measurement is the application of microelectronics and microcomputers to instrumentation. The art
of intelligent measurements as an isolated sophisticated
individual creative solution of the problem is changing into
the new quality of science and technology, that is into
systematically organized knowledge about intelligent measurements and its abundant practical application [3]. The
unification of contradictionary meanings on a higher level
of understanding concerning the nature of intelligent measurement is a contribution to streamline the application of
computer-aided systems in social and individual life on an
higher level of productivity.

The 5th International **IMEKO** Symposium Intelligent Measurement well reflects this situation. Plenary lectures
dealing with the impact of microelectronics, precision
mechanics and optoelectronics on intelligent measurements.
Main fields of application are laboratory, process and
manufacturing automation, medical and technical diagnostics.
New demands for education and training are stated. Session
papers illustrate new ideas in measurement theory mainly
concerned with sensors and software, describe new intelligent
systems for laboratory automation and testing using com-

puterized experimental facilities and local area networks, give first solutions for optical inspection and pattern recognition including knowledge engineering, speak about solved and unsolved problems in multi-coordinate measurement and position sensing due to recently developed standards and broaden the scope of signal analysis and information processing with analogue and digital intelligent instruments. Intelligent sensor systems and interfaces inaugurate a new era of on-line identification and quality testing as well as on-line monitoring and quality control on a higher level of productivity. Poster sessions give detailed information on new practical applications of intelligent measurements for mechanical, electrical, thermal, optical and analytical quantities.

INTELLIGENT MEASUREMENT AND SCIENTIFIC-TECHNOLOGICAL PROGRESS

Labour, entertainment and reproduction of working force are increasingly accompanied and determined by measurements.

Artificial matter like industrially produced textiles, housings and artificial energy sources like industrially produced food-stuffs, thermal and electrical power as well as artificial information acquisition and processing means like sensors and computers quickly satisfy growing demands depending on the fast increase of world population and fast increase of consumer habits.

Decisive prerequisite to social progress is at last the increase of labour productivity. Therefore it is necessary to apply intelligent measurements not only at the lowest level of automation that is computer-aided manufacturing CAM but step by step also to include the higher levels of the hierarchy that is computer-aided design CAD, computer-aided engineering CAE, computer-aided organization CAO, computer-integrated manufacturing CIM, computer-aided administration CAA and computer-aided management information systems CAMIS (Fig. 2).

For defective production usually the workers at the lowest level of hierarchy are blamed. Nevertheless most errors are caused by people from higher levels of the hierarchy. That is because information is to short in facts and data 4. Therefore a growing number of intelligent measurements must be carried out in agriculture, production, transport, trade, office work, research, health care, security systems, home work and education. Quality control has to be based on intelligent measurements 5.

NEW APPLICATIONS OF INTELLIGENT MEASUREMENTS

Recent developments in computer-aided measurement information acquisition and processing with reliable strong and cheap sensors and computers allowed:
- agriculture automation AA with automated breeding houses and combines
- production automation PA in flexible small and medium batch production with computerized numerically controlled CNC tool machines and industrial robots IR

Fig. 2 Fundamental conception for computer-aided production processes

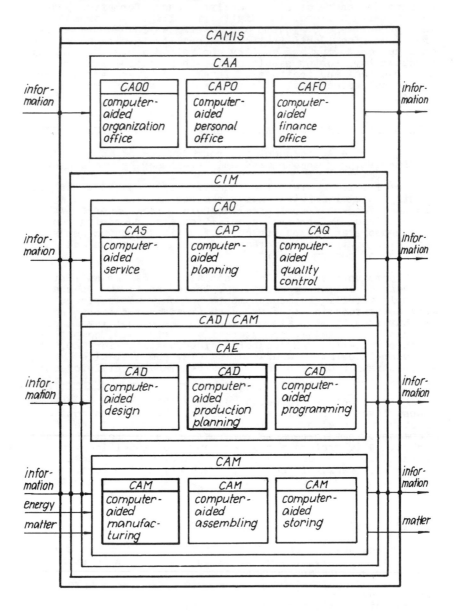

- transport automation TA with automated jet liners, public trains and workshop transport lines
- trade automation TA with automated computer scales, automated cashiers and money changing, automated bargaining, automated warehouses and vendor automata
- office automation OA with automated word processors, translating machines, handy-type sophisticated personal computers with standardized intelligent measurement interfaces, intelligent copiers with editing capabilities and automated filing on solid state memories or magnetic and optical discs
- research automation RA and laboratory automation LA with expert systems and knowledge processing units as well as computer-aided information and experimental systems
- health-care automation HCA with computerized diagnostical and therapeutical equipment, screening facilities and survival kits
- total security systems TSS for homes and plants through remote sensing and controlling with local area networks including sensors for gas, fire, temperature, door, window, light, air conditioning, laundring and visual out-door and inhouse detection
- home automation HA with computer-controlled radio, television, audio and video drives, teletext, telefax, telephone, heating, cooling, cleaning, cooking as well as electronic blood pressure meters and digital body thermometers
- computer-based education CBE with design and retrieval facilities in school and at home, training support with learning capabilities of computerized text books to meet the actual level of knowledge of pupil, student, apprentice or trainee, with examination kit and also adjustable for the individual choose of the level of difficulty.

Modern interfaces like the open systems interface OSI and the manufacturing automation protocol MAP realize an unified approach to data transfer [6]. Local and wide area networks together with intelligent measurements facilitate objective information acquisition and processing in near future (Fig. 3).

The deeper understanding of measurement procedures as well as the easier use of reliable and strong computational means till the end of our century leads to knowledge-based measurements. They are the natural basis for the next step to associative measurements (Fig. 4).

CONCLUSIONS

Modern products contain an increasing number of intelligent sensors and measuring circuits as a fundamental basis of their functional principle.

Modern design, production, maintenance, trade, service and office processes unthinkable without application of sophisticated intelligent measurements.

It is my sincere conviction that "Intelligent Measurement" can contribute to the qualification for design, development and application of reliable cheap and efficient intelligent measurement hardware and software to enhance

9

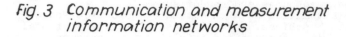

Fig. 3 Communication and measurement information networks

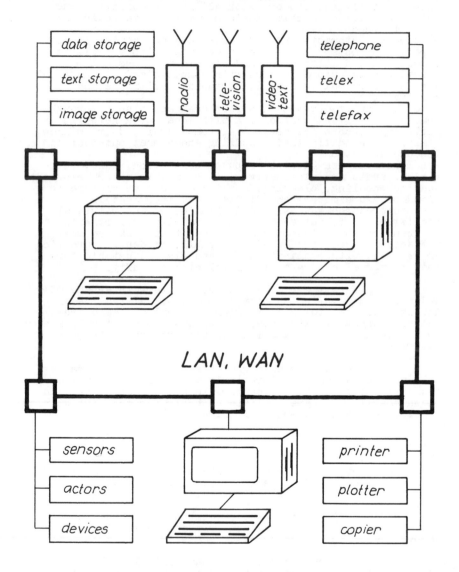

Fig.4 Computer-aided technologies and measurement

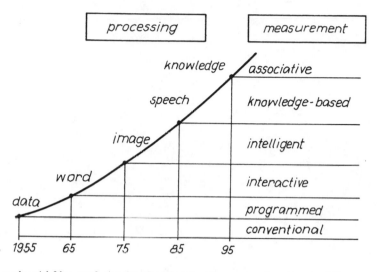

the scientific and technological progress to the welfare of man.

REFERENCES

1. Völz, H.: Information I and II. Berlin: Akademie Verlag 1982 and 1983
2. Hofmann, D.: Handbuch Meßtechnik und Qualitätssicherung (Handbook Measurement Engineering and Quality Assurance). 3rd ed. Berlin: VEB Verlag Technik 1986
3. Finkelstein, L.: State and advances of general principles of measurement and instrumentation science. Measurement 3 (1985) 1 pp. 2-6
4. Ishikawa, K.; Lu, D.J.: What is total quality control? The Japanese way. Englewood Cliffs, N.J.: Prentice-Hall, Inc. 1985
5. Hofmann, D.: Rechnergestützte Qualitätssicherung CAQ (Computer-aided quality assurance CAQ) Berlin: VEB Verlag Technik 1987
6. Pfeifer, T.; Rühle, W.: Derzeitige Situation und Chancen von MAP (Present situation and chances of MAP). Automatisierungstechnische Praxis atp 28 (1986) H. 3, S. 109-116

MODEL BUILDING AND PARAMETER ESTIMATION

AS MEANS FOR INTELLIGENT MEASUREMENT

A. van den Bos   P. Eykhoff
Physics Dept.   Electrical Eng. Dept.
Delft University   Eindhoven University
Delft   Eindhoven
                The Netherlands

Abstract - For the time being the notion of intelligent measurements is still rather vague. It is clear, however, that the implicit and explicit use of models in such measurements is quite pronounced. For that reason it is appropriate to explore the relations with system identification (i.e., model building + parameter estimation). That is the subject of this paper. Also discussed are the actual and potential contributions of identification, the availability of appropriate software, and the desirabilities with respect to further contributions to intelligent measurements.

1. WHAT IS INTELLIGENCE/WHAT ARE INTELLIGENT MEASUREMENTS?

'Intelligence is a hypothetical construct used to describe individual differences in an assumed latent variable that is, by any direct means, unobservable and unmeasurable. In its popular usage, the concept refers to variations in the ability to learn, to get along in society, and to behave according to contemporary social expectations. ...' (Encycl. Brittan., 1983)

After reading this 'definition' of intelligence, ... do we really want to speak about intelligent measurements? This adjective refers to something that is unobservable and unmeasurable by any direct means, and is related to 'learning', 'getting along' and 'expectations on behaviour'. These are all notions not usually associated with measurements! Nevertheless, with some caution we may conclude that intelligence is related to the intake and storage of information, leading to 'experience' and resulting in adaptive changes in behaviour. Highly intelligent systems might, therefore, distinghuish themselves from non-intelligent ones by properties like the following:

This is a condensed version only. The full text of the paper, including a list of references, is available form either author. The adresses are: A. van den Bos, Dept. of Applied Physics, Delft University of Technology, P.O. Box 5046, 2600 GA Delft, The Netherlands, and P. Eykhoff, Dept. of Electrical Engineering, Eindhoven University of Technology, P.O. Box 513, 5600 MB Eindhoven, The Netherlands.

13

- the ability to concatenate measurements into information that is less
  redundant and is better adapted to the nature of human perception.
- adaptive behaviour, i.e., 'learning' by the instrument from previous
  experience in order to improve the experimental design.
- the ability to estimate, i.e., to measure indirectly, through process
  identification.

Admittedly, the terminology is somewhat vague. For example: what is
the difference between 'adaptive' and 'learning'? An intuitive answer
might be that 'learning' and 'adaptive' refer respectively to a permanent
and a temporary record of experience. Although these notions remain
imprecise, they may suffice for the purpose of this paper.

2. THE AIMS AND USES OF MODELS

Among scientists/engineers there will be little disagreement that,
ultimately, the only way in which we may describe reality is in terms of
'models'. The model may be conceptual, physical or mathematical – in all
cases it should represent those aspects of reality in which we are
interested at that moment, and ... for a particular purpose in mind. The
range of models is virtually unlimited. It includes laws of nature:
physical and chemical processes; dynamic behaviour of airplanes, power
generators; economy; weather; etc. In what follows, all such entities
with dynamical aspects will be called processes. Process models can be
associated with the interpretation of the past behaviour, the monitoring
of the present behaviour or the prediction of the future behaviour. The
'fixed' a priori knowledge in a model can be considered as just a
condensation of existing information. Of particular interest, however, is
the increasing information, i.e., the a posteriori knowledge that is
added to the a priori one by the updating of the model. This might be
called "learning", particularly, if this knowledge is used for adaptation
of the measurements that will follow. Note that such updating can be
directed to the structure of the model as well as to its parameters. For
a proper understanding of the model concept it is desirable to specify
more clearly the terms 'process' and 'model':

- the real physical process is only partially known and is, probably, not
  even fully knowable
- the theoretical model should adequately represent the real process for
  the intended purpose of the identification activities
- the estimation model should approach the theoretical model in an
  'optimal' way. It is often assumed that the theoretical model is
  contained in a set of estimation models and that it can be approached
  through estimation of the 'order' and the parameter values.

In all these considerations the choice of model structure is
crucial. The terms 'black (box) model' and 'gray (box) model' reflect the
virtual absence, respectively, the presence of a priori knowledge about
the process. Examples of black boxes are impulse response and Volterra
kernel models; differential and difference equation models are gray ones.
System/process identification consists of model building and
order/parameter estimation. Through the concept of learning its technques
are very relevant for intelligent measurement indeed. Consequently, in
the next section some aspects of parameter estimation will be briefly
reviewed.

3. THE AIMS AND USES OF PARAMETER ESTIMATION

As examples of aims of model building and parameter estimation

14

techniques we mention the following: prediction (weather, economy); diagnosis (not directly measurable quantities in engineering or biomedial applications); monitoring (wear, aging), design (information for optimal properties). With respect to control it is observed that over the years both the processes to be controlled and their models have become more complex; the requirements that have to be met have become much more demanding. Consequently, information about the process itself becomes more and more crucial when designing control schemes. This requirement leads to an increasing number of applications of identification techniques in control engineering.

4. EXPERTISES REQUIRED

Prominent among the expertises required for intelligent measurement is experience in the modeling of physical observations. This includes the modeling of the system response as well as the systematic and random errors. The choice of model is, ideally, connected to the <u>purpose</u> of the measurement. Such models may also include constraints on and/or initial guesses of the parameters. This modeling of the observations is crucial for the success of the measurement procedure and requires close cooperation between experts in the field concerned on the one hand and experts in measurement/estimation on the other. Further indispensable ingredients of modeling and parameter estimation expertise are knowledge of procedures for the checking and preprocessing of the observations. Other prerequisites are familiarity with signal processing, linear system theory and time series analysis. Knowledge of parameter estimation methods is, of course, a necessary conditon for making a well-motivated choice. Once this choice has been made, the question arises how to implement the method numerically. If no relevant programs are available, one must either assemble general numerical routines or make a program oneself. In this respect, familiarity with numerical methods, those for minimization in particular, may be extremely helpful. Finally, if a method has been selected, the question arises as to whether or not software is available. This topic is discussed in the next section,

5. WHAT IS AVAILABLE?

To put the expertises described in the preceding section into practice, a substantial amount of software is required. Fortunately, the number of quality software packages and subroutines for implementing statistical, numerical, system modeling and signal processing techniques is gradually increasing. In this section, first a short description will be given of some of these packages and routines. Next, modern developments leading to intelligent computer based measurement procedures or instruments will briefly be discussed.

5.1 General Numerical Software

For general numerical purposes, comprehensive scientific software libraries as NAG and IMSL are available. Both also contain a number of routines of direct interest for parameter estimation. Other routines are suitable as building bricks. Recently, a NAG personal computer version has become available. Unlike the sources of NAG and IMSL routines, those of the much smaller and more specialized LINPACK and MINPACK libraries are freely available from the literature. They solve various linear and nonlinear problems respectively. The advantages of using high quality software as NAG, IMSL, LINPACK and MINPACK can hardly be exaggerated. The routines in these libraries have been tested on a variety of computers, they are clearly documented and their properties are well specified. The designers of these routines have combined or balanced a wide variety of properties as, for example, numerical stability, accuracy, speed and robustness with respect to users' requirements. Practice has shown that all these properties are hard to achieve by non-experts even for seemingly simple numerical problems. Moreover, these routines are portable. In addition, IMSL and NAG are regularly updated and thus reflect the state of the art in numerical analysis. They also enable the user to make a motivated choice.

5.2 Specializd Software

For modeling and parameter estimation purposes a number of special software packages have been developed which are now in use for a considerable time and are well-tried; see the full text of this paper. Being interactive these packages differ from collections of individual subroutines for system modeling and identification and for signal processing found in the literature. Nowadays there are several directions in which these special identification packages are evolving. Emphasis may be on the real-time interaction with the process under study, where the intention is to provide the user with more a priori information for his next decision. On the other hand, the purpose may be to make an existing identification package easier to use by different categories of users. Although these objectives differ, they both lead to the use of intelligent software to be discussed further below.

5.3 Intelligent Software

During the last twenty years identification has evolved rapidly. In addition, useful software has become available. As a result parameter estimation is increasingly applied in the laboratory environment. However, the use of it elsewhere has clearly lagged behind. This in spite of the fact that a surprisingly large number of measurement problems can very compactly be formulated as solvable parameter estimation problems. The reason for this discrepancy is that is does not suffice to have available the mathematical tools and the corresponding software. There is a duality between the 'science' tools available (theory, algorithms) and the 'art' aspects ('insight', expert knowledge). For example, the autofocusing of a transmission electron microscope may be formulated in terms of a simple parameter estimation problem. However, if the microscope is grossly out of focus, heuristic steps are needed to first roughly focus it. This may involve: starting from a low magnification, focusing, increasing the magnification, etc. So, the estimation procedure can only be successful if it is accompanied by heuristics in the form of control actions, logical decisions and checks. These actions are all in the domain of the expert of the particular application. He alone knows what steps have to be taken, in what order, and how to iterate. This example is characteristic for technical science in general: there is a growing need of intelligent software, i.e., software combining heuristic expert knowledge and, often standard, mathematical-numerical procedures. Fortunately, during the last twenty years software of this kind has been developed for widely varying purposes in the form of expert systems. The purpose of expert systems is to represent human expert knowledge and to use it for a restricted category of problems. The first examples of applications of expert systems in measurement instruments can now be found in the literature. Undoubtly, many of the future intelligent measurement devices will make use of the impressive variety of identification tools. Thus the existing discrepancy between what, in principle, is feasible using these tools and what is actually done in

16

practice may be expected to diminish in the coming years.

6. WHAT IS DESIRABLE?

The number of numerical methods and techniques developed during the last decade is impressive and this process is still in full progress. In addition, improvement of performance has made modern microcomputers increasingly suitable for carrying out demanding numerical procedures. In order to enable the measurement engineer to fully exploit these resources for increasing the intelligence of his instrument designs, two particular developments would be highly beneficial:

- In the first place, a more widespread use of standard, high quality software libraries would, for all active in the measurement field, alleviate the programming burden and drastically enhance the reliability, portability and standardization of their software.
- In the second place, the measurement engineer would greatly benefit from a further development of software packages for building expert systems. These so-called knowledge engineering tools should preferably be portable and user friendly.

17

INTELLIGENT MEASUREMENT IN CLINICAL MEDICINE

L. Finkelstein and E.R. Carson

The City University,
Northampton Square,
London EC1V OHB U.K.

INTRODUCTION

The advance of information technology has led to the embedding of machine information processing in measurement processes and instrumentation leading to so called intelligent measurement. There have been great advances in the use of intelligent measurement in clinical medicine.   In this paper a brief critical account is given of intelligent measurement in clinical medicine, using this also as a means of examining carefully the nature of intelligence in measurement.

THE NATURE OF INTELLIGENCE IN MEASUREMENT

The term intelligent instrumentation is used to describe instrumentation in which machine information processing is used to enhance measurand information/1/.   Intelligent measurement may be defined as a measurement process in which machine information processing, on- and off-line, is used to enhance and interpret measurement data.

The above definitions are unsatisfactory. Firstly, virtually all measurement instrumentation involves some machine information processing and is to that extent intelligent. Secondly, the above definitions of intelligent instrumentation are much wider than the concept of artificial intelligence.

The classic definition of artificial intelligence is the "Turing Test"/2/ in which a machine is allowed to interact freely with a human interrogator. An intelligent machine is one in which the interrogator cannot distinguish whether it is a human or a machine which provides the response. This definition fails to provide for limited machine competence. It is possible to develop it by defining artificial intelligence as the property of information processing systems which imitate the way in which humans function.   The most practical way of defining artificial intelligence is by noting what is included within it. Key areas are machine reasoning and learning, knowledge acquisition, representation, and elicitation, natural language understanding, machine vision and pattern recognition /3,4,5/.

We may distinguish a number of levels of intelligent instrumentation and measurement (i) **Measurand signal enhancement:** Here information processing is used to correct for imperfections of measurand signal acquisition, performing such tasks as linearisation, self-calibration, deconvolution and the like, as well as extraction of signal from noise. Although such digital machine information processing provides greatly increased capability it does not provide qualitatively different capability from what can be achieved by simple sensor construction or analog signal processing. It does not involve any real machine intelligence. (ii) **Measurand signal processing:** In this form of instrumentation embedded information processing is used to evaluate some characteristics of the measurand signal. This form of intelligent instrumentation does not again involve any real machine intelligence. (iii) **Inferential measurement:** Here the actual measurand is a variable or parameter of a model and is not directly observed but is identified from measurements of other directly observed variables of this model. This involves generally substantial on or off-line computing and may be said to involve some level of machine intelligence. (iv) **Pattern cognition:** In this form of measurement the machine induces the characterisation of an attribute from measurements of other attributes. This is true machine intelligence. (v) **Measurement as part of an integrated information system:** Measurement always forms part of a wider information system. Having been acquired, enhanced and processed into an appropriate form, measurement information forms input to decision and action or control processes and the like. Modern information technology enables much of this information handling to be carried out by machine. This may be viewed as the highest level of intelligent measurement.

THE CLINICAL MEASUREMENT INFORMATION SYSTEM

To see intelligent measurement and instrumentation in the overall clinical information system consider the patient/clinician interaction as shown in Fig. 1.

The clinician obtains information on the patient state from a variety of measurements and observations; physiological monitoring (e.g. ECG and heart rate recording); analysed blood and urine samples; measurements obtained from physical investigations (e.g. X-ray, ultrasound); and from direct clinical observation (quantitative and qualitative). From such data and the resultant information, the clinician makes decisions on diagnosis, prognosis and management.

Information extraction from such data needs machine information processing, followed by interpretation for clinical decision making (e.g. treatment selection). Examples of intelligent measurement in the form of inferential measurement and pattern cognition and application within the total information system are now considered. Other levels of intelligent measurement are not considered since their clinical use does not provide distinct lessons.

INFERENTIAL MEASUREMENT EXAMPLES

Many examples relate to chemical processes in man. Usually the only measurements that can be made readily are those of concentrations of naturally occurring chemical species and drugs in blood or urine samples. Whilst such measures provide useful clinical information, more valuable insight into the patient state or his response to treatment can be obtained from estimates of variables and parameters not directly observable; typically concentrations in the liver and

20

other organs, distribution volumes and rates of transfer of material from one site to another.

Examples drawn from a wide range of model-based studies in metabolism and endocrinology have been documented /6/. For liver disease, models have been developed enabling estimates to be made of liver distribution volume, rate of uptake by the liver and liver concentration of substances such as bilirubin /6,7/. This is a substance occurring naturally in the body (which can also be administered as a test signal) whose dynamics provide a good diagnostic indicator of certain forms of liver disease. Tracking the changes of such quantities by estimation enables patient response to treatment to be quantified.

Similarly inferential measurement techniques have been applied to the study of diabetic patients. /6,8/. For instance a mathematical model has been developed to describe the carbohydrate metabolic state from clinical data that can be routinely gathered. These are glucose and insulin concentrations which with model-based analysis can be used to estimate parameters defining the sensitivity of insulin secretion to changes in glucose concentration; parameters providing valuable clinical indices of the diabetic state.

Inferential measurement can also be used to plan treatment. Having estimated the parameters from observable input/output measures providing a representation of the particular patient the model can be used to assess alternative therapies by simulation, thus providing an additional clinical decision aid.

Inferential measurement can yield estimates of parameters of thyroid hormone dynamics which are altered in certain forms of thyroid disease. By recursive estimation parameter estimates can be updated, providing quantitative measures for the clinician of the patient's response to treatment /6,8,9/.

PATTERN COGNITION EXAMPLE

An example of pattern cognition arises in diagnosing certain forms of thyroid disease. Here an estimate of the state of the patient is sought from a number of clinical observations (including eye signs and the enlargement of the thyroid gland) and laboratory tests (measures of the thyroid hormones concentrations). The expert clinician undertaking diagnosis essentially without the aid of machine intelligence, takes a number of these signs and symptoms and laboratory tests and combines them into a single diagnostic measure, mapping the patient into one of a number of diagnostic classes. Such classes might typically correspond to "definitely", "probably" or "possibly" hypothyroid, euthyroid, and "possibly", "probably" or "definitely" hyperthyroid /10/.

Whilst the expert can achieve a correct diagnosis in the majority of cases, the diagnosis is more difficult for more junior clinicians since the underlying processes may not be fully understood and the high data dimensionality adds to difficulty in interpretation. Here machine intelligence can provide a useful function as an aid as evidenced by pilot study results /10,11/. Since the process models are not well understood, they have been induced through pattern cognition. Thus from the available measured variables those combinations of laboratory tests and clinical observations which are most dicriminative of the classes of thyroid patient can be identified. This use of intelligent measurement offers the benefit of increasing diagnostic efficiency and the prospect of being able to reduce the number of laboratory tests which are needed.

INTELLIGENT MEASUREMENT AND INSTRUMENTATION IN A TOTAL CLINCIAL SYSTEM

Intelligent measurement and instrumentation offer the greatest potential in critical care medicine where a high level of machine support is already available to the clinician and nurse in managing the critically ill patient. Fig. 2 shows the overall patient/clinician feedback loop in such a context. Commonly the balance of fluids and electrolytes in the body is deranged and treatment may include the infusion of potassium, saline or dextrose as indicated.

Conventionally the clinician as decision maker draws upon knowledge represented as a set of conceptual models to interpret and act upon patient data which take the form of physiological, physical and biochemical measurements and clinical observations. Machine intelligence is generally limited to signal extraction procedures applied to the monitored physiological signals together with subsequent signal processing. The complexity of fluid and electrolyte balance is such, however, that only a limited number of clinical experts fully understand the underlying processes and their control in quantitative terms. This coupled with an almost unmanageable volume and array of measurements (although many crucial variables are not directly observable) renders it difficult for the clinician to provide the best possible management for such critically ill patients.

Using higher levels of intelligent measurement, the clinician can be assisted in the decision making processes necessary for patient care as shown in Fig.2. Mathematical models in inferential measurement can be incorporated into an overall knowledge-based system which includes dimensions of artificial intelligence and pattern recognition. Thus estimates can be made of crucial quantities which are not directly measureable and such estimates used for more effective decision making. Equally the knowledge-based system can seek evaluations of alternative treatments for a given control strategy using a model of the relevant physiological processes. This model outputs its predictions for the given patient to the knowledge-based system, the parameters having been set to their patient-specific values using of available measurements and related observations /12/.

Initially such developments are being directed towards implementation in the clinical environment and assessment of requirements for user acceptability. It is expected, however, that subsequently the loop may be formally closed and aspects of fluid-electrolyte therapy such as the administration of potassium, saline or dextrose totally automated as shown in the lower section of Fig. 2. This is bringing about a transfer of intelligence from human to machine so that the clinician may be assisted by computer-based schemes to achieve a more effective level of patient care.

CONCLUSIONS

This paper provides a critical analysis of what constitutes intelligent measurement. Application of information processing within instruments contributes to a substantial increase in their capability. It is however such applications as inferential measurement, pattern cognition and integration of measurement with control which constitute significant machine intelligence in measurement. Intelligent

22

measurement has an important role to play in clinical medicine, particularly inferential measurement and the integration of measurement into the information machine based system of patient care.

REFERENCES

1. Barney, G. C. Intelligent Instrumentation: Microprocessor Applications in Measurement and Control, Englewood Cliffs:Prentice/Hall. 1985.
2. Turing A.M. Computing Machinery and Intelligence, Mind LIX (1976), p236.
3. Proceedings 8th IJCAI 1983.
4. Proceedings 9th IJCAI 1984.
5. Proceedings 10th IJCAI 1985.
6. Carson E.R., Cobelli C., Finkelstein L. The Mathematical Modelling of Metabolic and Endocrine Systems, New York; Wiley, 1983.
7. Evans N.D., Carson E.R., Cramp D.G., Jones E.A. Bilirubin dynamics: good modelling practice, in: Medinfo 83 eds. J.H. van Bemmel, M.J. Ball, O.Wigertz, Amsterdam:North Holland, 1983, p.863.
8. Carson E.R., Cobelli C. Model-based measurement in physiology and clinical medicine, Measurement 3(2) (1985), p.11.
9. de la Salle S., Leaning M.S., Carson E.R., Edwards P.R., Finkelstein L. Control system modelling of hormonal dynamics in the management of thyroid disease, in: A Bridge between Control Science and Technology, eds. J. Gertler, L. Keviczty, Oxford: Pergamon, 1985.
10. Edwards P.R., Britton K.E., Carson E.R., Ekins R.P., Finkelstein L. A Control system approach to the management of thyroid disease, in: Medinfo 77, eds. D.B.Shires and H. Wolf, Amsterdam:North Holland, 1977, p.541.
11. Edwards P.R. A systems approach to thyroid Health Care, PhD Thesis, London, The City University, 1983.
12. Cramp D.G., Carson E.R. Design of an intelligent system for closed-loop control of fluid balance, Biomedical Measurement Informatics and Control, 1 (1986) (in press).

A RELATIONAL - TOPOLOGICAL MODEL

OF THE MAN-MACHINE SYSTEM

Krzysztof Podlejski, Roman Myszkowski

Institute of Electrical Metrology,
Technical University of Wroclaw,
Wroclaw, Poland

INTRODUCTION

One of the main tasks of ergonomics, metrology and psychology is studying the properties of a man-machine system. It concerns both accepting an adequate model and selecting methods of studies as well as finding the rules of interpretation. There exists a conviction that the man-machine system is, because of the participation of man difficult to describe and to formalize. Many attempts have been made in order to find a formal description of the man-machine system.

In the attempts of searching the model of the man-machine system there prevails a system approach as an operational union of man or a group of people with one or numerous technical objects, in order to obtain definite results from the input data with regard to the limitations arising from the properties of the surrounding environment. On this basis one may operate with: morphological, functional or parametrical descriptions of the man-machine system. Analogically, a man-machine system is treated in terms of an anthropocentric approach which determines man's role in the system as an active one and fundamental especially in the area of decision making.

Most of the models is used for describing the functioning of man-machine system, whereas during studies its properties become dectptive. Thus, there exists a necessity of finding a relational - topological model of the man-machine system on the basis of which it would be possible to determine, what should be measured and in what way.

AN EMPIRICAL MODEL OF THE MAN-MACHINE SYSTEM

Let us consider a given system which has a subsystem, the man-machine system. This system may be approached in terms of a relational system determined on the basis of the relationships between man and machine. Machine is defined here as a set of devices used for information visualization and of the answering devices of the system. It is known that the man-machine system is determined by a set of features $c(u)$ which determine the relational system G. According to the measurement

theory determining the parameters of features $c(u)$is possible
when there exists a transformation n giving a homomorphical
projection of the relational system G to some relational system H determined in n-dimensional Euclidean space $\mathcal{E}^m$. So,
in order to enable the measurement of features $c(u)$ of the
man – machine system i.e. an ability to project a realization
of features $c(u)$ realization n of the system H is indispensable
for determing the model of the system α as well as the subsystem of features $c(u)$ in the man – machine system.
    Let U_α be a set of elements $u_{\alpha i}$ of certain system and
R_α be a set of relations between elements $u_{\alpha i}$. System α is
understood as a functional complex of machines, processes
and phenomena in which man-machine system can be distinguished.
In general, system α is formed by elements $u_{\alpha i} \in U_\alpha$ because
of set R_α having properties $v_{\alpha i}$, if relations $r_{\alpha i} \in R_\alpha$
with properties $v_{\alpha i} \in V_\alpha$ exist between the elements of set U_α.
In such a defined system, a sequence of states of the system
represent different realizations if relation $r_{\alpha i}$ can be
distinguished.
    System α will be in a different state when relation $r_{\alpha k}$
characterized by property $v_{\alpha k}$ occures between elements $u_{\alpha i}$
and $u_{\alpha j}$ than that in which it will be when relation with
property $v_{\alpha l}$ occurs between the same elements. States of
system α are called as situations S_α.
    Let the man-machine system be a subsystem of system α.
The man – machine system can be isolated from system α under
the condition, at the adopted criterion K, the remaining elements of the system are ordered in such a way, that the relations between system and man are reduced to machine-man relationship /Fig.1./

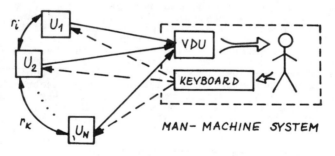

Fig.1. A schema of the system α

System α in the above form will be effective, due to criterion
K if the man-machine system is effective. Effective operation
of man-machine system is called as behaviour Z and it is a
global feature of the system. Practically speaking, man is
the element of man-machine system which performs the operation. This operation can be described by function of operation F determined by the following complex activities:
    – reception and processing of information,
    – formulation of responses,
–   – realizaton of responses.
Sequence of situations $S_{\alpha i} \in S_\alpha$ sets off a sequence of
realizations F^x of the operational function. Generally, behaviour Z is a family of realizations F^x of the operational
function. Realization F^k of the operational function may be
presented as the carrying out of instruction J . The instruction

26

describes activities q, usually essential ones, performed by
man and relations of the order in which these activities are
performed[2] Instruction J is defined in m-dimensional space
ε^m ; where m is the number of possible activities.

A TOPOLOGICAL MODEL OF THE MAN-MACHINE SYSTEM

The geometric representative of instruction J is a simplex
whose verices are points A_i representing activities q_1 i =
= 1,...,m and the walls are relations between the activities.
The simplex's dimension is m^3. It has been shown that the
global feature $c(u)$ of the man-machine system is behaviour Z.
Since Z is a family of realization F^x of the operational
function, Z is a simplifical complex. Therefore, in order to
get to know the man-machine system, one should evaluate its
behaviour Z. The evaluation in real conditions, i.e. in system
α is troublesome and often impossible. Hence, artificial
/laboratory/ conditions should be created, to make evaluation
Z possible. In general, new system β analogous to system α
should be formed. The system must be a relation system with
defined set of elements U_β and set of relations R_β having
properties V_β .
Let covariant functor $\overline{a}$ of system α to system β be given
which assigns element $u_{\alpha i}$ of system α to each element $u_{\beta i}$ of
system   and relation   to relation r . The transformation
of system $\beta\alpha$ into system β has, moreover two features:
- $u_\beta = \overline{a}(u_\alpha)$ for the man-machine system,
- $Z_\alpha = Z_\beta$ for the behaviour in the two systems.
It means that situations S_α which occur in system α and
β give rise to comparisons, due to functor $\overline{a}$, behaviour
Z_α and Z_β of the man-machine system.
The geometric model of behaviour Z is a simplified complex
defined in space ε^m. Therefore, such a simplified transformation ξ must exist which transforms behaviour Z_α in system
α into analogous behaviour Z_β in system β . If $Z_\alpha = Z_\beta$
the set of vertices of complex Z_α becomes the set of vertices
of complex Z_β and thus there is an identity of simplex J_α
and J_β . This means that the same characteristic of the
man-machine system as in system   should be evaluated in
the new system β .
Situation S_β which occurs in system β will be called
as tasks in contradiction to situations S_α . A task requires
that specified instruction J is to be carried out. If the
man-machine system is to perform sequence N of tasks S_β
(i = 1,...,N) a sequence of simplex J_i is obtained, i.e.
complex Z_s formed by the simplexes. The sum of all simplexes
J_i of complex Z_s is a polyhedron $|Z_s|$ of complex Z_s. A poly-
hedron $|Z_s|$ is compact metric space. The distance between two
points of polyhedron Z_s is assumed to be the distance of the
two points in the Euclidean space which contains $|Z_s|$
Let the principle of evaluation of the man-machine system
be a concept of comparing the standard behaviour to real
behaviour. The comparison can be defined by the distance between behaviours in certain metric space. If Z^x is standard
behaviour, the set of standard instructions J^x can be determined for a given set of tasks S_β . Set S_β is a test of system
β . If a man-machine system is subject to test β , a set
of real instructions $\overline{J}$ defining behaviour $\overline{Z}_s$ of the considered
system is obtained. Since Z_s^x and $\hat{Z}_s$ are simplified complexes
defined in the same metric space ε^m , there exists a

simplified transformation λ of Z_s' into $\hat{Z}_s$ and $|Z_s^x|$ into $|\hat{Z}_s|$

Analog considerations could be carried out for the pair of simplexes $\mathfrak{I}^x$ and $\hat{\mathfrak{I}}$. If simplexes are not identical, they differ in their barycentric coordinates b_i^x and b_i'. Therefore there is a nonzero measure of the distance between simplexes $\mathfrak{I}^x$ and $\hat{\mathfrak{I}}$ defined by barycentric coordinates. Thus, there is a function ψ which is the measure of the behaviour of the man machine system in system β. Function ψ is the measure of barycentric deformation of real polyhedron $|\hat{Z}_s|$ relative to standard polyhedron $|Z_s^x|$

CONCLUSIONS

A general feature, called behaviour, has been determined for the man-machine system. While observing the effects of the man-machine system functioning one may determine the behaviour of a system in the same large arrangement where it has a role of a subsystem. As the measurement of behaviour in real conditions is practically not possible, a transformation of the real system (system α)into artificial system (system β) ought to be done. System β is equal to system α.

It has been shown that we have found a functor $\bar{\bar{\varkappa}}$, which changes neither the man-machine system nor the behaviour represented by a simplified complex. The measurement of the same feature in laboratory conditions becomes possible. The measurement Z in a definite system β is limited to determining a resulting polyhedron which is a deformation of the standard polyhedron known before. A barycentric measure of this deformation is function ψ determining the behaviour of the man-machine system.

REFERENCES

1. Hempel,L., "Czlowiek i maszyna", WKL, Warsawa /1984/.
2. Hiler,G., Myszkowski R., Podlejski K., "A model of driver behaviour in the road traffic" Modelling, Simulation and control, vol 3, No.1. /1985/
3. Kuratowski K., "Wstep do teorii mnogosci i topologii", PWN, Warszawa, /1980/.

COMPUTERIZED EDUCATION IN ELECTRICAL ENGINEERING

M. D'Apuzzo   C. Savastano

Department of Electrical Engineering
University of Naples, Italy

1. INTRODUCTION

At present, in all Italian Universities the courses for obtaining a degree in Electrical Engineering are organized in two propedeutical years dedicated to Physics and Mathematical Sciences and three years where applied disciplines are taught according to the different curricula chosen by the students. The following courses must be included in all curricula: Electric Circuits, Applied Electronics, Electrical Measurements, Electrical Machines, Automatic Controls and Power Systems. Each of the above mentioned courses includes theoretical lectures and practical experiments; the latter pose serious problems arising from (i) the increasing number of students and consequently the necessity of repeating the same experiments many times in order to allow the students to operate in groups of not more than four, (ii) the sophisticated and expensive apparatus now available which makes the upgrading of the technical staff in charge of its operation and maintenance very difficult, (iii) severe safety standards which must be observed when using the apparatus which may be dangerous for the operator (i.e. high voltage testing equipment).

Aim of this project was to improve traditional theoretical lectures and to create interdisciplinary laboratories for practical experiments, as well as to reduce global costs and the number of technical staff required. The implementation of the project was possible thanks to the widespread diffusion of Personal Computers equipped with sufficiently large mass memories (10-30 Mbytes) and peripherals with good graphic performances on the one hand and the direct interrelation between the above mentioned courses on the other hand. The following supports were necessary: (i) working groups for analysing the contents of lessons and/or technical experiments; (ii) digital equipment for simulating laboratory experiments and measurement apparatus; (iii) software and hardware for teaching purposes; (iv) display systems in class-rooms and in working-areas; (v) video tape and magnetic disk libraries; (vi) a technical committee for coordinating all the above mentioned activities.

At the Department of Electrical Engineering of the University of Naples four working groups were created to deal with: measurement techniques, electrical machines, power transmission and distribution, control engineering. Due to the shortage of space, this paper will deal in detail only with points (iii) and (iv) which, in the Authors' opinion, are the most peculiar of the entire project. For the other points reference can be made to previous papers /1,2,3/ and to the oral presentation.

2. SOFTWARE AND HARDWARE TOOLS

It is well known that the major difficulty in developing computer aided education consists in text editing. The teacher can: (i) employ a programmer, (ii) implement by himself the software for the course, (iii) make use of standard software products. Solution (i) is expensive; solution (ii) is time consuming for the teacher who is usually a user of scientific programs but cannot be considered a skilled programmer. Solution (iii) is the most convenient because it minimizes the time of text editing and makes the hardware and the software easily accessible to users without specific experience.

In our case the courses are organized in theoretical lectures and in laboratory experiments. The number of lectures and experiments is determined by the teacher. The computer aided lecture may be fundamentally structured in three parts: (a) formation, (b) verification, (c) movement. In (a) the teacher gives the information related to the subject of the lecture and defines the progression criteria. In (b) the teacher defines the answers that the student must give at the end of each lecture and assigns, if necessary, a mark to each answer; in (c), according to the predetermined progression criteria, the student either passes to a successive theoretical lecture, or repeats the same lecture and/or a previous one, or he reviews, if necessary, the contents of other courses useful for understanding the lecture. For theoretical lectures, interactive tutorial languages, such as CSR Trainer 3000, were experimented.

During laboratory experiments the student operates on a portable IBM Personal Computer which can be used in two ways: as a simulator and as an emulator. In the first case the student must make some choices (i.e. dimensional parameters, physical quantities, planning constraints, etc.) and must operate according to the choices previously made in order to obtain a practical result (i.e. electrical machine or plant design, quality control, systems identification, etc.). In the second case the student acts as if he were performing a real measurement and the PC becomes a general purpose intelligent instrument by means of hardware and software implementation described in a previous paper /3/.

In view of the quality of the images which have to be displayed on the screen (i.e. electric circuits, machine sections, measurements instrumentation front-panel, control and regulation components, testing apparatus etc.) hardware systems which can drive also high resolution peripherals were opted for. Therefore IBM Personal Computers of XT and ES types were adopted: the second one is an XT implemented with a Professional Graphic Controller which permits to display simultaneously 256 colors from a palette of 4096 on a color screen with 640x480 pixel.For laboratory experiments and text editing, PILOT language and Graphical Kernel System together with Professional Fortran were successfully employed.

Standard curve-follower apparatus were used for diagram digitizing, whereas two devices, both interfaced with an IBM AT Personal Computer and characterized by high resolution, were considered for image acquisition:
-   the Sony DXC M3A color telecamera (750 lines horizontal resolution, 25 frames/s of 625 lines each) connected to a MATROX PIP-512 real time image digitizer;
-   the Personal Scanner (manufactured by Electronic Informatic Technology) which is similar to optical scanners employed in high quality facsimile transmission.

The first solution is obviously more expensive (its cost is almost three times that of the Personal Scanner) but it is more flexible and suitable for moving images recording. Other audio-video traditional supports are also available, such as:
-   slides and sheets for overhead projector prepared by means of special

software running on Personal Computers; for our purpose the EXECUVI-
SION software made by Visual Communications Network Inc. was found
to be more complete and easy to use;
- video-cassettes and/or video-disks containing scientific programs
previously recorded; this material is readily available from manufacturers and specialized libraries.

3. DISPLAY SYSTEMS

Theoretical lectures are presented in a different way as compared to
laboratory experiments. For the theoretical presentation in a classroom of
250 students, a high-definition (1000 vertical lines, horizontal scanning
up to 33 kHz with automatic synchronization) color video projection system
is installed: it replaces and does away with the need for other audio-video systems which are normally used (slides projector, movie projector,
overhead projector, etc.) because, by means of a direction-room, (i) video
signals from video-cassettes and video disks, (ii) alfa-numerical signals
from personal computers through interface boards, (iii) digital signals
from image codifiers are sent to the same apparatus. In addition two small
class-rooms are equipped with XT and ES Personal Computers in "open-shop"
operation to allow each student to: (i) get an aid for electrical lectures, (ii) repeat the simulation of the practical experiences, (iii) more
generally train himself in programming techniques.
In laboratory experiments the student will immediately receive the
output data both from the screen and from a low-cost graphic printer. In
each laboratory four work stations were set up linked together by means
of a local network which connects them also with other terminals located
in the Department of Electrical Engineering. In this way many peripheral
devices may be shared, for instance:
- low cost Laser Jet Printers (like HP 2686A)with 8 pages/min printing
speed and 150 dot/inch resolution;
- XY plotters with a variety of options necessary for highly accurate
technical drawings;
- high quality and high speed printers for text editing.
Finally it has to be pointed out that Personal Computers together
with dedicated work stations are widely used by teachers and students for
word processing: in this activity some of the most popular software (Easy
Writer, Microsoft Word, etc.) are successfully used. The problem of
storing all the documentation produced is solved locally by using off-line
low cost hard-disks and magnetic tapes and, in a centralized way, through
the connection via modem to the IBM 4331 Computer located in the main
building of the Engineering Faculty.

4. CONCLUSION

In the present paper the hardware and software system architecture
for computer aided education in Electrical Engineering are reported. The
project, initially carried out for Electric Engineering undergraduates,
was subsequently extended to the Electronic Engineering undergraduates who
pose more serious problems because of their higher number. The same
experiences can also be extended to High Schools where the present situation, above all for what concerns laboratory experiments, is particularly
heavy.
As far as the teacher is concerned, though a greater initial effort
is required of him, the following advantages are being obtained: (i) a
more careful control of the students' learning level, (ii) the possibility
of rapidly modifying the content of theoretical lectures and laboratory
experiments, (iii) a fast updating in compliance with new technical
developments and/or student requests.

As for the students, the following advantages, besides the ones previously mentioned, are found: (i) a better learning level, (ii) a more thorough preparation in the use of the Personal Computer, (iii) an appreciation of the interdisciplinarity of the different courses. For the success of the project the activity of the above mentioned technical committee was found to be of great importance because: (i) it avoids duplication of initiatives, (ii) it ensures a better use and proper maintenance of hardware and software, (iii) it takes charge of the updating of the technical staff, (iv) it manages the magnetic disk and video-cassette library.

Technical supports were given by the Computer Science and Electronic Engineering Departments. A special financial support was obtained from the Welfare Service for University Students and IBM contributed with a study contract concerning the automation of Departments at the University of Naples.

5. REFERENCES

1. A. Baccigalupi, M. Savastano: Simulation of an Electrical Measurement Laboratory - IMACS Congress, Oslo, August 1985.
2. N. Polese, G. Betta, C. Landi: Intelligent Measurement for Quality Testing of AC Motors - IMEKO Symposium on Intelligent Measurement, JENA, June 1986.
3. A. Baccigalupi, M. Savastano: Computer Aided Electrical Measurement Experiments - IMEKO Symposium on Intelligent Measurement, JENA, June 1986.

COMPUTERIZED EDUCATION and AN EXAMPLE ON MEASUREMENT —— As a Example of CAI aimed at Universal Education kept Harmony with Normal Lecture Essence of Education by used Low Cost Micro Computer

Komyo Kariya

Department of Electricity, Faculty of Science and Engineering, Ritsumeikan University
56-1, Tojiin-kita, kita-ku, Kyoto 603 Japan

Many styles of Applications to Education of Computer have been developing and there is a Computer Assisted Instruction(CAI) in one of them. Then it has been attempting in individual stages of School Education and in each field of Special Field Education according to development and diffusion of some kind of Computers. In this paper, after introduced the status of Applications to Education of Computer in Japan, an education method by using Handy Type Low Cost Micro Computer which is kept good harmony in Normal Lecture is offered and I give a Proposal to make project team internationally to propagate this education method.

INTRODUCTION

It is well-known that the computer is classified four grades, that is,
1. Large scale computer (Computer system),
2. Middle class computer (Mini computer),
3. Micro computer (include Handy type computer, Personal computer),
4. Calculator.
Even if which grade computer is used having how purposes on education problems it is said as "Computer Assisted Instruction". The essential thing is how grade up the education effect by setting the help of several kinds of computer functions. In that time, we must not forget the essence of education that is symbolized as a talk, a touch, comunication and reliance relation with a man and a man. In the first time, the computer was used for numerical calculation, numerical analysis, experimental data arrangement and experimental graph making on education problems. In the second time, the computer became to be used as answer formarity by conversation style. This is a method that the computer gives a problem and student inputs the answer then the computer announces or indicates "Correct" of "Incorrect". A trainning computer of foreign language (for example: TEXAS INSTRUMENT LANGUAGE TUTOR, CRAIG M-100 (CRAUG cooperation)) will be included also in this study. In the third time, the graphics functions of computer became to be applied to rise up educational effects. And generally the computers over middle class are using to make graphic picture. On the education effects by graphics are very remarkable at indication of moving and colour graphics to student.
National education budget is not so high cost in each country. Of course we can not forget the fact general saying "Percentage of national education budget is indicates the degree of cultural and science level of its

country". But it is not more good more use high cost.

Even if CAI has been abroading, the situation is still in a trial stage and there are many kinds of problems on the national education principle, the school system in each country, the national budget for education, the education fund to be pay by students and parents, the computer allergy of students and teachers not acceptable computer in personality, the countermeasure that the essencial essence of education is not lost and so on. When fundamental education system or method by CAI is concerned, it must be avoided to become in irregular essencially.

An education method which is offered in this paper is a method to solve above mentioned many kinds of problemes that lie down on the realization of CAI. And particularly the method is realized by low cost and by having well good harmony with normal lecture had originally education essence.

STATUS OF APPLICATIONS TO EDUCATION OF COMPUTER IN JAPAN

The CAI has been advancing in very wide range of education in japan. When talk about status of CAI, it is need to give a review about Applications to Education of Computer. Because CAI is one of them classified in three fields /1/.
    (1) Education of computer itself (Computer Literacy)
    (2) Result processing of education test and another management processings in education (CMI; Computer Managed Instruction)
    (3) Computer aided Education (CAI/CAL; Computer Assisted Instruction /Computer Assisted Learning)
And now it is possible to concern about three categories. The first category is the status in University, the second category is the status in Primary school (Elementary school), Junior high school (Middle school) and High school and then the third category is the status of sub-side Education (Home study, Private study room, --- etc.).

* Status in University:- In University, the education advances in parallel with several kinds of research works. Then the applications of computer are more prosperous in special fields on Experimental and Practice Educations, Planning and Design Education, Mathematical Education, Computer Education Itself (Hardware and Software), Normal Lecture Education, Introduction of Research Education and so on. Each University has one or two computer centers and the centers are connected with another University's centers or public information centers. Sometimes the centers are connected with another foreign University centers or information centers by using saterite system or cable system. And then the terminals are settled in some lecture rooms, experimental rooms, visual eduction rooms and laboratories. Over and above several kinds of micro computers are introduced in almost laboratories, particularily in Natural science laboratories. On the Experimental and Practice Education, the analysis and the arrangement of data and the experiments by modeling and simulation are advanced using the computer functions. On the Planning and Design Education, the use of CAD (Computer Aided Design), CAM (Computer Aided Manufacturing) is developing. On the Mathematical Education, the computers are working with calculation itself and understanding of mathematical structures. On the Computer Education itself (Computer Literacy), the understanding of computer operation is advancing through computer graphics and the learning of important software (FORTRAN, COBOL, BASIC, etc.) is practiced. Particularily the relation with human intelligence will becomes strongly and Logo which excellents as education language will be regarded in future. At normal lecture room, the rise up of education effects of foreign language, staistics and prediction problems on economics, mapping on geography and all problems that is possible to use of computer functions is discussing by using mainly pattern graphics. In the Introduction of Research Education of each field, several kind of computers are used as the tools. Especially analytical instruments assisted with computer are used on physical, chemical, biological and medical

fields, and then diagonostic instruments supported with computer are operating on the fields of mechanical and civil engineering. The result processing of education test and another management in University have normalized already. And the research work managements are done in each laboratories individually.

* Status in Primary school, Junior high school and High school:- Introduction of CAI(CAL) is remarkable in this category. Japanese trial of education assisted by computer had begun in 1967. In the first time, from fifty to one hundred terminals of a computer center had used. But, according to development of micro computer now almost schools have been testing about CAI(CAL) by using micro computer. And many types of learning methods are developing. The main types are next.
   1. Tutorial type to give exercises, conceptions and knowledges of drill type.
   2. Graphic type to give physical and chemical phenomena and mathematical principles as figure elements by using colour graphics and animations.
   3. Simullation type to do sham experiments and experiences by using simullation models.
   4. Game type to rise up education effect through enjoining education game.
   5. Solving type to go to solve problems in order.
Japanese Goveranment Education Ministry and Education Committee of Each Prefectures are having a strong plane to be advance and to be widly spread CAI(CAL). Here, there are two directions. One is the method to set micro computer in each education room of each school and another one is the method to connect between an infomation processing education center and each school. Particularly the latter is advancing on the Education of Computer itself (Computer Literacy) of Industrial schools and Commercial schools.

* Status of Sub-side Education:- Recently, the applications of computer on

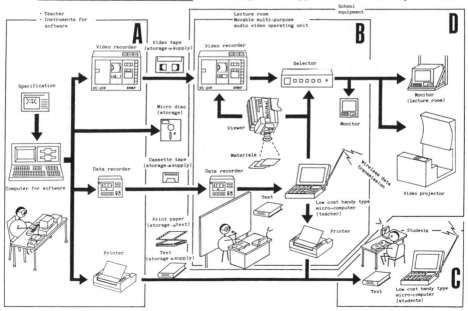

Fig.1　Education System by using Low-Cost Micro-Computer

home study and at private study rooms became prosperous (assisted by Tape or Floppy disc) are used. The softwares are constructed by above mentioned learning methods by as there are no standards so that the distribution of wrong contents of education is anxioused and any countermeasure becomes need.

LOW COST COMPUTER ASSISTED MEASUREMENT EDUCATION ON NORMAL LECTURE

We have been constructing a CAI(CAL) had well harmomy with normal lecture by used Low cost micro computer /2/,/3/,/4/. Fig.1 shows an education system.

Block A is a personal system of teacher and it is set in official private room (desk) (sometimes, private home). Block B, C and D are the system for lecture room. System B is set in teacher's desk, System C is handy type Low cost micro computer which is prepared individually by student himself and System D is an usual Visual system in lecture room used in Video monitor system.

At system A, the menu (softwares) which fit to the contents of lecture are made by teacher and supplied to students and another teachers by Video tape, Floppy disc (Micro disc) and Cassette tape. Usually, only the Cassette tape is supplied to students on the reason of Low cost supply. The Video tape and the Floppy disc (Micro disc) can supply to teachers. Then teachers are possible to have lecture by using the Video tape at lecture room only prepared Vido monitor system and are possible to reffer to construct education menus (graphics) by using the Floppy disc quickly. The hard copy by Printer paper becomes easy to construct the Text book or Preprint for students.

At system B, teachers use Low cost micro computer same one of students and Data recorder and have lecture effectively by using the education menus put in Cassette tape with usual visual system constructed by Video recorder, Viewer, Monitor, Video projector and Blackboard relate with System D. In this case Selector operates effectively for rise up of education effects. The Printer is used when the menus (softwares) are changed in the way of lecture and the hard copy is distributed to students by using Copy system connected with the Printer. Still more, the Low cost micro computer used at present is

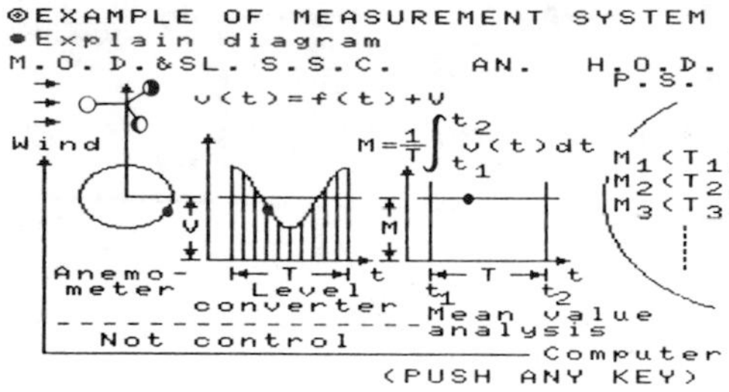

Fig.2 An example of Education menu to understand MEASUREMENT SYSTEM (hard copy)

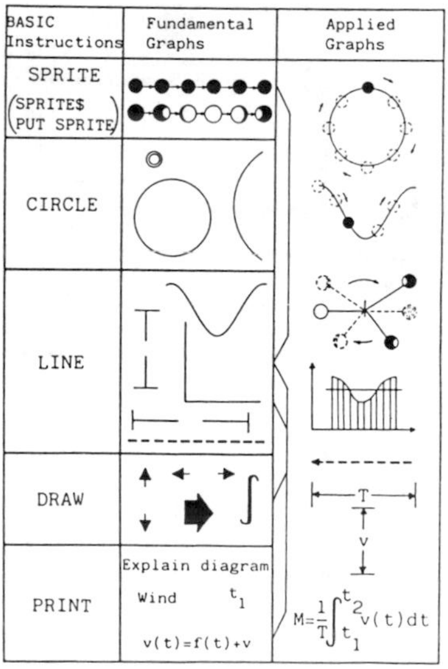

Fig.3 Basic Instruction and its Fundamental graphs and Application graphs of Fig.2

operated by MSX-BASIC and the computer is constructed by Z-80A as CPU. And then ROM is 32kB(for MSX-BASIC), RAM is 16kB and Video RAM is 16kB. The cost is less than US$400-(computer and colour monitor) at present.

Fig.2 is an example of education menu to understand MEASUREMENT SYSTEM (by hard copy). A wind speed gauge (anemometer) is used as a sensor and the wind speed is measured quantitatively by calculation of mean value. This menu is constructed by Basic Instruction and its Fundamental Graphs and Applied Graphs (see Fig.3). It is very important to construct an education menu (software) by simple instructions as possible, otherwise it takes large time to construct a menu. Two examples of Measurement education menus are shown in Fig.4 and Fig.5. Fig.4 is an understanding moving graphics of WAVEFORMS MIXING, TIME DOMAIN and FREQUENCY DOMAIN of SIGNAL(by photograph of display (a) and hard copy (b)). In this menu, a signal in time domain S(0) is indicated as the sum of signal S(1), S(2), S(3) and S(4) and the state that a frequency spectrum (frequency domain) goes to be constructed is shown by using moving graphics. Fig.5 is an understanding menu of WHAT IS INFORMATION (by hard copy). In this menu, the relation between "Information" and "Intelligences" and an expression of information by a train of numerical value are indecated in order by using moving graphics.

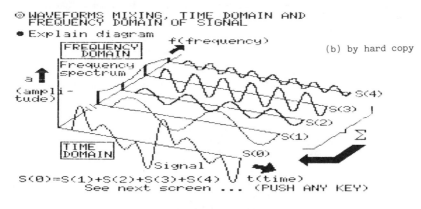

(a) by photograph
of display

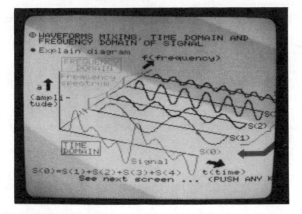

(b) by hard copy

Fig.4 An understanding moving graph of WAVEFORM MIXING,
TIME DOMAIN and FREQUENCY DOMAIN of SIGNAL

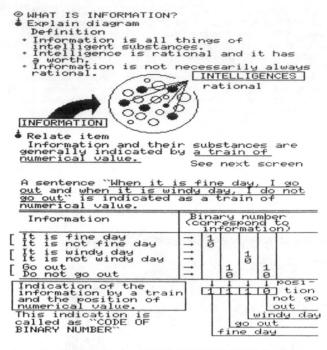

⊕ WHAT IS INFORMATION?
♣ Explain diagram
  Definition
  * Information is all things of
    intelligent substances.
  * Intelligence is rational and it has
    a worth.
  * Information is not necessarily always
    rational.

INTELLIGENCES
rational

INFORMATION

♣ Relate item
  Information and their substances are
  generally indicated by a train of
  numerical value.
                          See next screen

A sentence "When it is fine day, I go
out and when it is windy day, I do not
go out" is indicated as a train of
numerical value.

Information	Binary number (correspond to information)		
[It is fine day →	1		
[It is not fine day →	0		
[It is windy day →		1	
[It is not windy day →		0	
[Go out →			1
[Do not go out →			0

Indication of the
information by a train
and the position of
numerical value.

`[1 1 1 1 0]` posi-
tion
not go
out

This indication is
called as "CODE OF
BINARY NUMBER"

windy day
go out
fine day

Fig.5 An understanding menu of WHAT IS INFORMATION (by hard copy)

CONCLUSION

In the status of development of CAI(CAL), we saw a direction to advance
an education method had good harmony with application of low cost micro
computer and normal lecture. The method indicated in this paper is very
effective to spread widly computer assisted education in lecture room. To
propagate this method, it is need to construct a good team work with computer
hardware makers, software makers, book company and teachers. Then I give a
proposal to make project team in internationally.

REFERENCES

/1/ S.Takahashi: Learning by Personal Computer -in center of CAL and Logo,
     The Journal of the Institute of Electrical Engineers of Japan, Vol.105,
     No.10, pp.959-962, 1985
/2/ K.Kariya: Intuitive and Image Learning Method of Fundamental Points on
     Measurment, Electric and Electronics by using Low-Cost-Micro Computer,
     IMEKO TC-1 Colloquim Preprint pp.205-222, Graz-Austria, 1984
/3/ K.Kariya: A proposal of Intuitive Learning Method on Special Fundamental
     Education of Electric Couurse in University and A proposal of Low-Cost
     Personal Handy Type Micro-Computer for Education, Journal of JSEE,
     Vol.33, No.3 pp.18-25, 1985
/4/ K.Kariya: -Having a Purpose of Universal Validity on Micro-Computer
     Aided Education -Intuitive and Image Learning Method by Using Graphic
     Function of Low Cost Micro-Computer, Memoirs of The Research Institute
     of Science and Engineering, Ritsumeikan University, No.43, pp.119-131,
     1984

UNITARY BLOCK SCHEME FOR MEASURING SYSTEMS

Hans-Joachim Dubrau

College for information engineering of Dresden

Department of information electronics
DDR-8019 Dresden, Hans-Grundig-Straße 25

INTRODUCTION

The theory of measurement develops into a new quality in the last tens of years /1-3/. A great number of measuring devices and systems appears in this time. But there is a dividing line between theory and practice in the understanding of measurement. Therefore we are proposing an unitary block scheme for all meters, measuring devices and systems, joining the theoretical base with the practical systems.

COMPARISON - THE FUNDAMENTAL OPERATION OF MEASUREMENT

Each measurement, every yield up of a measuring value includes a comparison between the quantity to be measured A_{mx} or its imaging A'_{mx} and one or more comparison quantities A_{cj} ore their imagings A'_{cj} with $j = 1,2,\ldots,n$, independent of the measuring method /4/. There are three possible results of a comparison

$$A_{mx} < A_{cj} \qquad (1)$$
$$A_{mx} =_q A_{cj} \qquad (2)$$
$$A_{mx} > A_{cj} \qquad (3)$$

$=_q$ - analogic quantization comparison gives "true"

The measuring comparison is a quantization comparison between analogic quantities. The set of analogic quantities to be measured $\{A_m\}$ is infinite, whereas the set of (known) analogic comparison quantities

$$\{A_c\} = A_{c1}, A_{c2}, \ldots , A_{cn}$$

is finite. Because the set $\{A_m\}$ is infinite and the set $\{A_c\}$ is finite, there gets an infinit number of quantities to be measured to each comparison quantity. All quantities to be measured in a range

$$A_{mx} = A_{cj} \pm \Delta A_c =_q A_{cj} \qquad (4)$$

are leveled to the quantity A_{cj}. This operation is indicated by the special sign of analogic equality $=_q$ (speak: quantizated equal). At linear graduation of sca-les yields

$$\pm\Delta A_c = \pm 0,5 \text{ lsb} .$$

There are the parallel and the serial comparisons. Parallel comparisons requires the simultaneous comparison of A_{mx} with all elements of the set $\{A_c\}$, for instance by measuring a length with a linear measure. By serial comparison the elements of the set $\{A_c\}$ are compared one after the other with the quantity to be measured A_{mx}, until yields the relation (2), for instance by weighing with weights.

THE MEASURING COMPARATOR

The comparing element between the quantity to be measured or its image and the comparison quantities or their images is named measuring comparator. It points out, which of the relations (1),(2) or (3) is true. The measuring comparator consists either of two comparator devices, one for the beginning (start) and one for the end (stop) of A_{mx}, or one comparator device serves serial two comparisons.

There are two types of measuring comparators: Type A and Type B. Type A consists of one reference comparator c_r for the beginning and one quantizating comparator c_q for the end of the quantity to be measured. The Type A is working only if there is a common beginning of A_{mx} and all A_c (Fig. 1).

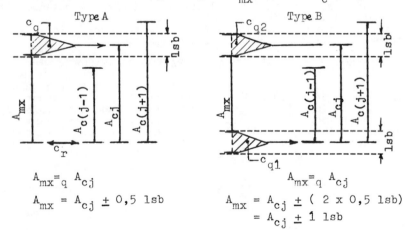

Type A
$$A_{mx} =_q A_{cj}$$
$$A_{mx} = A_{cj} \pm 0,5 \text{ lsb}$$

Type B
$$A_{mx} =_q A_{cj}$$
$$A_{mx} = A_{cj} \pm (2 \times 0,5 \text{ lsb})$$
$$= A_{cj} \pm 1 \text{ lsb}$$

Fig. 1   Measuring comparators

Typical reference comparators are the straighting plate for linear measuring, the short circuit (interconnection) for electrical measuring or the common start for time meter and measuring time.

Sometimes it is not possible to find a reference comparator, for instance crystal oszillators can't work at start-stop-conditions. In this cases it is necessary to use a measuring comparator of the Type B. It consists of two quantizating comparators c_{q1} and c_{q2}, one for the beginning and one for the end of A_{mx}. The quantization error amounts to $\pm$ 0,5 lsb at the Type A and to 1 lsb at the Type B (Fig.1).

At analogic measuring devices the operator serves the function of the measuring comparator, in digital measuring devices there are technical comparators. That is the fundamental difference between analogic and digital measuring devices.

MEASURE REPRESENTATION /5/

Each measuring device includes one or several (n) comparison quantities $A_{c1}, \ldots A_{cn}$. After the analogic comparison, which is performed not only by analogic measuring, but also by digital measuring, it is necessary to read the value S_{cj}, correspondending to the comparison quantity A_{cj}, which is satisfying the relation (2). The set $\{S_c\}$ is a scale. At analogic measuring devices the scale is legible for the operator, at digital measuring devices for the control unit. A set of comparison quantities $\{A_c\}$ and the attached set of comparison values $\{S_c\}$ are forming a measure representation. There are single-valued and multi-valued measure representations.

GENERAL SCHEME OF ALL MEASURING DEVICES

The general scheme for measuring devices is shown in Fig.2. At minimum a measuring device consists of a measure representation. The operator performs the comparison. Additional there are transducers for the measuring and/or the comparison quantities. In digital measuring devices there are moreover a technical measuring comparator and a control and reading unit. The result of analogic and digital measuring is always a numerical measuring value S_{mx}.

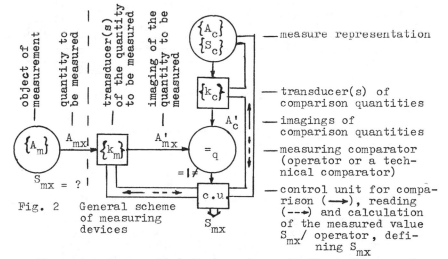

Fig. 2   General scheme
of measuring
devices

Measuring is provided in two steps. At first the operator (analogic measurement) or the control unit (digital measurement) by variation or choice of A_c, k_c and k_m attains comparison:

without transducers	with transducers

$$A_{mx} =_q A_{cj} \qquad\qquad A_{mx} \cdot k_{mi} =_q A_{cj} \cdot k_{cr} \qquad (5)$$

After comparison by reading the values of k_{mi}, k_{cr} and S_{cj}
we obtain
$$S_{mx} =_q S_{cj} \qquad\qquad S_{mx} =_q \frac{k_{cr}}{k_{mi}} S_{cj} \qquad (6)$$

The quantization error of analog as well as of digital measuring, in the equations (5) and (6) indicated by the sign $=_q$, is depending only on the typ of scaling (linear, quadratic or other) and the typ of measuring comparator (A or B). At linear scaling it amounts for the typ A to $\pm$ 0,5 lsb and for the typ B to $\pm$ 1 lsb.

UNITARY BLOCK SCHEME FOR MEASURING DEVICES

Fig. 3 shows the unitary block scheme for all analogic and digital measuring devices. At analogic measuring there are working the sections A and B, at digital measuring the sections A and C. The blocks 1 and 2 are performing a measuring generator, the blocks 3 and 4 an analogic recorder.

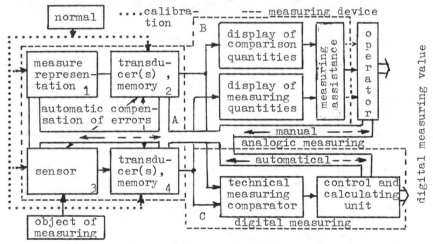

Fig. 3 Unitary block scheme for measuring devices

REFERENCES

1. Pfanzagl,J. Theory of measurement. New York, Wiley. 1968
2. Leaning,M.S.;Finkelstein,L. A propabilistic Treatment of Measurement uncertainty in the formal Theory of measurement. Acta Imeco 1979,V.1.p.73-81
3. Piwowarczyk,T.;Sluszkiewicz,T. Practical and theoretical Consequences resulting from the Set-theory Description of the fundamental metrological Notions.Acta Imeko 1979.V.1,p.93-101
4. Dubrau,H.-J.Messen und seine technische Realisierung.Wissenschaftliche Beiträge.Ingenieurhochschule Dresden.1983,5,p.51
5. Dubrau,H.-J.Hybrid memories of measures in meters and measuring devices.3rd Imeko TC-8-Symp.Berlin October 8-10 1986, Proceedings of the 3rd Symposium on Theoretical Metrology,185
6. Dubrau,H.-J. Grundlagen der Meßtechnik.Hochschulfolienreihe HFR 849 (in German and in English)Inst.f.Film,Bild u.Ton Berlin 1986

USE OF A-PRIORI INFORMATION FOR INTELLIGENT TESTING
OF MULTIRESONANCE ELEMENTS

Stanisław Żmudzin, Adam J. Fiok, Jacek Cichocki

Institute of Radioelectronics
Warsaw Technical University
Nowowiejska 15/19, PL-00665 Warsaw, Poland

1. INTRODUCTION

One of the new possibilities offered by intelligent
testing systems is better utilization of a-priori information
about the measured object. Many mechanical, electrical and
electromechanical objects have multiresonance character. Main
difficulties in measurement of their resonances are caused by
large values of corresponding Q-factors. The resonance bandwidth is usually very small in comparison to the frequency
range to be scanned and the inertia of DUT is high. If the
speed of tuning is too high, then the danger of skipping over
the resonance occurs. If it is too low, the searching time
becomes too long for industrial application. An intelligent
testing system can solve above problem using adaptive control
of the tuning speed based on combining a-priori information
about DUT with that acquired during the measuring process.
    In the paper ways of solution of thas and related
problems in the case of measurements of resonances of quartz
crystal units (QCU) have been considered.

2. METHOD OF MEASUREMENT OF QUARTZ CRYSTAL UNIT RESONANCES

An equivalent circuit of QCU is shown in Fig.1. The
most important part of it is the series resonant circuit

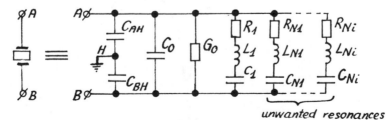

Fig.1 Equivalent circuit of quartz resonator

$L_1 C_1 R_1$ - an electrical equivalent of the main mechanical
resonance. Its frequency f_s is called the DUT series resonant
frequency. Other resonant branches represent mechanical
resonances connected with other modes of mechanical vibrations
which may be excited in the DUT.

In many cases, the resonant frequencies f_N of some modes
($L_N C_N R_N$) are near to the frequency f_s. These unwanted resonances (UR´s) can cause much trouble in application circuits.
In the paper [1] basic problems in measurement of QCU resonances as well as the proposed measuring method have been
discussed in more detailed manner.

In the proposed transmission measuring method QCU is
inserted into a special T-type transmission three-port.(Fig.2)

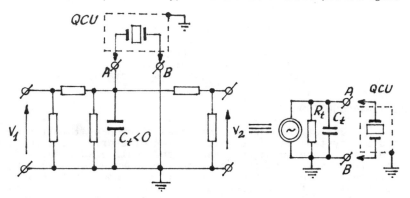

Fig.2 T-type transmission three-port with partial
       compensation of C_o

The original construction proposed by Smolarski and Wójcicki
[7] enables one to obtain a negative internal capacitance
C_+ (parallel to the measuring terminals). The C_+ is used for
partial compensation of QCU parallel capacitance C_o in order
to diminish systematic errors of resonator frequency and
resistance measurements.

Determination of parameters of main and unwanted resonances requires precise tuning to each resonance. Too fast

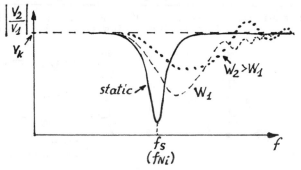

Fig.3 T-network responces for signal of linearly growing
       frequency (W-tuning speed)

tuning can cause dynamic effects shown in Fig.3. To find a
compromise between contradictory requirements of short tuning
time and high accuracy an original tuning method has been
proposed [5] . The method is based on a kind of adaptive
control of tuning speed. An illustration of this method is
shown in Fig.4.

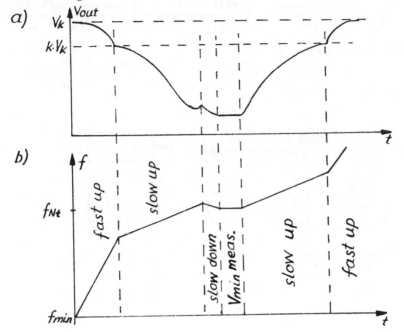

Fig.4   Illustration of UR measuring process
        a) dynamic changes of V_{out}
        b) dynamic changes of signal frequency

For each resonance a minimum of transfer function magnitude
occurs. The measuring process is accomplished in 3 steps:
- calibration of transfer function magnitude, i.e. measuring
  and storing of the V_k value,
- searching for and measuring of main resonance (MR),
- searching for and measuring of unwanted resonaces (URs).
Three values of tuning speed are used in searching for UR. The
highest is applied if the measuring receiver output voltage
V_{out} is greater than kV_k, i.e. when the instantaneous frequency
is far from resonance. In a vicinity of the resonance it is
reduced to diminish dynamic effects. The final precise tuning
to an UR frequency is performed at the lowest speed.

3. CONTROL OF PARAMETERS OF THE MEASUREMENT PROCESS

        In industrial conditions batches of products with usually
the same nominal parameters are to be measured. Then three
kinds of a-priori information about the measured object may be
available statistical information about parameters of products,
results of measurement of the first object in the tested batch

and results obtained at previous stages of measurement process
of a just tested object. To obtain reliable performance of
measurement task, required accuracy and high throughput, all
these kinds of information should be utilized in controling
the measurement process.

In general a-priori information may be used by: designer
of the measuring system, operator of the system and the system
itself (Fig.5), if it is an intelligent one. Designer can use
only the first kind of information, operator - up to certain
degree - also the second one. Intelligent testing system is
the only one which can utilize the whole a-priori information.

Let us consider exemplary ways of the use of a-priori
information in measurements based on the method described
above (QCU´s designed for oscilators working in frequency
range 4 to 125 MHz).

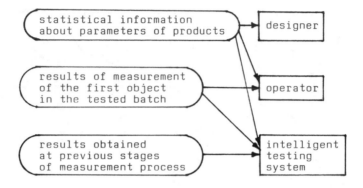

Fig.5   The kinds of a-priori information and their users

A very simplified functional diagram of the measuring system
is shown in Fig.6. It is essential that parameters of the
measuring signal generated by the signal source are controlled
by the Tuning Control System, which can be a simple analog
unit, operator itself or a complex network with microcomputer
assuring sophisticated adaptive control.
According to the results of statistical analysis of QCU parameters the T-network and system parameters have been chosen:
C_t = -6 pF (typical values of C_0 are 4 - 8 pF), R_t = 50 Ω .
The measuring range of R_1 is 3 $\stackrel{\circ}{-}$ 150 Ω (typical values of
R_1 are 5 - 100 Ω) and measuring range of R_N to 500 Ω (typical
tolerable minimal value of R_N is about 3 times R_1).
Minimum a-priori information needed for the tested batch is
the value of MR nominal frequency f_{nom}. The operator can also
enter other information which may change parameters of measurement process: tolerable minimal value of R_N (or of R_N/R_1
ratio), kind of QCU construction, range of searching for
UR in which the low-resistance resonances is probable, etc.
An intelligent system uses all a-priori information to

46

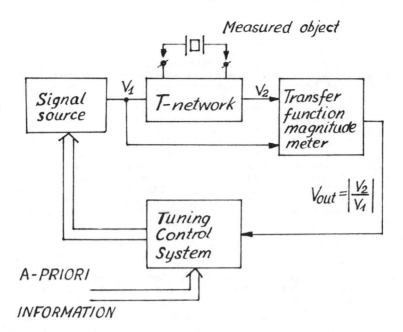

Fig.6 Very simplified functional diagram of the measuring system

accomplish measurements in the shortest possible time.
A kind of calibration (without QCU for lower f_{nom} or with QCU at frequency equal to $0.99\ f_{nom}$ for higher f_{nom}) is chosen. Calibration is made only for the first QCU in a production batch, and its results are utilized in measurements of other QCU´s in the same production batch.

Measurement time depends mostly on tuning speed W and comparison threshold k. If the value of W increases and the value of k decreases, then the measuring time becomes shorter but it may cause skipping over a weak UR. Optimization of W and k values is rather complex and must be done on the basis of information about: f_{nom} (W can be higher for higher f_{nom}), typical R_1 value (k can be lower for lower R_1), tolerable R_N (or R_N/R_1) value (for lower R_N tuning process may be faster) etc..

Moreover, if the parameters of the DUT are far from the mean values, an intelligent system can try to modify the parameters of tuning. For example if the searching for MR is not effective then it may be due to much higher value of Q-factor or of R_1 then has been assumed so the system can repeat searching for MR but at the smaller tuning speed and greater value of k. For smaller driving power the intelligent system can use the narrow er bandwidth and smaller speed to diminish the influence of interferences.

Most of above possibilities have been used in the testing system CMS-3 worked out in our Institute [1,6] . In that system 32 values of k and 15 values of tuning speed W can be

automatically chosen using all existing kind of a-priori
information.

4. CONCLUSION

Use of a-priori information in intelligent testing system,
as it can be seen from the considered example, enables to obtain
much better metrological and exploatation features of the system. Necessary algorithms for the adaptive control of measuring
process parameters can be implemented using simple microprocessor or microcomputer. But development of such algorithms requires statistical investigations of values of tested object parameters (and correlations between these parameters), detailed
analysis of utilized measuring method and good knowledge of
unwanted phenomena in utilized instrumentation. First version
of algorithms can be based on theoretical considerations, but
the final version should be fixed after a few months of normal exploitation of the system in the industrial environment.
Changes in the algorithm can be based on measurement
data acquired by the system itself.

5. REFERENCES

1. Fiok A.J., Żmudzin S., Cichocki J.: System for industrial
   measurements of main and unwanted resonances of HF Quartz
   Crystal Units, Xth IMEKO World Congress 1985, Preprint
   Vol.9., p.167.

2. Żmudzin S.: Piezoresonatoren Messmethode mit Vorabs-
   timmung, Digest of Conference "Elektronicke a piezoelektricke prvky a obvody" - Liberec, 1983, pp. 68 a-c.

3. Cichocki J.: Application of transmission methods for
   measurement of low frequency quartz crystal units,
   Digest of Conferency "Elektronicke a piezoelektricke
   prvky a obvody" - Liberec, 1983, pp.7a-d.

4. Fiok A.J.: Problems connected with identification and
   measurement of h.f. quartz crystal unit parameters,
   IMEKO-Symposium on Measurement and Estimation - Bressanone, 1984.

5. Żmudzin S.: Method of automated measurements of unwanted
   resonances in quartz resonators. Prace Instytutu Tele-
   i Radiotechnicznego, No 99-100, 1985, pp.46-49 (in
   Polish).

6. Żmudzin S., Cichocki J., Fiok A., Królak S.,
   Słowikowski A.: Automated system for measurements of
   quartz resonator parameters. Prace Instytutu Tele- i
   Radiotechnicznego, No 99-100, 1985, pp.50-54 (in Polish).

7. Smolarski A., Wójcicki M.: T-type twoport for measurement of quartz crystal resonators in frequency range up
   to 200 MHz. Elektronika 11-12, 1985, pp.21-23 (in Polish).

MEASUREMENTS OF THE FUTURE IN MICROELECTRONICS

Andrzej Sowinski

Industrial Institute of Electronics

44-50 Długa St., Warsaw, Poland

INTRODUCTION

If the future of measurements in microelectronics is to be
discussed, we should talk not about concrete systems solutions
but rather about the philosophy of measurements of the future.
It results from three observed and significant trends:
- vast development of computer graphics,
- creating numerous dedicated test systems,
- the popularity and universality in using a computer,
  and microprocessor and its systems are the base.
The factory of the future is characterized by a minimum role
of man, and the system approach to the whole. It also concerns
measurements in which we may assume that today's circuit
solutions will survive in a generally unchanged form for a
couple of years.

PHILOSOPHY OF THE MEASUREMENT OF THE FUTURE

We may define three prevailing directions in measurement,
control and testing:
- testing process at subsequent stages of product manufacturing,
- control and test in-line of production processes; here
  we can see the closed loop of computer integration of
  testing and calibrating,
- computer system of management data, including materials,
  connected to the network of control, testing and production processes.
Nowadays, it is above all testing and computerized testers,
defined as CAT /computer aided test/. CAI /computer aided
inspection/ is the "automatic sensitivity" which eliminates
man's subjectiveness and improves /so it optimizes, and does
not increase/ the testing frequency and states perception,
leading to the objective organization of patterns quality.
Sensors react very precisely, fast, in a linear way, and
repeatedly on the most of physical phenomena. Such far-going
and just complex integration - is just the future of measurements in microelectronics.

Two questions arise at the base of considerations of this philosophy:
- what is the optimum set of test points at the maximum economy of costs and efficiency?
- how many tests are enough?

"Quality people" prefer fast, automatic and so-called developed /sealtered/ testing, manipulating higher test level. "Process people" prefer to have, for the same money, more precise, automated technological equipment allowing for less test elements. The test solution lays, as usual, in the middle. The use of full 100 per cent testing protects from putting wrong materials to the production process, and full 100 per cent testing of critical technological operations protects from adding errors to already defined semiproducts. Simultaneous monitoring identifies anomalities, critical operations, and data gathered from the former may lead on-line, to the process modification and correction. Thus testing becomes the integral element of the factory process control. A "young" notion of product testability becomes more and more popular. In order to utilize this, we apply CAD to the testing process, which simplifies the design of logics of product testability as well as suitable patterns of tests for given products. The development of sophisticated sensor system and progress in AI /artificial intelligence/ are another aspects of the future. We have functional integration between CAD and CAM, CAT and CAD, AI and MIS / management information systems/.

UNION OF MEASUREMENT AND ASSEMBLY

The testing of processes, and not products draws some consequences. One of many of them is the union of measurement and essential technology operation of process. The assembly techniques are connected with it in microelectronics. There are two basic assembly techniques:
- constant or hard automation assembly, in which there is dedicated equipment assigned for high production of difficult products,
- flexible automation assembly, served with programmed equipment; also industrial robots are included in this group.

We do not assume the hard automation in the future because, should there be an error in a process, the equipment will produce a defective product with the same speed. Re-programming assembly equipment seems the most advantageous, according to the assumption that it is more advantageous to protect from a defective product than to identify errors. Changes in a product, revealed in the result of testing, generate correction data to the process whilst improving productivity in effect. So we deal in a process control in a closed loop.

Here we have to mention also new technologies of contactless sensor of an image.

From the features of microelectronic industry there arises a possibility of using equipments, which are, to a large extent, standardized, in order to fulfil substantial productivity. A question is connected with the latter:
- shall we buy fast and cheap equipment yet just for the production of a one-type product /and the one that needs to be set for each production line/?
- shall we buy slower yet more expensive equipment, which may be used for various types of products, with little or even no time needed for setting?

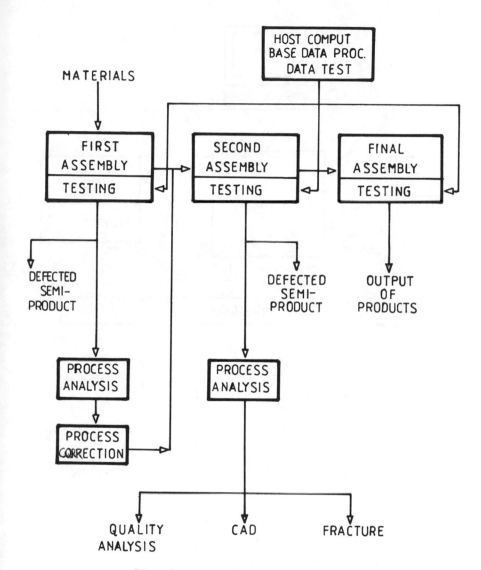

Flow diagram of CAD, CAT, CAM

The element moving one to accept the second concept is the
development of vision system equipment - image recognition.
There are flexible devices, yet they need fairly rich software. The future belongs to the automation of two factory
functions: process and testing, controlled from one source.
Displaying in real time is the basic requirement from CAD to
CAM and CIM. Here software is built according to MIS and MRP,
basing on the product identification common throughout all the
production operations /number, input data, time/.

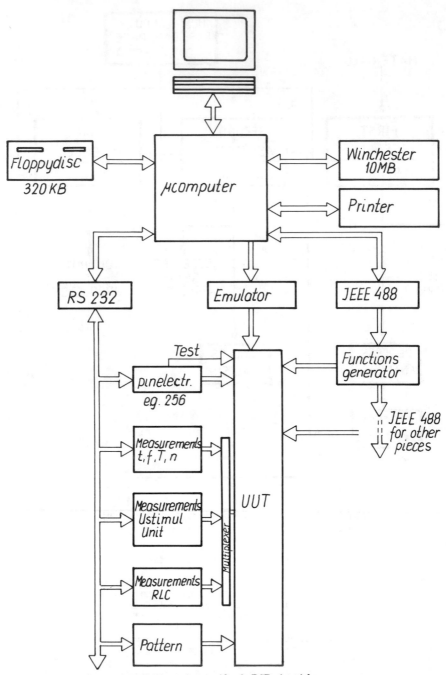

Block diagram of Multi-mode method PCB testing

AUTOMATED TEST EQUIPMENT /SYSTEMS/ ATE

Three-dimensional, colourful and animated image will become common. A notion of Konfutius "vision is worth one thousand words" takes in here as well a new expression in data collecting. It means the end of interminable printents, of histograms not always legible, and of tons of paper.

The basic matter of ATE part is the ability of coupling test systems with themselves, of these systems with superior computers, of these computers with host ones, and of test systems with technological equipments. We may put a question how far the closed loop of test and production will reach. The ways of connecting equipment are to be flexible in a maximum way, easily adaptable to cooperate with themselves, by interfaces.

A monitor is gaining a substantial role, yet multiplicity of applications creates the problem of software, is it to be kernel software, or free application software? Answer is - the latter. Test hardware is getting closer and closer to the physical limits of speed, precision, etc. Software solves these problems only partially. Hardware should exist, though, just for creating signals generators, simulators, words generators, A-C and C.A converters, etc.

A network joins components with themselves, and in other words, it is carried out by various interfaces.

Three interfaces amongst computer systems, peripherals and measurement equipment are recommended for the future: RS232C /in the USA, V.24 outside the USA/. ETHERNET and GPJB, beside IEC-625 used nowadays /IEEE 488 in the USA/.

Each of these interfaces has its own characteristic which makes one solution for optimum multiple applications. In such a way we can make the whole production process testable, starting from the product design.

And the last but not least question!

An electronic product is, above all, the PCB. The set of these boards, smaller or bigger in quantity and assortment, forms more or less complex electronic product, or other in which electronics has been applied. That is the reason why whenever speaking about the future in microelectronics, one should not omit the future of PCB testing.

Nowadays, we apply three known ways of PCB testing:
- in-circuit test,
- function test,
- emulation testprogramme.

The possibility of joining these three methods is being seeked for. It results from two trends, easy to be noticed. One aims at cheap test systems yet of high output. The second one looks for. It results from two trends, easy to be noticed. One aims at cheap test systems yet of high output. The second one looks for possibilities of integrating more and more methods and ways in one system. So a new test system has been created, called the Multimode Testing. It has, as it seems, a future before it. It is an endeavour to combine these three test methods mentioned. The figure shows the examplary block scheme of Multimode testsystem. Here as well, yet, the basis of the method is the in-circuit testing together with troublesome yet still developing bed of nails. A distinct trend is the creation of analogue in-circuit test with functional test.

As it concerns function test, one can notice a distinct growing trend. On one hand it results from the full supply of in-circuit testers on the market, yet especially from the obtained higher speeds of fault detection as well as from growing programming and diagnostics comfort. On the other hand, emulation shows clear advantages at real time testing, in comparison

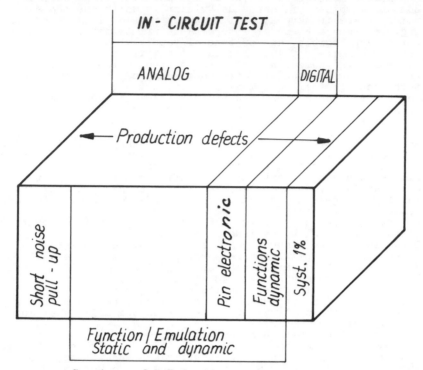

IN-CIRCUIT TEST

ANALOG

DIGITAL

← Production defects →

Short noise pull-up

Pin electronic

Functions dynamic

Syst. 1%

Function / Emulation
Static and dynamic

Spectrum of PCB testing methods

with real-time pinelectronic. There emerge the growing prognosis of emulation in the future. Yet according to the today's state of knowledge, the in-circuit method is economically optimum both for testing PCB as well as fault diagnostics of microelectronic elements on PCB.
One should remember that constant growth of the scale of IC integration causes the subsequent increase of fault of circuits applied. Several reasons cause that:
- non-testing of circuits by input inspection because the former are: too complex, too expensive, comprise too many pins,
- producer's problems, because of too short time for innovations, too short production series,
- user's problems, because of too fast conversion and design of IC's.
The solutions of circuit testability are a great help here. The circuit design makes it testable, up to obtaining selftest. Selftest solves, to great extent, the inspection of circuits itself and, later, PCB designed from them. In this way we are consequently setting from homo faber via homo sapiens, for homo ludens.

REFERENCE

Test and measurement: Factory of Future, February 1984

THE PROBLEMS OF INTEGRATED VISUALIZATION

OF THE MEASUREMENT INFORMATION

Grzegor Hiler, Roman Myszkowski

Institute of Electrical Metrology
Technical University of Wrocław
Wrocław, Poland

INTRODUCTION

We can observe a widespread increase of the number of measurement paramters in modern control - measurement systems. Furthermore, the measurements become more and more precise. This involves rapid growth of the amount of information which must be processed and transmitted to man who has a role of an operator. At the same time the demands concerning system reliability and security have grown. The effects of possible failures become more and more dangerous for man and his environment, and besides, they are more and more expensive.

This problem is found in manufacturing laboratory and medical measuring systems. The growing number of information, application of computers cause the change of manner of visualization of measuring information. Two basic trends may be distinguished in the development of the methods of visualization constructing integrated graphic images and constructing animated graphic images. It is proposed to name complex images produced by the readout devices as hypertexts. The first group of hypertexts is composed of simple geometrical figures[3] such as: lines, columns, stars, etc. Each figure is a graphical model of the measured object or its element. The change of size, colour or position of a figure represents quantitative changes of the parameter or the set of parameters of the object. A man reading such a hypertext may estimate quantitative and qualitative changes of parameters of the object.

In measurements of complex objects with numerous parameters one may consider the measurement of general features, e.g. object behaviour. In such a case the use of situational, graphical images is proposed. These images are built from animational graphic images in which elements /fragments/ of an object are pictured as well as the relations between these elements. Moreover, the image may present time relationships of the past /history of the object behaviour/ and the future /forecast of the object behaviour/. In relationship with the increasing importance of visual methods it is necessary to consider the problem of building hypertexts and their relations with the measured objects and also the criteria of hypertest estimation.

HYPERTEXT - GRAPHICAL MODEL OF THE OBJECT BEHAVIOUR

A basic function of any hypertext is presenting to man, in an optimal way, the object behaviour i.e. man studies the object behaviour by means of a hypertext. Thus, there occurs the following dependency:

object - hypertext - man

A hypertext is here an agent mediating between an object and man.

It is well-known that in order to get acquainted with the object we ought to define what state it is in. It is possible only, when a set of parameters describing the object is distinguished and their measurement is made. As a result of this measurement an information about the present state of the object is obtained: whereas making the measurements for a longer time enables us to determine the object behaviour.

This information ought to be transmitted to man in such a form that it could be received and understood easily. In classical methods or in the situation when there are few parameters, an individual device, visualizing single parameters, is used. One reads indications from each device separately and after receiving all accessible information may process this information and then one can imagine in what state the given object is in. Then we may say that we have generated a subjective image of the object and in the course of time we may say that we have a subjective model of the object behaviour and knowledge of the tasks demanded from the object makes it possible to undertake activities by man /e.g. controlling a power station/.

With a constantly growing amount of information delivered to man and the use of individual devices visualizing information it is more and more difficult to generate a correct subjective model of the object behaviour. Particularly in cases of failures. Generating an integrated image in which an object behaviour is visualized in a complex way enables us to make a proper estimation of the real object behaviour. In such terms a hypertext becomes a graphical model of the object behaviour. The hypertext not only transmits the information but its graphical form may simplify and hasten the generation of the subjective model of the object behaviour /Fig.1./

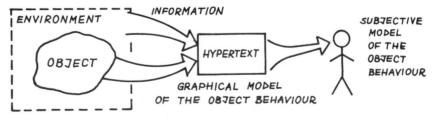

Fig.1. A scheme of an object - man - hypertext system

56

Obviously, a hypertext should be the only source of information but is the most essential means of delivering information. This is why it is very important to determine what are
the properties of a definite hypertext realization assuring
optimal work with the object.

The described system: object - hypertext - man may be
divided in two examples: the object - hypertext system and the
hypertext - man system. Additional sources of information
about the object have little importance. It is assumed that
the subsystems are independent[4] and various factors have an
influence on them. The object - hypertext is influenced by
the informational and metrological factors, whereas the man
- hypertext system is influenced by the ergonomical factors
and the properties of man-hypertext communication. These
factors make it possible to determine the criteria of hypertext estimation.

HYPERTEXT ESTIMATION CRITERIA

The theory of information and the queuing theory can be
applied for the description of estimation of hypertexts. The
basis of the informational estimation of the hypertexts is
checking the condition of the complete destination of object
state Y by the hypertext X. It is obtained,when the conditional
enthropy H Y/X of the hypertext in relation to the object
tends to be zero[1].

$$H (Y/X) = - \sum_i p (x_i) \sum_j p (y_j/x_i) \cdot \log_2 (y_j/x_i) \to 0 \qquad (1)$$

where $p(x_i)$ - probability of i-th hypertext x
$p(y_j/x_i)$- conditional probability of i-th object
state by i-th hypertext x

In the metrological estimation it must include additional
conditions carried by inaccuracy of measuring devices as
systematic errors and probabilistic errors. The probability
of an appearancy of systematic errors is similar for digital
and analog devices. Conditional probability p y_j/x_j is not
a number, it is a section between the values e, where e is
the relative measure of quantum error. It increses the conditional enthropy about ΔH_s. Similarly, it can be demonstrate
that the probabilistic errors of Gaussian type cause the
scattering K of information and lead to the increase of enthropy
by about ΔH_p. The qualification valve of the enthropy with
systematic errors and probabilistic errors is difficult,
because up to now, there is not a strong methodological approach to information J in measuring signals.

These studies are usually concentrated on one parameter
which assumes n-states and the conditions are considered by
the assumption that the measurements of subsequent parameters
are mutually independent. In such conditions one may determine
the enthropy matrix and determine the value called the hypertext usefulness G[2,4].

$$G = \min_{i = 1,\ldots,N} \left(1 - \frac{\Delta H_i}{H_i} \right) \qquad (2)$$

where: ΔH_i – i-th matrix element of the errors enthropy
$\quad\quad\quad H_i$ – i-th matrix element of the hypertext for i-th
$\quad\quad\quad\quad\quad$ object state.

An enthropy cannot be an only criterion of estimating the hypertext because it concentrates on one type of hypertexts and measurement errors of the parameters.

The man-hypertext communication will function correctly, when man reads the hypertext correctly. The readout device is a transmitter of the hypertexts and man is an observer. The observer ought to read all hypertexts, but a part of hypertexts and enthropy get lost. The readout device shows the hypertext sequence with the intensity L, because this sequence is of Poisson type. The observer potentials are characterized by the parameter t_r mean readout time.

Let S denote the coefficient of loss of the hypertext /Fig.2./ and that is equal:

$$S = \frac{r^2}{(1+r)^2} \quad\quad\quad\quad (3)$$

for readout without interruption, when $r = 1.t_r$ is the coefficient of load.

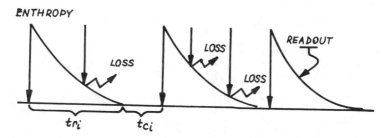

Fig.2. A scheme of readouting without interruption

Let M denote the coefficient of halt-time:

$$M = \frac{t_c}{t_c + t_r} \qu\quad\quad\quad (4)$$

where: t_c – waiting time for changing the hypertext.

Calculation of parameters of the man – hypertext communication must be preceded by ergonomical experiments.

CONCLUSIONS

An analysis of the object – hypertext – man system enabled us to determine an influence of informatic and metrological factors on hypertext. By including the probability distributions of systematic and probabilistic errors we can determine the hypertext enthropy loss. On the basis of the knowledge of the hypertext usefulness G it is possible to compare the hypertext of various types with no regard to graphical form.

This criterion takes into consideration both the complexity
of the object and the number of parameters and states and the
influence of error occurence.

The coefficient of loss of hypertext enthropy S is the
measure of the advantage of the man - hypertext communication.
With an increase of the values of the coefficient r the value
of the loss enthropy is growing but at the same time the
coefficient of halt - time is decreasing i.e. man spends less
and less time on waiting for the hypertext. The demands of the
least enthropy losses and the shortest waiting time are contradictory and ought to be reflected in the proper work organization
of a man working with the object.

REFERENCES

1. Hiler G.:"Wyhrane zagadnienia ergonomiczne, metrologiczne
   i konstrukcyjne dotyczace wskaznikow odczytowych", IME
   Wroclaw /1976/

2. Hiler G., Myszkowski R., Podlejski K.: "Hypertexts - a
   problem of visualization of measuring information",
   IMEKO Conference, Graz /1984/

3. Hiler G., Myszkowski R.: "Integrated visualization of
   measuring information", IMEKO Conference, Jena /1986/

4. Myszkowski R.: "Wybrane zagadnienia ergonomiczne i metrologiczne oceny hipertekstów" Archiwum Automatyki i Tele-
   mechaniki. 1, /1984/.

WHY CONSIDER TRANSITION MEASUREMENT OF

AMPLITUDE PROBABILITY DENSITY DISTRIBUTION(APDD)

Shigeru Takayama, Satoshi Nishikawa,
Etsushi Sugiura and Komyo Kariya

Department of Electricity, Faculty of Science and Engineering,
Ritsumeikan University
56-1, Tojiin-kita, Kita-ku, Kyoto 603 Japan

Already, some examples of "Time Transition of Amplitude Probability Density Distribution(TT of APDD) of Detected Signal and State or Situation Measurement" had reported on 10th IMEKO Congress and 4th TC-7 IMEKO Symposium/1/. On these papers, the transition state of amplitude probability density distribution of detected signal by a sensor settled in the measurement object had reported at relation of the parameters in measurement object and the some prediction problems of the amplitude probability density distribution(APDD) on future time had treated. In this time, we show about the effectiveness of this measurement method for the state and situation of measurement object.

INTRODUCTION

Generally, the consideration that a time series signal (Physical data) detected by a sensor settled in a measurement object is whether stationaly or non-stationaly is treated as follows. In the case of stationaly, it is possible to formulate mathematically. Becourse the knowledge about the object which generates the signal is given. And in the case of non-stationaly, it is not possible to formulate mathematically. Becourse the knowledge about the object which generates the signal is not given. Namely, there is the infinite probability that the uncountable events happen in future. Normally, almost statistical analysis have a big hypothesis that the time series signal accords to stationary process. And moreover the time series signal is by carrying into stationaly process even if the time series is non-stationaly. There is a big effort in here. And the procss becames ergodic process in the special case. Authors have been setting a strong standpoint that the time series signal obtained from measurement object is non-stationaly, and on the reason of that, authors have been studying to grasp to the change of state and situation of measurement object by using TT of APDD of detected signal. Now authors are thinking that this measurement method becomes to possible to apply to a Quality Control Measurement. Because it is possible to monitor the quality as a measurement object by using TT of APDD of the signal detected by a sensor settled on the quality point and by using the quantity on a production process of manufacture goods.

In this paper, in the first, the conception of measurement method of TT of APDD is introduced by using some our laboratory's models. In the second, the possibility of this measurement method on quality control is suggested.

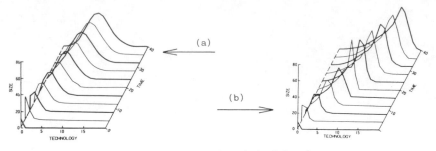

Fig.3 Transition of technical level

and (b), when the technical imitation is possible, it is obvious that the growing and declining process of new technical level is too fast. Here, each transition of distributions is formed by the limitation law for new technical innovation. Fig.4 shows a TT of APDD made by a model equipment/3/. This equipment is constructed by ten lanes which are possible to change the slope. Many small balls roll on the lanes, and the speed of balls are measured together in same time on the bottom of each slope. So that, when each lanes have up-down motion individually, this equipment is possible to make up the random flow situation of material having many kinds of speed. Namely, it is possible to consider this equipment as a measurement object that the same sort of parameter takes many states. Fig.4 is a measurement result of TT of APDD that the amplitude corresponds to ball speed when the speed of up-down motion of each lanes was set to same speed and the motion was done independently. In this time, the relation of starting position of each lane was given under a condition that a distribution of APDD becomes to a Normal distribution. And over then, some lanes (two lanes) were moved with special lower speed. By observation of this TT of APDD, it is evident that a peak (peculiar peak) goes to shift according to the time change of a slope. Namely, the appearance of peak and the shift show the change of the situation of measurement parametar in the measurement object. So that, it becomes possible to know the structure of transition matrix by realization of measurement of TT of APDD.

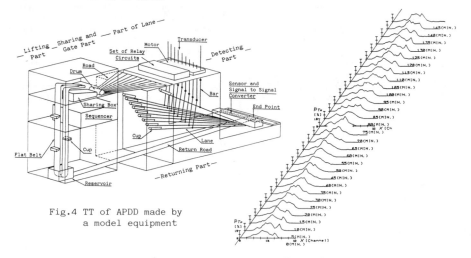

Fig.4 TT of APDD made by
a model equipment

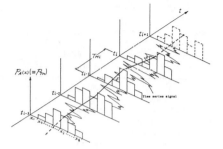

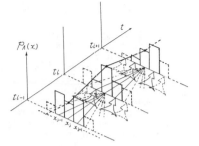

Fig.1 Conception of measurement
    method of TT of APDD

Fig.2 Relation of each APDD

FUNDAMENTAL CONCEPTION EXAMPLE OF MEASUREMENT OF TT OF APDD

Fig.1 shows a simple explanation of fundamental conception of measurement method of TT of APDD. By setting a measurement time, an APDD of time series signal detected by a sensor set in a measurement object is measured. And according to advance the measurement time T_{Mi} (i $=$---,-2,-1,0,+1,+2,---), a series of TT of APDD is formed. In this figure, x is the amplitude value devided to k channels and $P_A(x)$ is the probability value taking each x. Here, the transition of each $P_A(x_j)$ (j =1,2,---,k) can also be treated just as the discrete stochastic process. In this process, the time series signal is shown as $P_A(x_j)(t;\omega)$.
Then, it must pay attension that the transition of each $P_A(t_i, x_j)$ does not occur individually but to depend on every previous $P_A(t_{i-1}, x_j)$ (j=1,2,---,k) as shown in Fig.2. The relation in Fig.2 are given as a following matrix indication. By setting the reletion of a matrix like this, an APDD is indecated as a matrix $P_A(t)$, then the change of state or situation of a measurement object from time t_i to time t_{i+1} is indicated as a transition matrix $P_T(T_{Mi+1})$. In equation(1), the element $P_{T,ij}$ indicates the degree

$$P_A(t_{i+1})=P_T(T_{Mi+1})\cdot P_A(t_i) \quad\text{————————}(1)$$

$$\begin{pmatrix} P_A(x_1,t_{i+1}) \\ P_A(x_2,t_{i+1}) \\ \vdots \\ P_A(x_i,t_{i+1}) \\ \vdots \\ P_A(x_k,t_{i+1}) \end{pmatrix} = \begin{pmatrix} P_{T,11} & P_{T,12}\cdots P_{T,1j}\cdots P_{T,1k} \\ P_{T,21} & \\ \vdots & \\ P_{T,i1} \text{------} P_{T,ij} \text{---} \\ \vdots & \\ P_{T,k1} \text{--------} P_{T,kk} \end{pmatrix} \cdot \begin{pmatrix} P_A(x_1,t_i) \\ P_A(x_2,t_i) \\ \vdots \\ P_A(x_i,t_i) \\ \vdots \\ P_A(x_k,t_i) \end{pmatrix}$$

of influence from the j th channel of an APDD to the i th channel of next APDD. It is shown in the measurement object and it includes the information of changes of the state and situation of measurement object. Therefore, the state and situation of a measurement object and their change are grasped by the observation of some changing parameters (for example, mean value, variance, moment, distribution form, position where peaks appear,---etc.) appear in the TT of APDD. Two examples of consideration about TT of APDD at any rate the measurment sethod are shown in following figure. Fig.3 shows how the transition of technical level is changing with progress of time/2/. The vertical axis "SIZE" indicates the "scale of companies". The difference between (a) and (b) is caused by whether the imitation of technology is possible or not. Each company has grown up with the rate of growth according to the present technical level itself. Moreover, each company studies toward higher technical level and try to occupy a profitable position. But the technical innovation occures in a statistical manner, not according to the effort of a company. And most of company grows up thier technical level by the imitation of technical innovation by other company. Comparing Fig.3 (a)

APPLICATION OF TT OF APDD TO DISCHARGE PHENOMENON MEASUREMENT

In Fig.5, one example where the discharge phenomenon is set as a measurement object is shown. The aim of this study is to get the information of metal material on discharging electrodes by using TT of APDD. In Fig.5(a), by generating a discharge at the gap of one pair of sample metal electrodes in the discharge power source unit, the unique discharge current as a metal is induced. That discharge current i_s is detected as a Arc discharge voltage v_s generated on detective resistance R_s. And when the discharge is repeated in same interval, the continuous periodic signal is observed. But the peridic signal's waveform(Arc discharge voltage waveform) changes according to the state of discharge conditions(air, atmosphere, electrode temperature and so on). Then if the periodic signal is sampled by enough short time to one periodic time, a APDD $D(a)$ is got in each one constant measurement time. For analyzing this APDD, the characteristic function $\bar{\Phi}(\lambda)$ is calculated with the following equation;

$$\Phi(\lambda) = \int e^{i\lambda a} D(a)\, da \quad\quad\quad (2)$$

Fig.5(b) shows the correspondence between APDD $D(a)$ and that characteristic function $\bar{\Phi}(\lambda)$. The sum of all harmonic waves indicated by characteristic function constructs APDD $D(a)$. By regarding the characteristic function obtained from APDD $D(a)$ of the discharge periodic signal, it is confirmed that the APDD $D(a)$ is occupied mainly by smooth fluctuating component of probability density distribution which the Arc discharge voltage signal appears. And that thing shows that the element influenced strongly to discharge current is not so much. Because, if many kinds of element relates to discharge current, the form of APDD becomes more complex and higher component of characteristic function become more larger. From this consideration, the correlation coefficient r between the low component area $S(\lambda_f)$ of characteristic function $\bar{\Phi}(\lambda)$ and the unique property of metal material(melting point, boiling point, atomic number, work function, thermal conductivity and so on) is calculated with equation (3). The result of that correlation is shown in Table.1. The coefficient of Debye Temperature is well relatively. Debye temperature relates to the fluctuating frequency of atomic lattice that shows the strength of material. So the Debye Temperature is complex, we hope to consider more detail discussion.

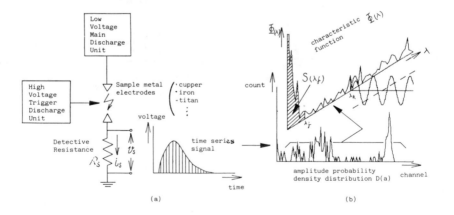

(a)   (b)

Fig.5 Detection of Characteristic of Discharge Current appeared on APDD

Table 1 The result of correlation coefficient

Property of Material	Correlation coefficient	Property of Material	Correlation coefficient
Atomic number	0.509	Work function	0.375
Atomic weight	0.554	Melting point	0.217
Specific heat	0.375	Boiling point	0.106
Specific gravity	0.444	Thermal conductivity	0.059
Debye temperature	0.693	Ionization potential	0.056

$$ \gamma = \left\{ \sum_{i=1}^{n} (\chi_i - \bar{\chi})(S_i(\lambda_j) - \overline{S(\lambda_j)}) \right\} \Big/ \left\{ \sum_{i=1}^{n} (\chi_i - \bar{\chi})^2 \sum_{i=1}^{n} (S_i(\lambda_j) - \overline{S(\lambda_j)})^2 \right\}^{\frac{1}{2}} \quad (3) $$

γ : coefficient of sample correlation
χ : value of unique property of metals
i : sort of meterials (iron, cupper, ...)
$S(\lambda_j)$: square of characteristic function from λ_1 to λ_j

$$ S(\lambda_j) = \sum_{m=1}^{j} \phi(\lambda_m) \quad\quad\quad\quad\quad (4) $$

APPLICATION OF TT OF APDD TO QUALITY CONTROL MEASUREMENT

To increase the quality of manufacture goods is very important to keep the stabilize economy of the country and to develop the trade among each country. The measurement assists very much to grade up of quality of manufacture goods, then the measurement methods fitable to the quality control must be study. In nowadays, as almost production process of manufacture goods are constructed automatically so the setting of quality points on the flow of the goods and the measurement methods in the points become to important factors to amend the drop down of quality by an expected accidents. Fig.6 shows an example of the system monitoring quality (length, width, roughness, form and others of metal boards) using this measurement method of TT of APDD in the automated mass-production process. After passed the process K, the elements indicating quality of metal board, they are length, width, roughness, form etc., are measured for one measurement time T_M under a sensor gate K set between the process K and the Process K+1. The APDD of sensor's signal at t_i is formed for one measurement time from at t_i to at t_{i+1} according to each element (in Fig.6, the length is selected as monitoring quality). In this TT of APDD, the bias of APDD like distribution at t_{i+1} shown in right side figure is predicted from the transition state of APDDs at t_{i-1} and t_i. Here, the transition matrix $P_T(T_M)$ is used to this prediction. By this prediction of APDD at t_{i+1}, in practice the estimation of transition matrix $P_T(T_M)$ by the previous distributions, the control to the previous processes 1,2,---,K is carried out and the bias of APDD is corrected. When the measured APDD becomes coincidence with standard distribution (Ds) and is supperposed, the transition matrix $P_T(T_M)$ will becomes to a unit matrix. But, when the measured APDD has a bias from Ds, it is considered that the factor based on deviation of the 1-3rd order moment, the differnce area, the form and others of APDD comes appear in the comporents of transition matrix $P_T(T_M)$. Authors are considering this correspondence between the parameters of distribution and the comporents of transition matrix $P_T(T_M)$ the estimation method of trasition matix $P_T(T_M)$ and the control method of the production system.

65

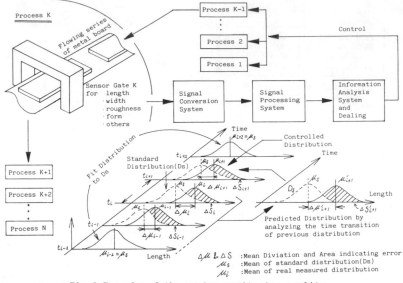

Fig.6 Example of the system monitoring quality

CONCLUSION

In this paper, the conception of the measurement method of TT of APDD, three examples of this method and the application method to quality control measurement were reported. This measurement method is an effective measurement method to grasp the state and situation of measurement object and also it becomes to a good method to estimate the state and situation of measurement object in near future on the natural and social phenomenon. As mentioned above, especially, this method has strong adaptability in the quality control measurement. In operation of the production system, it is difficult to measure analitically the dynamic state and situation. Because,
- the production system is almost operated in the non-stationary condition,
- the relation of function among each system are much complex,
- the phenomenon of dynamic state and situation are impossible to indicate by simple physical quantity,
- comparatively the decision of system character is depend on the human sense and experience.

The acquisition of operation situation of production system and the systematization of quality measurement are important subject. A method to solve the subject is to realize quality control by an excellent measurement method. Also from this sense, the transition measurement method of TT of APDD will become important key.

REFERENCES

1. K.Kariya, T.Nishiki and S.Takayama:"TT of APDD of Detected Signal and Situation Measurement", IMEKO Xth World Congress, Praque 1985
2. K.Nishiyama:"The Flicker and Developement of The Social System", Journal "ELAN" published from BUNKA HOUSOU BRAIN, No.49 pp.28-31 Japan 1983
3. T.Nishiki and K.Kariya:"Construction of Fundamental Experimental Device to consider -TT of APDD-", Memoirs of Research Institute of Science and Engineering, Ritsumeikan University, No.43 1984

A NEW IMAGE PROCESSING TECHNIQUE FOR THE HIGH-SPEED MEASUREMENT OF PARTICLE-SIZE DISTRIBUTIONS

Koichiro DEGUCHI

Department of Information Engineering
Yamagata University
Yonezawa 992 Japan

INTRODUCTION

The automatic measurement of particle size distributions from the object image has been developed in many fields such as microscopic analysis in metallography, crystallography, biology, etc., and reported in the literature[1,2]. Even for an image of many small particles distributed on a plane, it may be not difficult to determine automatically the size of each particles by measuring the area of each particle regions. However, since we have usually a large number of particles on an image, this conventional technique requires considerable processing time. When real-time processing is essential, a new high-speed processing technique must be developed. In steel industries, for example, we must measure the size distribution of raw materials moving on high-speed conveyers in real-time.

The technique proposed in this paper is one based on sampling. We use a number of parallel lines to obtain a distribution of "run-length". By employing a TV camera for image pick-up, we can use its scanning lines as the sampling line. When a scanning line is put across a particle region, a run-length is obtained as the length crossing over the particle. After the scanning of one image frame, we obtain a distribution of run-length as a histogram of its occurrence counts for the image frame.

Next, this run-length distribution is converted to the particle size distribution. It will be shown that the conversion is given by solving a linear equation.

In the followings, the basic principle of the technique is described, first. Then the problems for taking place practical applications of the proposed method are discussed. Finally, some experimental results are shown.

BASIC PRINCIPLES

The schematics of the measurement of particle size distribution proposed in this paper is shown in **Fig. 1**.

We assume the shape of object particle image to be measured its size is circular. When a sampling line runs across the circle, we get a length crossing over it, which is called run-length. If the sampling line crosses randomly over the circle of radius R (**Fig.2**), the probability of a half of the run-length lying between r and $r+dr$ is given as

$$\frac{\mid 2dx \mid}{2R} = \frac{\mid d(\sqrt{R^2-r^2}) \mid}{R}$$

$$= \frac{r\ dr}{R\sqrt{R^2-r^2}} \tag{1}$$

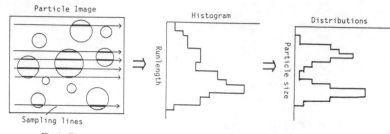

Fig.1 The schematics of the high-speed measurement procedures of particle size distribution using run-length histogram of the particle image

Let p_{ij} denote the probability with which the run-length lies between integers $i-1$ and i for a circle having a diameter of integer j. By integrating (1), this p_{ij} can be given as

$$p_{ij} = \begin{cases} \sqrt{1-(\dfrac{i-1}{j})^2} - \sqrt{1-(\dfrac{i}{j})^2} & ; i \leq j \\ 0 & ; i > j \end{cases} \qquad (2)$$

To obtain the distribution of the run-length from a given image frame of parti-cles, a number of sampling lines must be used. By employing TV camera for an image input device, its scanning raster lines can be used as the sampling lines at the same time of particle image pick-up. The run-lengths of the object particle images are measured whenever the raster line crosses over them as shown in Fig.1. Then we obtain the distribution as a histogram of the run-length $\{ h_i \}$, where h_i is the number of occurrence count of run-length i. If there exist in the given image frame N circles with diameter of j only, for example, the h_i results in

$$h_i = N j p_{ij} \qquad (3)$$

Fig. 3 shows the theoretical forms of this h_i with respect to various j's.

Now, let a vector $\boldsymbol{h}_n$ express the run-length histogram obtained from an image having various size particles, i.e.,

$$\boldsymbol{h}_n = (h_1, h_2, \cdots h_i, \cdots h_n) \qquad (4)$$

where n is the possible maximum run-length for the image frame. It must be noted that each h_i is the summation of the contributions of various size particles. Then, let a vector $\boldsymbol{r}_n$ express the number of particles of each size to be measured, i.e.,

$$\boldsymbol{r}_n = (r_1, r_2, \cdots r_i, \cdots r_n) \qquad (5)$$

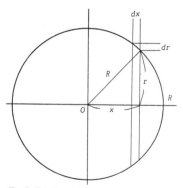

Fig.2 Relation between location of sam-pling line crossing a circle of radius R and the length running over it

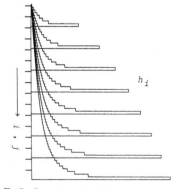

Fig.3 The theoretical forms of run-length histogram h_i for various par-ticle size j's

where r_j is the number of the particle having diameter j. Using these vectors and the matrices

$$P_n = (p_{ij}) \quad \text{and} \quad K_n = \begin{pmatrix} 1 & & & \\ & 2 & & 0 \\ & & 3 & \\ & 0 & & \ddots \\ & & & & n \end{pmatrix} \tag{6}$$

(3) can be expressed as

$$r_n \, K_n \, P_n = h_n \tag{7}$$

Therefore, the particle size distribution can be obtained by

$$r_n = h_n \, K_n^{-1} \, P_n^{-1} \tag{8}$$

From (1) the matrix P_n has the form

$$P_n = \begin{pmatrix} p_{11} & & & & \\ p_{12} & p_{22} & & 0 & \\ p_{13} & p_{23} & p_{33} & & \\ \vdots & & & \ddots & \\ p_{1n} & \cdots & & & p_{nn} \end{pmatrix} \tag{9}$$

so that, P_n^{-1} of (8) is given as

$$P_n^{-1} = \begin{pmatrix} & & & \vline & \\ & P_{n-1}^{-1} & & \vline & 0 \\ \rule{2cm}{0.4pt} & & \rule{0.6cm}{0.4pt} + \rule{0.6cm}{0.4pt} \\ -p_{nn}^{-1} & p_n \, P_n^{-1} & \vline & p_{nn}^{-1} \end{pmatrix} \tag{10}$$

where

$$p_n = (p_{1n}, p_{2n}, \cdots p_{in}, \cdots p_{nn}) \tag{11}$$

This means that P_n^{-1} can be obtained by recursively, and once P_m^{-1} has been obtained for a large m, P_n^{-1}'s $(n < m)$ were obtained simultaneously as the parts of it.

MEASUREMENT SYSTEM OF PARTICLE-SIZE DISTRIBUTION

In this section building-up the measurement system of particle-size distribution based on the principle described in the previous section will be discussed.

As shown in Fig. 1 the sampling of the run-length from a object particles can be took place by thresholding the TV camera scanning output of the particle image frame into black-and-white binary values and counting the duration time of the output signal indicating that the scanning line crossing a particle image. Then the histogram is made up by counting occurrences of each run-lengths. Therefore, the run-length histogram can be obtained with on-line and real-time measurement. The calculation (8) of the conversion from the obtained run-length histogram to the distribution of particle-size can be also took place with real-time processing using already obtained P_n^{-1}, even with a microprocessor.

The problems which must be noted and considered for real application may be followings ;

(i) the run-length histogram has statistical error from the ideal expected form when the sampling image frame has not sufficient number of particles,

(ii) the form of the object particles is not necessarily circular, and,

(iii) the images of some object particles are overlapped in the image frame.

These problems will be examined, respectively, in the followings.

On the Error of the Run-length Histogram

A sampled histogram of run-length will meet its theoretical form of (3) only when there are sufficient number of particles distributed randomly in the given image frame. If it does not hold, the obtained histogram would have systematical error because a set of parallel sampling lines is employed. To reduce this error, two methods may be utilized.

The one method is a non-parametric estimation technique using so-called Parzen kernel, in which the sampled histogram is smoothed by moving average operation with some weight function. This method is well-known as a unbiased estimation of distribution when only a little number of samples can be available.

The another is to add sampling by further more independent sampling lines than the TV raster-scanning lines. When locational distribution of the particles in the given image frame is random, we can utilize the count of run-length along sets of sample lines of different directions, such as diagonal and/or vertical, in addition to the horizontal scanning lines (Fig. 4). This can be took place with a simple optional hardware, and results in increasing the number of independent trial of sampling for the same image, so that the error of sampled histogram can be reduced without additional processing time.

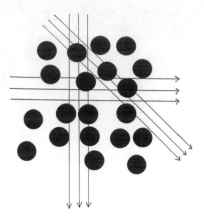

Fig.4 Addition of sets of different directional sampling lines to increase the number of independent trials of sampling the run-lengths

On the Form of Object Particles

We have assumed that the object particles are circular. However, it does not means that this method can be available only for circular particle images. But, in fact, for non-circular particles the definition of their size is not distinct. It may shown that, applying this proposed method to a non-circular particle image, we obtain the distribution of diameters of circles which have almost equivalent areas to each object particles in the image, if the object particles have not regular allocations nor directions. Fig. 5 shows examples of simulation results for non-circular particles. When the form of each object particle are square, triangular, hexagonal and elliptic as shown in Fig.5, the diameters of the circles over-drawn on each figures are obtained as their average particle sizes by applying proposed method, respectively.

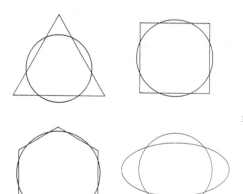

Fig.5 The measurement results for non-circular particles. For the square, triangular, hexagonal and elliptic particles, the diameters of the circles over-drawn on each figures are obtained as their average particle sizes, respectively.

Effects of Object Image Overlapping

When images of some particles have mutually overlapping area, longer false run-lengths than each their individual diameters are sampled, as shown in **Fig. 6**. However, the counts of their run-lengths are much fewer than those when particles having such longer sizes actually exist in the image. If we employ multi-directional sampling lines described before, their effects can be reduced much further. Therefore, except that almost all particles in the image frame are overlapping each other, the effects can be reduced within small errors and the distribution of principal particle sizes can be obtained in spite of some object image overlappings.

EXPERIMENTAL RESULTS

Figs. 7, 8 and **9** show the experimental results of the proposed method. Original particle images used in preliminary experiments and the final results of particle size distributions obtained are shown. In fig. 7 the sampled run-length histogram is also shown. Fig. 9 is the result of an application, where the particle size distribution of a gravel stone image was measured. These results shows that the distribution of principal sizes can be obtained in spite of object particle image overlappings.

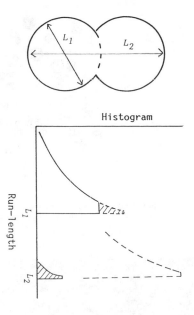

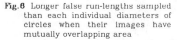

Fig.6 Longer false run-lengths sampled than each individual diameters of circles when their images have mutually overlapping area

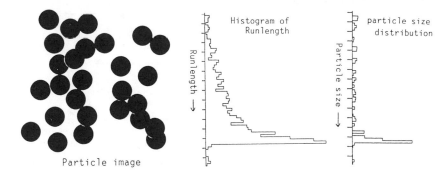

Fig.7 Experimental result of measurement of particle size distribution when the image has almost same size particles

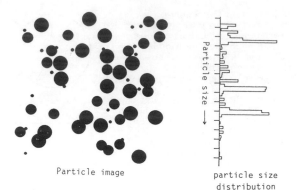

Particle image   particle size
                              distribution

Fig.8 Experimental result of measurement of particle size distribution for the image in which there has three sizes of particles

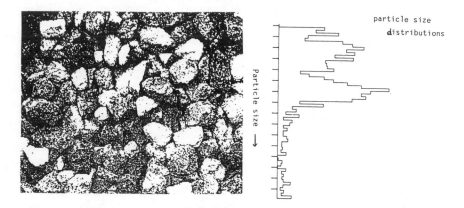

Fig.9 An image of gravel stones and its particle size distribution measured by the proposed method

CONCLUSIONS

A new image processing technique for a high-speed measurement of particle-size distributions is described. It is shown that this proposed method does not require complex hardware implementation and the calculations can be done with microprocessors.

Theoretically, the conversion from the run-length histogram to the particle size distribution employed in this method becomes unstable if the size resolution is selected too small. However, it works for many applications without matters, practically. But stable and high-speed conversion algorithm is thought to be the next problem.

REFERENCES

1. E.Underwood, *Quantitative Stereology*, Adison-Wesley (1970)
2. R.DeHoff & H.Rhines, *Quantitative Microscopy*, MacGraw-Hill (1968)

ERROR CHARACTERISTICS OF
INTELLIGENT MEASURING INSTRUMENTS

E.I.Tsvetkov

All-Union Research Institute of Electrical
Measuring Instruments
195267 Leningrad, USSR

INTRODUCTION

In an effort to determine the intelligence level of a processor measuring instruments /PMI/ the following statement should be taken as the basic one: the intelligence level of PMI is dependent on the capability of using a priori /knowledge and data banks/ and current data for the synthesis and the performance of measuring algorithms characterized by the best metrological quality of measurement results obtained for the specified sets of conditions and constraints. To ensure the adequate functionability PMI is complete with processor, memory and systematic software. Intelligence of PMI consists in synthesis /selection/ of rational measuring algorithms as applied to particular cases. In other words the main control character at the above approach is the possibility of usage a priori and current data for the selection /synthesis/ of measuring algorithms as well as for the evaluation and correction of parameters and characteristics to be realized during the performance of conversion algorithms.

ERROR CHARACTERISTICS

The following a priori and current data should be considered:

$\{\mathcal{P}_{m_i}\}$ — measuring functions, types of measurement problems to be solved by means of a specially designed measuring instrument;

$\{\alpha_i, f_\rho(t, z)\}$ — characteristics of input signals /input actions/

$\{\beta_i, \psi_\rho(t, z)\}$ — characteristics of operating conditions;

$\{\gamma_i, \psi_f(t, z)\}$ — characteristics of the requirements and constraints.

The following software is taken into account:

$\{A_{r i}\}$ — algorithms to be realized during measurement performance /available in software/ ;

$\{A_{ai}\}$ — auxiliary algorithmic software necessary for
selection /synthesis/ of a rational measuring
algorithm from the $\{A_{mi}\}$ set.

Based on the above information the following five levels
of PMI intelligence are introduced:

1. PMI makes it possible to perform the specified measuring
function in accordance with wired-in program. Considered here
is a unifunctional (φ_m) measuring instrument equippped with
a processor providing a means only for realization of predetermined algorithm (A_m) of a random complexity.

2. PMI enables the performing of the specified set of
measuring functions $(\{\varphi_{mi}\})$ each of them being performed
according to an appropriate wired-in program. In this case
we have to deal with a multifunctional measuring instrument
which allows the program to be chosen in compliance with an
identified function. Thus, memory of PMI should contain software necessary for the realization of any alsogithm from the
$\{A_{mi}\}$ set and auxiliary software $\{A_{ai}\}$ for the selection
of the A_{mi} algorithm in compliance with an identified
function φ_{mi}.

The procedure $\mathfrak{J}$ of the A_{mi} algorithm selection is
expressed by: $\mathfrak{J}: \varphi_{mi} \to A_{mi}$.

3. PMI makes it possible to perform the specified set of
measuring functions $\{\varphi_{mi}\}$. For each of these functions the
corresponding algorithm, numerical values of characteristics
and parameters can be selected using the information on properties of the object of measurements, conditions of measurements, requirements and constraints as the base. In this case
considered is a multifunctional measuring instrument which
provides for each function performance the set of algorithms
with varied characteristics and parameters selected in
compliance with the established rules $\{A_{a \rho}\}$ using data
files $\{\alpha_{i\rho}, f_{i\rho}(t,\tau)\}$, $\{\beta_{i\rho}, \varphi_{i\rho}(t,\tau)\}$ and $\{\gamma_{i\rho}, \psi_{i\rho}(t,\tau)\}$.
Measuring instruments of this type are noniterative, adaptive
The memory of these instruments stores data necessary for the
realization of all $\{A_{mi}\}$ algorithms, data files $\{\alpha_{i\rho}, f_{i\rho}(t,\tau)\}$,
$\{\beta_{i\rho}, \varphi_{i\rho}(t,\tau)\}, \kappa_{i\rho}, \psi_{i\rho}(t,\tau)$ and auxiliary algorithmic software
$A_{a\rho}$. As the $A_{mi j}$ algorithm is selected two pro-
cedures should be performed:

a/ $\mathfrak{J}_1$ — to specify the set of permissible measuring
algorithms in conformity with the identified
function $\varphi_{mi j}$

b/ $\mathfrak{J}_2$ — to select a particular algorithm from the set
accounting for the properties of the object of
measurements, conditions of measurements,
requirements and constraints, namely: $\mathfrak{J}_1: \varphi_{mi} \to \{A_{mi\tau}\}^e$;
$\mathfrak{J}_2: \{A_{mi\tau}\}^e / \{\alpha_{i\rho}, f_{i\rho}(t,\tau)\}, \{\beta_{i\rho}, \varphi_{i\rho}(t,\tau)\}, \{\gamma_{i\rho}, \psi_{i\rho}(t,\tau)\} \to A_{mij}$

4. PMI makes it possible to perform the specified set of
the measuring functions $\{\varphi_{mi}\}$. For each of these functions
one can select the corresponding iterative measuring algorithm
whose characteristics and parameters vary as a single measurement result is being obtained. Measuring instruments of this

type are iterative, adaptive. The memory of the instruments stores data structurally similar to those mentioned in the foregoing point with the difference that measuring and auxiliary algorithmic software $\{A_{mi}\}$ and $\{A_{a\rho}\}$ respectively involve procedures providing the realization of iterative measurement results with account of the intermediate results of measurements   /m-number of iterations/.

When selecting the A_{mij} algorithm three procedures should be performed:

a/ $\mathfrak{I}_1$, $\mathfrak{I}_2$ are similar to those described in Point 3 with the difference that $\mathfrak{I}_2$ is in the selection of the A_{mij1} algorithm to be realized at the first iteration;

b/ $\mathfrak{I}_3$ involves the algorithm evaluation at the succeeding steps m-1 of the iterative measuring process with regard to intermediate results of measurements. Thus, we have: $\mathfrak{I}_1 : \varphi_{mi} \rightarrow \{A_{mi\tau}\}^e_1$

$$\mathfrak{I}_2 : \{A_{mi\tau}\}^e_1 / \{\alpha_{i\rho}, f_{i\rho}(t,\tau)\}, \{\beta_{i\rho}, \psi_{i\rho}(t,\tau)\}, \{\gamma_{i\rho}, \psi_{i\rho}(t,\tau)\} \rightarrow A_{mij}$$

$$\mathfrak{I}_3 : \{A_{mi\tau}\}^e / \{\alpha_{ije1}, f_{ije-1}(t,\tau)\}, \{\beta_{ije-1}, \varphi_{ije-1}(t,\tau)\},$$

$$\{\gamma_{ije-1}, \psi_{ije-1}(t,\tau)\}, \lambda^*_{ije-1} \rightarrow A_{mije}, \quad e = \overline{1,m}$$

5. PMI of this intelligence level would operate in much the same manner as the previous one, while allowing not only for iteration to be used but also algorithmic software to be automatically developed, thanks to self-education. Consequently, we have to deal with PMI characterized by an open set of $\{A_{a\rho}\}, \{A_{mi}\}$ algorithms. Widening the $\{A_{mi}\}, \{A_{a\rho}\}$ sets is possible by the use of the developed algorithmic software $A_{a\rho}$ which enables us to realize the data obtained during the instrument operation with the aim of improving the metrological quality of measurement results. Hence, apart from the selection and realization of a corresponding measuring algorithm, A_{mij} an iterative, adaptive PMI capable of self-eduation enables the widening of the set of measuring and auxiliary algorithms with the help of $A_{a\rho}$ and the obtained measurement results using data files $\{\alpha_{i\rho}, f_{i\rho}(t,\tau)\}$, $\{\beta_{i\rho}, \varphi_{i\rho}(t,\tau)\}, \{\gamma_{i\rho}, \psi_{i\rho}(t,\tau)\}$ and algorithmic software $\{A_{mi}\}^j, \{A_{a\rho}\}^j$ for the i-th measuring problem (φ_{mi}) to be solved. As a result of the j-th run completion /the j-th test measurement/ we have $\{A_{mi}\}^{j+1}, \{A_{a\rho}\}^{j+1}$ where $I_{j+1} \geq I_j$ and $\rho_{j+1} \geq \rho_j$. The j-th test measurement involves five procedures:

a/ $\mathfrak{I}_1$, $\mathfrak{I}_2$, $\mathfrak{I}_3$ are similar to those listed in Point 4 but their performance is based on the usage of the developed algorithmic software;

b/ $\mathfrak{I}_4$, $\mathfrak{I}_5$ are included for the development of $\{A_{mi}\}$ and $\{A_{a\rho}\}$ respectively. This can be expressed in the following way:

$$\mathcal{Y}_1 : \varphi_{mi} \to \{A_{mi1}\}_i^{c_i},$$

$$\mathcal{Y}_2 : \{A_{mi1}\}_i^{R_i} / \{\alpha_{i\vartheta}, f_{i\vartheta}(t,\tau)\}, \{\beta_{i\vartheta}, \varphi_{i\vartheta}(t,\tau)\}, \{\gamma_{i\vartheta}, \psi_{i\vartheta}(t,\tau)\} \to A_{mij1},$$

$$\mathcal{Y}_3 : \{A_{mij\ell-1}\} / \{\alpha_{ij\ell-1}, f_{ij\ell-1}(t,\tau)\}, \{\beta_{ij\ell-1}, \varphi_{ij\ell-1}(t,\tau)\},$$
$$\{\gamma_{ij\ell-1}, \psi_{ij\ell-1}(t,\tau)\}, \lambda_{ij\ell-1} \to A_{mij\ell}, \quad \ell = \overline{1, m},$$

$$\mathcal{Y}_4 : \{A_{mi}\}_i^{\overline{1,j}} / \{\alpha_{i\vartheta}, f_{i\vartheta}(t,\tau)\}, \{\beta_{i\vartheta}, \varphi_{i\vartheta}(t,\tau)\}, \{\gamma_{i\vartheta}, \psi_{i\vartheta}(t,\tau)\}, \{\lambda_{ij\ell}^*\}_{\ell=1}^m \to \{A_{mi}\}_i^{\overline{1,j+1}},$$

$$\mathcal{Y}_5 : \{A_{ap}\}_i^{R_i} / \{\alpha_{i\vartheta}, f_{i\vartheta}(t,\tau)\}, \{\beta_{i\vartheta}, \varphi_{i\vartheta}(t,\tau)\}, \{\gamma_{i\vartheta}, \psi_{i\vartheta}(t,\tau)\}, \{\lambda_{ij\ell}^*\}_{\ell=1}^m \to \{A_{ap}\}_i^{R_i+1}$$

The author's intention is to consider only PMI characterized
by the five levels of intelligence mentioned above. However,
it is clear, that even present-day possibilities are not
exhausted. It can be said with confidence that this will be
followed by the creation of self-educated and self-organized
PMI, PMI characterized by autowidening of not only algorithmic software but also $\{A_{a\bullet}\}$ and others.

PMI of all types /unichannel, multichannel, multiplex,
multiprocessor, etc/ designed for direct, indirect, statistic
and other measurements can be approached similarly because
of adequate equipment are taken into consideration as determining factors.

It is easily shown that in metrological sense the above
levels characterize PMI yielding measurement results of
gradually improved accuracy. Assuming that the metrological
level is expressed by the following formula:

$$\mathcal{Y}_I = \frac{1}{N_1 N_2} \sum_{i=1}^{N_1} \sum_{m=1}^{N_2} M_{im} \left[\mathcal{Y} \left[\Delta \lambda_j^* \right] / \{\alpha_m, f_m (t,\tau)\}, \{\beta_m, \varphi_m (t,\tau)\}, \{\gamma_m, \psi_m (t,\tau)\} \right]$$

where N_1 - number of measurement problems /function/;

N_2 - number of permissible sets of conditions,
requirements, constraints and numerical values
of characteristics of the object of measurements

$I = \overline{1, 5}$ - levels of PMI intelligence under consideration.

Let us compare the results obtained by means of PMI of
various intelligence level for the same set of measurement
problems $\{\varphi_{mi}\}^{N_1}$ and for the same set of permissible conditions, requirements, constraints and evaluated characteris-
tic of the object of measurements $\{\{\alpha_m, f_m (t,\tau)\}, \{\beta_m, \varphi_m (t,\tau)\},$
$\{\gamma_m, \psi_m (t,\tau)\}\}$.

Proceeding from the assumption that change from nonadaptive,
noniterative procedures to iterative procedures of adaption
/self-education and self-organization/ is benefitial for
equally /in all aspects/ improving metrological quality of

measurement results obtained for complete N_1, N_2 sets, we derive $y_{\Sigma_1} \geq y_{\Sigma_2}$ at $I_1 > I_2$.

In other words, PMI of higher intelligence makes it possible to obtain measurement results of higher accuracy. As this takes place, the additional functional potentialities of PMI contribute greatly to the increase of data output.

NEW EFFICIENT STATISTICAL PROCEDURES
FOR DETECTING THE POSSIBLE CHANGES
IN A SEQUENCE OF INTELLIGENT MEASUREMENTS

Nikolai A. Nechval

Department of Control Systems
Civil Aviation Engineers Institute
226019 Riga, USSR

INTRODUCTION

The purpose of this paper is to view the statistical models of a sequence of intelligent measurements and the efficient procedures for detecting the possible changes /i.e.
the changes of functional form of the underlying probability
distribution or its parameters/ in the sequence, concentrating
on mathematical or statistical details as well as on application. Supposing one is able to consider sequentially a
series of independent observations /intelligent measurements/
whose distribution possibly changes from F_0 to F_1 at an unknown point in time. Formally, X_1 , X_2 ,... are independent
random variables such that X_1 ,... , $X_{\nu-1}$ are each distrubit-
ed according to a distribution F_0 and $X_\nu, X_{\nu+1}$ ... are each
distributed according to a distribution F_1, where $1 \leq \nu \leq \infty$
is unknown. The ovjective is to detect that a change took
place "as soon as possible" after its occurence subject to
a restriction in the rate of false detections /or false
reactions/. The problem originally arose out of considerations
of quality control. When a process is "in control"   observations are distributed according to F_0. At the unknown point
ν , the process jumps   out of control   and ensuing
observations are distributed according to F_1. The aim is to
raise an alarm "as soon as possible" after the process jumps
"out of control". The early literature on the problem deals
mainly with a change in the mean of normal random variables
having known, fixed variance, or a change in theprobability
of success in Bernoully trials. Early solutions grouped observations and proposed testing each group individually for
an indication of whether or not the process is "in control".
Shewhart charts /Shewhart, 1931/ were standard procedure for
20-30 years. Procedures that are in current use were initiated by Page /1954/. In order to detect a change in a normal
mean from M_0 to $M_1 > M_0$ he proposed stopping and declaring the
process to be "out of control" as soon as $S_n - \min_{1 \leq \nu \leq n} S_\nu$
gets large, where $S_\nu = \sum_{i=1}^{\nu}(X_i - M^*)$ and $M_0 < M^* \leq M_1$ is suitably
chosen. This and related procedures are known as CUSUM /cu-
mulative sum/ procedures. Shiryayev /1963, 1987/ solved the
problem in a Bayesian framework. He considered a loss function whereby one loses one unit if stopping rule $N < \nu$ and
one loses c units for each observation taken after ν if

$N \geq \nu$. The prior on ν is assumed to be geometric. Shiryayev showd that the Bayes solution prescribes stopping as soon as the posterior probability of a change having occured exceeds a fixed level. He also solved a non-Bayesian version, minimizing the mean detection time after the onset of a stationary regime. In the non-Bayesian setting of the problem, the only other optimality result is that of Lorden /1971/. He showed that a certain class of stopping rules is asymptotically optimal and that Page's aforementioned procedure belongs to this class. In this paper, the problem is also treated in a non-Bayesian setting. The proposed statistical procedures for detecting the possible changes in a sequence of intelligent measurements are based on repeated application of a sequential probability ratio test /SPRT/ of Wald to a transformed sequence of measurements ordered in time.

FORMULATION OF THE PROBLEM

Let's suppose that x_1 , ... , x_n is a sequence of independent measurements ordered in time. Consider the hypothesis

$$H_0: \quad x_i \sim F_0(x|\theta_0) \quad , \quad i = 1(1)n \quad , \tag{1}$$

against the alternative

$$H_1: \quad x_i \sim F_0(x|\theta_0) \quad , \quad i = 1(1)\nu - 1 , \tag{2}$$
$$\sim F_1(x|\theta_1) \quad , \quad i = \nu(1)n \quad ,$$

where $F_0(x|\theta_0)$ and $F_1(x|\theta_1)$ are the underlying distribution functions, the functional forms of which are known; θ_0 and θ_1 are the parameters of these distribution functions, respectively. Under H_1 the parameter ν represents a point in time at which a change in an above sequence occours. We wish to test the null hypothesis H_0 against the alternative hypothesis H_1, i.e. to decide between two hypotheses H_0 and H_1. We will suppose that if H_0 is true we wish to decide for H^1 with probability at least $(1-\alpha)$, while if H_1 is true, we wish to decide for H_1 with probability of at least $(1-\beta)$ where α, β are the prespecified false and miss-alarm probabilities, respectively.

TEST PROCEDURES

Complete Information

Main test procedure. Let us assume that F_0 and F_1 are completely known. Then the main test procedure for decision taking is equivalent to performing a SPRT of H_0 against H_1 and repeating the test on successive new measurements until a decision in favour of either H_0 of H_1 is reached. The test operates as follows. Continue sampling as long as

$$B < R_n < A. \tag{3}$$

Stop sampling and decide for H_0 as soon as $R_n \leq B$, and stop sampling and decide for H_1 as soon as $R_n \geq A$. The constants A and B can be chosen to obtain approximately the prespecified false and miss-alarm probabilities α and β , respectively, e.i.

$$A \simeq (1-\beta)/\alpha, \qquad B \simeq \beta/(1-\alpha). \tag{4}$$

The likelihood ration

$$R_n = arg \max_{\{R_n^{H_0}, R_n^{H_1}\}} \left(B - R_n^{H_0}, R_n^{H_1} - A\right) \tag{5}$$

where

$$R_n^{H_0} = \prod_{i=1}^{n} \frac{f_1(x_i|\theta_1)}{f_0(x_i|\theta_0)}, \qquad R_n^{H_1} = \max_{\nu \leq n} \prod_{i=\nu}^{n} \frac{f_1(x_i|\theta_1)}{f_0(x_i|\theta_0)}, \tag{6}$$

f_0 and f_1 are densities of F_0 and F_1, respectively. $R_n = R_{H_1} \geq A$ can be ragerded as a "maximum likelihood" treatment of the unknown change point, i.e. stop when for some ν the measurements x_ν ... , x_n are "significant". For practical reasons, it is useful to have available a procedure using statistics based on the empirical distribution function /EDF/.

Test procedure based on EDF statistics. Putting $u_1 = F(x_i)$, i = 1,...,n, the hypothesis that x_1 ..., x_n with distribution function F is equivalent to the hypothesis that $u_1,...,u_n$ are $U(0,1)$ variables, i.e. are uniformly distributed in the interval $(0,1)$. Let $u_1,..., u_n$ be ordered to give the ordered values $u_{[1]} \leq u_{[2]} \leq ... \leq u_{[n]}$. For measuring a distance from univofmity, one of the following EDF statistics can be used:
the Kolmogorov statistics D^+, D^-, D:

$$D^+ = \max_{1 \leq i \leq n} \left[(i/n) - u_{[i]}\right] ;$$
$$D^- = \max_{1 \leq i \leq n} \left[u_{[i]} - (i-1)/n\right]; \tag{7}$$
$$D = \max (D^+, D^-);$$

the Cramér-von Mises statistic W^2:
$$W^2 = \sum_{i=1}^{n} \left[u_{[i]} - (2i-1)/2n\right]^2 + (1/12n); \tag{8}$$
the Kuiper statistic V:
$$V = D^+ + D^-; \tag{9}$$
tne Watson statistic U^2:
$$U^2 = W^2 - n(\bar{u} - \tfrac{1}{2})^2, \qquad \bar{u} = \sum_{i=1}^{n} u_{[i]}/n; \tag{10}$$
the Anderson-Darling statistic A^2:
$$A^2 = -\left(\sum_{i=1}^{n} (2i-1)\left[\ln u_{[i]} + \ln(1-u_{[n+1-i]})\right]\right)\Big/n - n. \tag{11}$$

To make the test procedure /3/ using one of the preceding statistics /call it T/, it is necessary to present /6/ in the following form:

$$R_n^{H_0} = \frac{f_1(T|x_1,\ldots,x_n)}{f_0(T|x_1,\ldots,x_n)}, \quad R_n^{H_1} = \max_{\nu \leq n} \frac{f_1(T|x_\nu,\ldots,x_n)}{f_0(T|x_\nu,\ldots,x_n)}, \quad (12)$$

where $f_0(T|\cdot)$ and $f_1(T|\cdot)$ are the probability density functions of T under the hypotheses H_0 and H_1, respectively. Which test procedure is best for a particular problem of change detecting depends to a great extent on F_0, F_1 and the preferences of the user.

Incomplete Information

Let us suppose that the functional forms of the underlying distribution functions $F_0(x|\Theta_0)$ and $F_1(x|\Theta_1)$ are known, while their parameters Θ_0 and Θ_1, respectively, are unknown, i.e. these parameters enter into the problem of change detecting as the nuisance parameters. An efficient way for circumventing the difficulty of forming sequential tests in this case is to consider a sequence formed by transforming the original measurements, the transformation so chosen that the new sequence does not depend on nuisance parameters.

Eliminating the nuisance parameters. A sequential transformation, by means of which the nuisance parameter can be eliminated from the problem, is given by the following theorems.

Theorem 1. /Invariant Factorization Theorem/. Let $(R^n, \mathcal{B}^n, \mathcal{P}^n)$ be a probability model /space/, which is invariant under a group G of one-to-one /measurable/ transformations $g \in G$ in the sample space R^n /from R^n into itself/, and $x^n = (x_1,\ldots,x_n) \in R^n$ be a sample of independent observations of random variable x from the Borel probability model $(R, \mathcal{B}, \mathcal{P})$ which are identically distributed with probability measure P_Θ on $\mathcal{B}$, where P_Θ is a member of a parametric family $\mathcal{P} = \{P_\Theta, \Theta \in \Theta\}$ of probability measures with absolutely continuous distribution function $F(x|\Theta)$ and a parameter Θ /in general, vector/ belonging to parameter space Θ. Each $g \in G$ induces a transformation $\bar{g} \in \bar{G}$ from Θ into itself defined by $P_{\bar{g}\Theta}(gx \in L) = P_\Theta(x \in L)$, $L \in \mathcal{B}$, $\Theta \in \Theta$. Let us assume that there exists a nontrivial sufficient statistic $t(x^n) \equiv t_n$ for Θ /in general, vector/ on R^n which has the property that G induces a group G_t of one-to-one transformations $g_t \in G_t$ in the sample space R^{kt} of $t(x^n), k_t \in \{1,\ldots,n\}$. Then the probability distribution element of a sample x^n /or, in other words, the probability distribution element of the likelihood function of x^n / can be presented as

$$dF^{(n)}(x^n|\Theta) = \prod_{i=1}^{n} dF(x_i|\Theta) = \prod_{j=j_0}^{n} dF_j(z_j)\,dQ(t_n|\Theta) \quad (13)$$

where $\prod_{j=j_0}^{n} dF_j(z_j)$ is the probability distribution element of a maximum invariant statistic /Nechval, 1984/ $z(x^n) = (z_{j_0},\ldots,z_n)$ on R^n under G, i.e. $z(x^n) = z(gx^n)$ for all $g \in G$ and $x^n \in R^n$, $n-j_0+1$ is the rank of $z(x^n)$ whose distribution does not depend on the nuisance components of the parameter Θ, j_0-1 is the number of the nuisance components of Θ, $\Theta \in \Theta$, $dQ(t_n|\Theta)$ is the probability distribution element of a sufficient statistic t_n for $\Theta \in \Theta$

<u>Proof</u>. The result follows by applying Theorem 2 given below.

<u>Corollary</u>. If either an ordered sample of observations $x^{[n]} = x_{(1)}, \ldots, x_{(n)}$ is the case, where $\ldots \le x_{(n)}$, of a sufficient statistic t_n includes an order statistic of $x^{[n]}$ then the result

$$dF^{[n]}(x^{[n]}|\theta) = n! \prod_{i=1}^{n} dF(x_{(i)}|\theta) = \prod_{j=j_0}^{n} dF_j(z_j) dQ(t_n|\theta) \quad (14)$$

holds.

<u>Theorem 2</u>. /Conditional Factorization Theorem/ Under assumptions of the Neyman Factorization Theorem, the probability distribution element of the likelihood function of a sample $x^n = (x_1, \ldots, x_n)$ can be presented as

$$dF^{(n)}(x^n|\theta) = \prod_{i=1}^{n} dF(x_i|\theta) \Longrightarrow \prod_{j=j_0}^{n} dF_j(x_j|t_j) dQ(t_n|\theta). \quad (15)$$

<u>Proof</u>. The proof being straightforward is omitted.

Note, that the result of Theorem 2 /as well as Theorem 1/ is true also in cases when x is a multidimensional continuous random variable. Moreover, the discrete version of Theorem 2 can be obtained, i.e., when x is a discrete / one-dimensional or multidimensional/ random variable.

<u>Theorem 3</u>. /Probability Integral Transformation Theorem/ The random variables

$$u_j = F_j(x_j|t_j), \quad j = j_0(1)n, \quad (16)$$

are independently and identically distributed uniform random variables in the interval $0, 1$

<u>Proof</u>. The result follows by applying the remarks on a multivariate transformation given by Rosenblatt /1952/.

<u>Corollary</u>. The random variables

$$u_j = F_j(z_j), \quad j = j_0(1)n, \quad (17)$$

are independently and uniformly distributed in the unit interval.

<u>Example 1</u>. Supposing that $x_1, \ldots, x_n$ are indenpendent random variables, each with $N(\mu, \sigma^2)$ where μ and σ^2 are unknown, then /in 13/

$$dF_j(z_j) = \frac{1}{B(\frac{j-2}{2}, \frac{1}{2})} (1-z_j^2)^{\frac{j-2}{2}-1} dz_j, \quad z_j \in (-1, 1), \quad (18)$$

$$z_j = \frac{x_j - t_j^{(1)}}{\sqrt{\frac{j-1}{j} t_j^{(2)}}}, \quad t_j^{(1)} = \frac{\sum_{i=1}^{j} x_i}{j}, \quad t_j^{(2)} = \sum_{i=1}^{j} (x_i - t_j^{(1)})^2, \quad j = 3(1)n, \quad (19)$$

$$dQ(t_n|\theta) = \frac{\sqrt{n}}{\sigma\sqrt{2\pi}} e^{-\frac{n(t_n^{(1)}-\mu)^2}{2\sigma^2}} \frac{(\frac{1}{2\sigma^2})^{\frac{n-1}{2}}}{\Gamma(\frac{n-1}{2})} (t_n^{(2)})^{\frac{n-1}{2}-1} e^{-\frac{t_n^{(2)}}{2\sigma^2}} dt_n^{(1)} dt_n^{(2)}, \quad (20)$$

$$\theta = (\mu, \sigma^2), \quad t_n = (t_n^{(1)}, t_n^{(2)}), \quad t_n^{(1)} \in (-\infty, \infty), \quad t_n^{(2)} \in (0, \infty) .$$

Example 2. Let's suppose, that $x_1, \ldots, x_n$ are independent random variables, each with the density function of the exponential distribution

$$f(x|\theta) = \tfrac{1}{6} e^{-(x-\mu)/6}, \quad x \in (\mu, \infty), \quad \theta = (\mu, 6). \tag{21}$$

The parameter θ is the nuissance. Then /in /14/

$$dF_j(z_j) = (j-2)(1-z_j)^{j-3} dz_j, \quad z_j \in (0,1), \tag{22}$$

$$z_j = \frac{(n-j+1)(x_{[j]} - x_{[j-1]})}{\sum\limits_{i=2}^{j}(n-i+1)(x_{[i]} - x_{[i-1]})}, \quad j = 3(1)n, \tag{23}$$

$$dQ(t_n|\theta) = \tfrac{n}{6} e^{-\frac{n(x_{[1]} - \mu)}{6}} \frac{1}{\Gamma(n-1)6^{n-1}} (t_n^{(2)})^{n-2} e^{-\frac{t_n^{(2)}}{6}} dx_{[1]} dt_n^{(2)}, \tag{24}$$

$$t_n = \left(x_{[1]}, \; t_n^{(2)} = \sum\limits_{i=2}^{n}(x_{[i]} - x_{[1]}) \right), \quad x_{[1]} \in (\mu, \infty), \; t_n^{(2)} \in (0, \infty). \tag{25}$$

Example 3. Let's suppose, that $x_1, \ldots, x_n$ are independent random variables, each with the density function of the gamma distribution

$$f(x|\theta) = \frac{1}{\Gamma(a)6^a} x^{a-1} e^{-\frac{x}{6}}, \quad x \in (0, \infty), \quad \theta = (a, 6). \tag{26}$$

The parameter θ is the nuisance. Then /in /15/

$$dF_j(x_j|t_j) = \tag{27}$$

$$= \frac{\int\limits_{(z_3, \ldots, z_{j-1})} \left[\frac{1}{4} - \frac{\left(\frac{x_j}{t_j^{(1)}} \left(1 - \frac{x_j}{t_j^{(1)}} \right)^{j-1} \right)^{-1}}{\prod\limits_{k=3}^{j-1} z_k (1-z_k)^{k-1}} t_j^{(2)} \right]^{-\frac{1}{2} \frac{j-1}{2}} \prod\limits_{k=3}^{j-1} \frac{dz_k}{z_k(1-z_k)} \frac{dx_j}{x_j \left(1 - \frac{x_j}{t_j^{(1)}} \right)}}{\int\limits_{(z_3, \ldots, z_j)} \left[\frac{1}{4} - \frac{1}{\prod\limits_{k=3}^{j} z_k (1-z_k)^{k-1}} t_j^{(2)} \right]^{-\frac{1}{2}} \prod\limits_{k=3}^{j} \frac{dz_k}{z_k(1-z_k)}}, \quad j = 3(1)n,$$

$$t_j = \left(t_j^{(2)} = \prod\limits_{i=1}^{j} x_i \Big/ \big(t_j^{(1)} \big)^j, \; t_j^{(1)} = \sum\limits_{i=1}^{j} x_i \right), \quad z_j = x_j / t_j^{(1)}. \tag{28}$$

The obtained results are completely confirmed by computer simulation and can be used in following fields of research and application: measurement and quality control, diagnosing the aircraft systems in real-time, manufacturing and testing, environment inspection, etc.

REFERENCES

1. Lorden, G.: 1971, Procedures for reacting to a change in distribution, Ann.Math.Statist., 42:1987
2. Nechval,N.A.: 1984, "Theory and Methods of Adaptive Control of Stochastic Processes", RCAEI, Riga
3. Page, E.S.: 1954, Continuous inspection schemes, Biometrika, 41:100.
4. Rosenblatt,M.: 1952, Remarks on a multivariate transformation, Ann.Math.Statist., 23:470
5. Shewhart,W.A.: 1931, "The Economic Control of the Quality of Manufactured Product" Macmillan, New York.
6. Shiryayev, A.N.: 1963, On optimum methods in quickest detection problems, Theory Probab.Appl., 8:22.
7. Shiryayev,A.N.: 1978, "Optimal Stopping Rules", Springer-Verlag, New York.

PROBLEMS OF REAL-TIME INTELLIGENT MEASUREMENTS

Eugen-Georg Woschni

Department of Information-Engineering
Technische Hochschule Karl-Marx-Stadt
9010 Karl-Marx-Stadt, GDR

INTRODUCTION

The beginning of intelligent measurement is marked by the integration of microcomputers in measurement devices.The development of single-chip microcomputers beyond it opens the possiblity of decentral processing of measurement data and especially of information reduction within the sensor itself. These "intelligent sensors" make possible the introduction of the principles of hierarchical structures in measurement,too.

This new technique requires some suppositions because of the digital processing of the measurement data: Analog-to-digital conversion and the processing itself need a certain time, which depends on the accuracy demanded. This fact leads to restrictions of the dynamic behaviour. Furthermore Shannons sampling theorem in practice often especially with direct digital sensors is not fulfilled, leading to several kinds of errors.

The paper deals with these problems in which the mutual influence of the several effects will be taken into consideration.

PROBLEMS OF SAMPLING

As well known so-called "aliasing-errors" arise if the sampling theorem /f_c critical frequency of the signal; t_s sampling time/

$$f_c \leqq 1/(2t_s) \tag{1}$$

is violated[1]. The following drastic example shows the effect: In Fig.1.a,b a shaft with an undulatory surface is represented. If instead of 24 samples due to the sampling theorem /1/ only 14 /Fig.1.a/ or 13 /Fig.1.b/ sampling points are used a wrong surface is counterfeited; an ellipsoid one with 14 and a derivation of the whole shaft with 13 samples.

To calculate the errors due to aliasing the results of signal-, system- and modulation-theory are used[2,3], leading to the error in the frequency domain with $\underline{\hat{X}}(j\omega) =$ spectral amplitude density

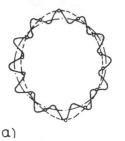
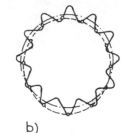

a)   b)

Fig.1. Testing the roundness

$$e(t) = F^{-1}\left\{\sum_{r=-\infty}^{\infty}\left[\underline{\hat{X}}j(\omega - r\omega_s) + \underline{\hat{X}}j(\omega + r\omega_s)\right]\right\}$$

$$= \frac{1}{2\pi}\sum_{r=-\infty}^{\infty}\int_{-\omega_s/2}^{+\omega_s/2}\left[\hat{\underline{X}}j(\omega - r\omega_s) + \hat{\underline{X}}j(\omega + r\omega_s)\right]e^{j\omega t}\,d\omega,$$

(2a)

or

$$e(t) = \frac{1}{2\pi}\sum_{r=-\infty}^{\infty}\left\{e^{jr\omega_s}\int_{-(2r+1)\omega_s/2}^{-(2r-1)\omega_s/2}\underline{\hat{X}}(j\omega)\,e^{j\omega t}\,d\omega\right.$$

$$\left. + e^{-jr\omega_s t}\int_{(2r-1)\omega_s/2}^{(2r+1)\omega_s/2}\underline{\hat{X}}(j\omega)\,e^{j\omega t}\,d\omega\right\}.$$

(2b)

An approximation is given by Churkin et al[4]: If

$$\left|\underline{\hat{X}}(j\omega)\right| < \frac{c}{\omega^{(1+\alpha)}}\quad;\quad c;\alpha > 0$$

(3a)

the maximal amplitude of the aliasing error is

$$|e(t)| \leq \frac{1}{\pi}\int_{\omega_s/2}^{\infty}\frac{c}{\omega^{(1+\alpha)}}\,d\omega = \frac{c}{\pi}\frac{1}{\alpha\left(\frac{\omega_s}{2}\right)^{\alpha}}$$

(3b)

For $\alpha = 1$ e.g. follows

$$|e(t)| \leq \frac{c}{\pi}\frac{1}{\omega_s/2}$$

(3c)

or

$$\omega_s \geqslant \frac{2c}{\pi |e(t)|} \qquad (3d)$$

Using the spectral power density of the signal $S_{xx}(\omega)$ the meansquare value of the cut-off error because of the ideal anti-aliasing filtering to fulfill the sampling theorem is

$$\overline{e^2(t)} = 2 \int_{\omega_s/2}^{\infty} S_{xx}(\omega)\, d\omega. \qquad (4a)$$

Because of the mirroring effect twofold this error results if there is no anti-aliasing filtering before sampling

$$\overline{e^2(t)} = 4 \int_{\omega_s/2}^{\infty} S_{xx}(\omega)\, d\omega \qquad (4b)$$

as Fig.2. shows

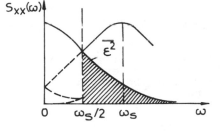

Fig.2. Calculation of the mean-square error,
        equations /4a,b/

Real-time processing needs a certain time depending on the accuracy demanded as explained in chapter 3 in detail. Therefore the sampling time t_s should be as long as possible. Before this background methods to reduce sampling errors are of great importance in the field of intelligent measurement:

a/ The best way to reduce the error is an ideal low-pass filtering before sampling. If this is not possible – as with some classes of direct-digital sensors – the pulse-shape of the sampling pulses should be choosen such a form, that the spectral amplitudes are decreasing with increasing frequencies as much as possible /cos^2 – or bell-shaped/ to reduce the aliasing error components of higher order.

b/ In the case of a mean-value operation after sampling – that means an extreme low-pass filtering – a sampling time twice the value due to the sampling theorem /1/ is convenient[5].

c/ If the signal has a band-limiter frequency range $mf_s/2 \leq f \leq (m + 1) f_s/2$ the so-called "reduced sampling" may be applied. The sampling time then may be enlarged by the factor $(m - 1)$; so-called ZOOM-effect[6].

Last not least it may be pointed out that the limited number of samples-instead of an infinite one due to the sampling theorem-yields supplementary errors as investigated in Herold et al[7].

PROBLEMS OF PROCESSING

A lot of tasks of processing especially most of all filter problems in measurement including those with variable or adaptive parameters as typical for intelligent measurement are described by linear algorithms. Examples of great importance for signal processing in this field are mean-value-, PI-, PD-, PID- or FFT-algorithms. In all these cases the procedure desired in the time domain

$$y = Op \; \{x(t)\} \qquad (5a)$$

can be written equivalent in the frequency domain[2,3]

$$\hat{\underline{Y}}(j\omega) = G(j\omega) \; \hat{\underline{X}}(j\omega) \qquad (5b)$$

with the complex amplitude density of the output $\hat{\underline{Y}}(j\omega)$ or the input $\hat{\underline{X}}(j\omega)$ and the frequency response $G(j\omega)$. The adequate relation may be written in the Z-domain too as well-known[2]. For linear operations one obtains

$$G(j\omega) = F \{Op\} \quad or \quad G(z) = Z \; \{Op\}. \qquad (5c,d)$$

Now all results of linear system theory may be applied, for instance to solve the error problems[2,3]. With the characteristic functions of the ideal system or algorithm (model) $G_{id}(j\omega)$ and of the real one $G_{real}(j\omega)$ the mean-square error is given by[2,3]

$$\overline{e^2(t)} = \int_{-\infty}^{+\infty} S_{xx}(\omega) |G_{id}(j\omega) - G_{real}(j\omega)|^2 \, d\omega. \qquad (6)$$

To optimize the error several effects are to be taken into consideration:

a) In general we have to pay with increasing calculation time for the processing the better the real algorithm is approximated to the ideal one as the example of a mean value operation shows. In figure 3 both the frequency response of the rectangular (---) and the trapezoidal approximation (-.-.-) are given[2]. The trapezoidal approximation is the better one because here only an amplitude error appeares. On the other hand the calculation time of the trapezoidal approximation is about twice the time of one.

b/ There is a relation between the quantization error given
by the word length and the calculation time. Using for
example an 8-bit microcomputer the extension to 16-bit
by means of software leads to increasing calculation
time by the factor 3 ... 8.

c/ The better the approximation the more the errors due to
the effects of parameter sensitivity are momentous as shown
in[8] for the example of a correction program.

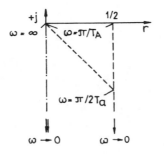

Fig.3. Approximations of the mean-value operation

CONCLUSION

The new techniwue of integration of microcomputers in
measurement devices and sensors for characteristics of intelligence requires some suppositions because of the digital
processing. In the paper at first the problems of sampling
are treated especially the aliasing errors and their reduction. The second part deals with the general problem of error
models for linear processing. The application of this method
shows the mutual influence of several effects. Here the limitations caused by these effects as calculation time needed,
accuracy of approximation and parameter sensitivity were put
together.

REFERENCES

1. E.G.Woschni: "Estimation of Errors caused by Sampling"
   Proc.IMEKO-Symposium on Measurement and Estimation
   Bressanone /Italy/ P.267-299 /1984/
2. E.G.Woschni: "Informationstechnik" 2nd Ed.VEB Verlag
   Technik, Berlin /1981/
3. E.G.Woschni: "Signals and Systems in the Time and
   Frequency Domains", in: Handbook of Measurement Science,
   P.H.Sydenham, John Wiley, Chichester, New York /1982/
4. I.J.Churkin, C.P. Jakowlew, G.Wunsch: "Theorie und
   Anwendung der Signalabtastung" VEB Verlag Technik, Berlin
   /1966/
5. E.G.Woschni: "Limitations of Sensor Dynamics caused by
   Aliasing" Preprints X.IMEKO, Praha S 5, P-32-39 /1985/
6. N.Thrane, ZOOM FFT, Brüel Kjaer Technical Review, No2.
   /1980/

7. H.Herold, E.G.Woschni: "Fehler bei der Abtastung von
   Signalen endlicher Dauer" Zeitschr. messen steuern
   regeln, No.11, P.485-509 /1985/
8. E.G.Woschni: "Parameterempfindlichkeit in der Messtechnik,
   dargestellt an einigen typischen Beispielen" Zeitschr.
   messten steuern regeln, No.4., P.124-130 /1967/.

APPLICATION OF THE ADAPTIVE ANALOG - DIGITAL CONVERSION PROCEDURE

Hans-Bernhard Bemmann

University of Electrical Engineering Mittweida
DDR-9250 Mittweida, German Democratic Republic

1. INTRODUCTION

Digital measurement of unknown amplitude-time functions, especially of the characteristics: effective value, mean value and peak value represents an important problem. In this paper there will be discussed the application principle of the adaptive analog-digital conversion procedure. This principle provides an improvement of efficiency and accuracy.

2. THE PRINCIPLE OF THE ADAPTIVE ANALOG-DIGITAL CONVERTER

The realization of the adaptive principle is based on the following idea. In signal measurement it is important to get many samples from unknown amplitude-time functions with fast temporary changing speed and thus there is to be found a compromise concerning the resolution. On the contrary, for signals with slow changing speed it is possible to get few but exact samples.

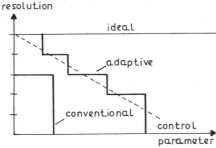

Figure 1

Dependence of the resolution upon the control parameter

The solution of this problem can be realized on principle with a parallel analog-digital converter (ADC) with a high resolution. This parallel ADC's are sometimes not available or too expensive. Therefore, it was taken into to create a converter with "controllable resolution and converting time". This fact is explained in figure 1 and 2. In figure 1 there is shown the resolution of the ADC in dependence upon the control parameters and, therefore, on the signal course for convential, ideal and adaptive conversion procedures. In figure 2 there is shown the reciprocal value of the conversion time as a function of the control parameter.

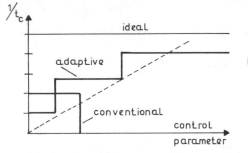

Figure 2

Dependence of the converting time upon the control parameters

The essential of the adaptive ADC consists in the adaption to the momentary signal shape. From the unknown amplitude-time function u(t) there is acquired the control parameter. This parameter can be the signal changing speed or the peak value if the signal function is periodical. The fundamental principle of an analog-digital converter is shown in figure 3.

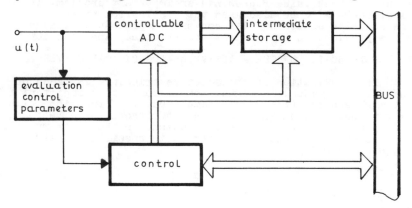

Figure 3   Fundamental principle of an analog-digital converter

3. ACQUISITION AND EVALUATION OF THE SIGNAL CHANGING SPEED

The signal changing speed $v = du(t)/dt$ depends on the following values of the amplitude-time function:

- curve shape
- frequency and cycle duration, respectively
- peak value
- crest factor

The task of a suitable hardware for the acquisition of the
signal changing speed consists in:

- the evaluation of the momentary signal changing speed
- the quantization of this value for generating the control
  signal for the adaptive ADC

The solution is possible on the base of an analog realization or a realization with quantized samples.
The analog realization is described in /1/. The digital realization is performed in the following way. The amplitude-
time function is sampled with a high sampling frequency. The
samples are quantized coarsely. By evaluation of the ADC temporary output changing we obtain the control signal for the
adaptive ADC device, see figure 4.

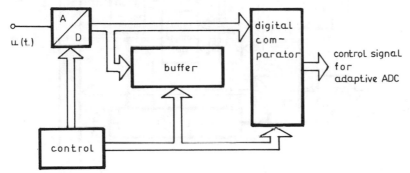

Figure 4   Acquisition of the signal changing speed from quantized samples

4. REALIZATION OF AN ADAPTIVE ADC

Now there is given a proposal of a circuit configuration
for an adaptive ADC with the signal changing speed as control
parameter. The fundamental idea is realized exemplarily by
application of two types:

- ADC 1 is a 4 Bit converter, conversion time t_{c1}
- ADC 2 is a 10 Bit converter, conversion time t_{c2}

Condition of realization for the ADC's: $t_{c1} \ll t_{c2}$

In figure 5 you can see the dependence between resolution
and signal changing speed for the created hardware configuration.

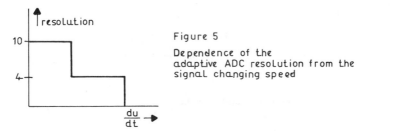

Figure 5

Dependence of the
adaptive ADC resolution from the
signal changing speed

In figure 6 there is shown the diagram of a practical realization /1/. ADC 1 takes samples from the amplitude-time function, the samples are quantized with small resolution. The
digitized samples are given into a high speed memory. In a
digital comparator there are simultaneously compared the momentary and the preceding samples to find out the signal speed
of u(t). Depending on this comparison there are attained
higher quantized samples from ADC 2. This is possible because
the smaller signal changing speed allows a higher converting
time and therefore a better resolution.

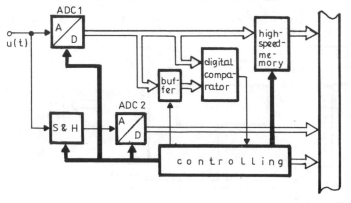

system BUS

Figure 6   Realization of an adaptive converter structure

The samples from ADC 2 are given in a further storage, the
equidistance of the samples is realized. Under this condition
the signal changing speed is defined as: $v = \Delta D / \Delta t$. ΔD
represents the changing of the quantized samples from
ADC 1; Δt corresponds the conversion time of ADC 2.

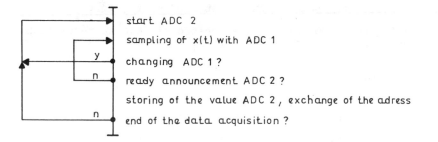

start ADC 2

sampling of x(t) with ADC 1

changing ADC 1 ?

ready announcement ADC 2 ?

storing of the value ADC 2, exchange of the adress

end of the data acquisition ?

Figure 7   Functional principle of the adaptive ADC

Conditions for finishing the ADC 2 conversion:

- during the conversion time of ADC 2 the output of ADC 1 does not change
- or changes only within a small value (multiple of the quantizing value)

5. CONCLUSION

The sampling method is useful for the calculation of the characteristics of the unknown amplitude-time functions, /2/. The main error sources in conventional ADC are /3/:

- static and dynamic characteristics of the sample and hold circuit
- sampling by the ADC with application of approximation methods

By application of the adaptive ADC there can be reduced the sampling and the total error. The total error is dependent on:

- the kind of the test function
- the amplitude (refered to the processing area, especially with ADC 1)
- the total number of the samples
- the ratio of the portions of the samples from ADC 1 and ADC 2 refered to the total number of samples

References

/1/ Rabe, J.; Ein Beitrag zur digitalen Kenngrößenberechnung unbekannter Amplituden-Zeitfunktionen. Diss. A IHM 1985

/2/ Bemmann, H.-B.; Wiebel, F.; Numerische Integration periodischer Größen - angewandt in der digitalen Wechselspannungsmessung. Nachr./Elektronik 33 (1983) H. 9, pp.361-363

/3/ Rabe, J.; Digitale Wechselspannungsmessung - Realisierungsprobleme von Hard- und Software. Wiss. Zeitschr. IHM 3/1984, pp. 47 - 57

INTELLIGENT MEASUREMENT AND EEG RECORDING

B.Bago, T.Dobrowiecki, Z.Papp, B.Pataki

Technical University of Budapest
Department of Measurement and Instrument Eng.
H-1521 Budapest, Hungary

INTRODUCTION

In the paper some considerations are given on how the device intelligence could be increased using the knowledge available of the given modelling and measurement process and in what way this knowledge-base could be stored and manipulated in the real-time measurement systems by means of nowadays obtainable adequate technology. As a possible solution to this problem a small-sized expert system is shown which is built in the measurement process and which can be successfully implemented even in the 8-bit microprocessor environment. A device structure of this kind makes it possible to design systems with an intelligence level significantly higher than those existing so far.

In the paper an intelligent EEG recorder with some built-in artificial intelligence is introduced. The EEG recorder supplied with user-friendly front panel and self-test facilities is controlled by a real-time rule-based expert system. This system by means of continuous analysis of the recorded data stream is expected to perform tasks normally administered by the supervising assistant, with unchanging attention even during a sleep analysis as well. Furthermore the attention of the evaluating physician may be drawn to the more noteworthy parts of the enormous amount of the recorded observations.

ARTIFICIAL INTELLIGENCE AND THE MEASUREMENT

In order to perform successful measurements the detailed knowledge regarding the specific application field and the general measurement procedures is always required. This special knowledge (expertise) consists of:
- selecting the quantities to be measured (model formation),
- temporal and spatial configuration of the measuring instruments required (experiment planning),
- measurement data acquisition, information processing (measurement),
- error and utilization analysis of the measurement results (result interpretation) [1].

A possible solution to the multi-purpose measurement system implementation is to connect different measurement peripherals with the universal processing and controlling units. For the measurement systems produced in this way a natural implementation emerges employing the expertise, both in information processing and in control.

The two basic constituents of the measurement expertise are the FUNCTIONAL (measurement theoretical) and IMPLEMENTATIONAL (measurement technical) EXPERTISE [1]. The background for the functional expertise lies in the measurement theory which exposes the general properties of the measurement process. This kind of expertise covers the specification of the measurement environment, the information processing and the evaluation of the results. The implementational expertise relies on the metrology and the measurement technique valid for the specific field of application and supports the proper configuration of the instruments and control of the measurement process.

The key issue of the intelligent measurement system design is the formulation and efficient implementation of the expertise required. One part of this knowledge (i.e. analytical mathematical algorithms - estimation, decision schemes etc.) is theoretically easy to implement. The steps of algorithms, their validity limits are given by mathematical formalism and can be translated into the traditional languages by well-developed methods. The second component is the heuristical knowledge, that is the knowledge originating mainly in the professional intuition and experience, frequently not cleared up enough but nevertheless very useful for the expert in avoiding solutions with excessive demand for operational time and resources. Putting it more plainly, a part of the error analysis and the information processing are in general analytical, all the rest of the expertise is of heuristic nature.

Regarding the implementation of heuristic knowledge the following three problems are to be considered:
- representation of knowledge,
- manipulation of the knowledge-base,
- integration of the knowledge-base with the numerical database and the algorithms within the unified real-time control structure of the measurement system [2].

The first two problems traditionally belong to the research field of artificial intelligence [3], while the third one is a new emerging chapter of the measurement system design.

Expert systems are computer programs solving tasks demanding expertise and belonging to the work area of the human experts. Their structure depends upon the knowledge representation being implemented, which depends again on the specific field of application. According to the experience so far the most suitable representation formalism for the mainly heuristic expert knowledge is the so called rule-based systems built from the elementary IF-THEN rules. The manipulation of the rule system is the goal-driven (backward) or the data-driven (forward) rule inference mechanism [4,5].

The implementation and manipulation of the knowledge-base in general is a task with significant time and resource requirements. Therefore in real-time measurement systems its usage is limited by the processing capability of the acceptable cost for the application field in question. For that very reason the expertise is usually implemented only partially e.g. into the information processing during measurement as well as into the final evaluating phase of the measurement process.

Measurement systems with the expert system support were designed till now mostly for the exclusive increase of the information processing intelligence. In the following part a different kind of intelligent measurement system is presented which by means of the built-in heuristic knowledge permits a multi-level multi-purpose data evalution and accommodates itself to the users with different demands.

EEG AND ARTIFICIAL INTELLIGENCE

The experiments concerned with the support of the EEG measurement technique by means of artificial intelligence have been going on for some years. The course of development is determined by the following factors:
- There are no striking results yet regarding the model of the EEG signals (clarification of their "genesis"). The numerous attempts in progress have not left the research laboratories so far. So it seems to be a little premature to think of the analytical EEG signal models.
- The evaluation of the EEG signals - at the present level of knowledge - is based merely on the properties of the "surface structure" (i.e. essentially there is no knowledge about the origin of the signals, the relationship between the measured signals and the symptoms is expressed in the numerous rules of thumb). Thereupon the most useful expert systems are modelling the EEG record reading activity performed by the physicians.
- Because of the enormous amount of the heuristic knowledge involved even the programs running on the main-frame computers give possibilities for no more than the evaluation of the well-defined restricted group of diseases only.

Up to the present the EEG evaluation meeting diagnostical demands and performed under real-time operational conditions (i.e. where the results of the automatic signal processing by some kind of feedback would have a direct effect on the analysis process itself) was not accomplished. The reason is that for the full diagnosis a very large knowledge-base is usually required which would demand on the other hand an enormous computing capacity. The diagnostical EEG devices reported in the literature are restricted to the limited fields of application, e.g. [6].

It can be concluded from the above that it is premature to make efforts toward the design of universal EEG evaluating system either from the technical or the medical/biological point of view.

As the first step it is of the greatest importance in the practice to obtain good EEG records, since they are the very basis for the evaluating staff. For this reason as the preliminary aim it was to provide the EEG processing with a support guaranteeing reliable EEG records.

EEG RECORDING AND ARTIFICIAL INTELLIGENCE

Our aim was to build an EEG recorder having "measurement technical intelligence" which is comparable with the knowledge of a well-trained assistant. In consequence the following aspects are to be realized:
- Automatization of the recording (the recording goes on without supervising). It is especially advantageous during the long-time (e.g. sleep) analysis periods.
- Increase in the realibility of the evaluation by means of good quality recording.
- Supporting the physician in the long record evaluation (by drawing his attention to the possibly abnormal record intervals).

The developed device is composed of two basic units. One of them is a traditional, although microprocessor-controlled programmable EEG recorder with remote-control possibility, while the other is the supervising analyzer on the signals coming from the recorder (Figure 1.). During the analysis the following items are closely watched:
- amplitude conditions of the channels (average intensity, trends),
- power density spectra in the common EEG bands (δ, γ, α, $\beta 1$, $\beta 2$),
- noise spectrum of the channels.

The measurement results are transformed into the set of bi-valued facts, for example representing high spectral component at line frequency, low average intensity etc. in certain measurement channels. The factual database can be extended with the information resulting from the other sources: non-analogue signals are also generated by the recorder and are directly convertable into facts (e.g. shortage of paper, some channels overdriven etc.).
These facts indicate the specific situational data stucture which the inference system handling the heuristic knowledge operates on. The operation of the analyzer inference system is based upon the heuristic knowledge constituted from facts and IF-THEN rules. It results in possible interference into the functioning of the recorder and/or messages intended for the operator and printed on the matrix printer connected to the recorder.

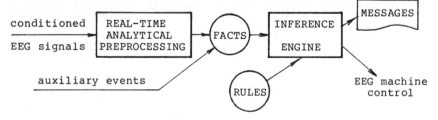

Fig. 1 Signal flow diagram of the analyzer

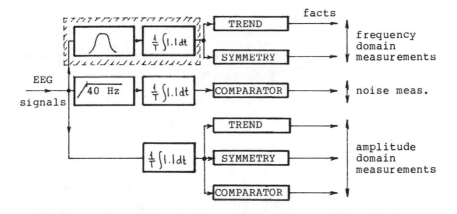

Fig. 2   Real-time preprocessing

REAL-TIME PROCESSING

The purpose of the real-time preprocessing is to transform the signals coming from the EEG channels into facts necessary to the operation of the expert system. For this the implementation of the processing steps shown in Figure 2. is required.
From the detailed analysis of the processing algorithms it was recognized that the required accuracy and real-time specification can be fulfilled only if the units marked with lines are implemented by hardware, and furthermore if the noise measuring channels are time-multiplexed (the channels are examined in each 15 seconds). The bandpass filters are digital filters implemented using Intel 2920 processor chips. All the other processing is performed by the Z80 processor of the analyzer itself.

THE INFERENCE SYSTEM

The rule manipulating strategy of the inference system is the usual forward chaining method [3]. The inference process is induced by the appearence of the fact and it continues until running out of the applicable rules. If no more is found the system stops and will start again only if some new facts appear. For the sake of the run-time being held at the minimum-level the direct code implementation was chosen for the system instead of the customary interpretative one. The forward chaining of the rules is accomplished by the special run-time system operating on the machine code from the translated IF-THEN rules.
In certain situations it is necessary to delay the inference process (e.g. to wait until the filter transients settle). The problem was solved by the introduction of the "time-dependent" facts giving exactly the same event-controlled structure of the inference system as in case of the facts coming directly from measurement results.

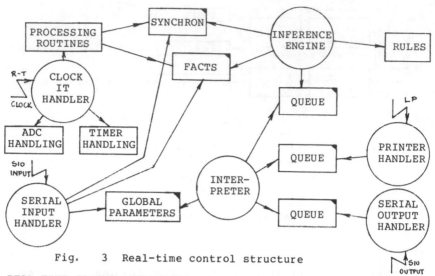

Fig.   3   Real-time control structure

REAL-TIME CONTROL STRUCTURE

The main aspects of the real-time control structure [7] implementation were:
- the separation of the strictly real-time scheduled processes from the other ones and
- the time-optimization of the event-driven inference process   (Figure 3.).

The real-time preprocessing is clock-synchronized and controlled by the process handling the clock interrupts, with the highest level of priority. The compiled facts are processed then by the inference engine process with the event-driven control implemented by means of the synchron module. The inference results are fed into a FIFO and are taken and evaluated by the interpreter process which sends suitable messages to the printer or to the serial output (i.e. the recorder). The recorder is connected to the analyzer by a serial line, in this way commands are sent by the analyzer as well as digital signals and parameter tables are passed describing the operational conditions of the recorder. By the FIFO-s the long-time operational balance and short-time overload tolerance of the system are provided.

THE TECHNOLOGY USED

The software of the devices was written by means of the real-time programming technology developed at the Department of Measurement and Instrument Engineering, Technical University of Budapest [8]. Approximately the 80% of the program was written in real-time Pascal language, the time-critical parts were written in structured assembly language. The cross development technology was performed on the DEC PDP 11 series computer under the RSTS/E operating system. The real-time debugging was supported by an integrated real-time/sequential debugger program being run on the end-product device.

CONCLUSIONS

For the intelligent EEG recording design problem presented in this paper a new approach was employed in terms of
built-in expert system. Although this expert system is limited
in its abilities, its size and facilities are well behind
those of the customary expert systems (e.g. the explanation
generation and the run-time rule system editor are not provided), the proposed solution is nevertheless a noteworthy
one, since it gives a method for the device implementation and
simple modification of the heuristic knowledge represented by
an expert physician.

The facilities of the implemented expert system are limited however only by the low level abilities of the microprocessor used with its origin in the low cost implementational
demands. At the same time it turned out to be sufficient to
satisfy the requirements of the recording itself. Considering
however the decreasing prices of the greater capacity (16 and
32 bit) microprocessor elements in the near future the possibility arises already to develop systems supplied with
expert systems or expert system teams able to cope with the
more sophisticated measurement-modelling tasks.

On the base of the presented device structure some further generalizing remarks are possible regarding expert systems working in the real-time measurement environment. First,
the time-optimal implementations of the expert systems are
emphasized because of the limited reply time specified for the
real-time data acquisition and operator interference. Second,
the presented measurement technical structure turns out to be
EEG recording independent. The system can be generalized in
parts or as a whole without filling it with any specific
knowledge-base. In this way it will be well suited to perform
similar signal processing task with subsequent re-filling its
knowledge-base with the knowledge specific to the given field
of applications.

REFERENCES

1. J.Sztipánovits and J.R.Bourne: Design of Intelligent
      Instrumentation. Proc. of the 1st Conf. on Artificial
      Intelligence Applicatons. 490-495 (1984).
2. J.Sztipánovits and J.R.Bourne: Architecture of Intelligent
      Medical Instruments. Proc. of the 7th Annual Conf. of
      the IEEE EMBS. 1132-1136 (1985).
3. E.Rich: Artificial Intelligence. McGraw-Hill. (1985).
4. F.Hayes-Roth, D.A.Waterman and D.B.Lenat: Building Expert
      Systems. Reading. Addison-Wesley. (1983).
5. F.Hayes-Roth: Rule-Based Systems. Comm. of the ACM.
      28/9:921-932 (1985).
6. L.Baas and J.R.Bourne: A Rule-Based Microcomputer System
      for Electroencephalogram Evaluation. IEEE Trans. on
      Biomedical Engineering. 31/10:660-664 (1984).
7. Z.Papp,B.Bago,T.Dobrowiecki and G.Peceli: Software Archi-
      tecture of Real-Time Medical Instruments. Proc. of the
      7th Annual Conf. of the IEEE EMBS. 1137-1142 (1985).
8. B.Bago,T.Gerhardt,G.Karsai,Z.Papp and G. Peceli: Software
      Tools for Medical Instruments. Proc. of the 7th Annual
      Conf. of the IEEE EMBS. 1143-1147 (1985).

KNOWLEDGE BASED PROGRAMMING ENVIRONMENT

FOR AUTOMATED PICTURE ANALYSIS

Rainer Hesse, Ruediger Klein, Reinhad Klette

Department of Artificial Intelligence
Central Institute for Cybernetics
and Information Processes
Academy of Sciences of GDR
Kurstrasse 33, Berlin 1086, GDR

INTRODUCTION

The analysis and interpretation of measured three-dimensional data can be considered as transformation of completely unstructured input data /some Mbit/ into a small set of diagnostic characteritics /some bit/ as shoen in Fig.1. This complex solution process is to be decomposed into two subtaks of different problem spaces:
/1/ transformation of the measured unstructured input data into structured object-oriented data by means of picture processing algorithms and
/2/ problem-oriented interpretation of these data by the human user to get the required parameters.

For example, in medical application like analysis of microscopical slides the solution of task /1/ must supply the physician with information about the cell structure, number and types of cell components etc. which will be used in task /2/ to characterize the slide as normal or abnormal.

KNOWLEDGE-BASED PICTURE PROCESSING

Nowadays computer-aided picture processing systems in general provide the user with more or less complex packages of tools and procedures for performing single picture processing operations /contour finding, local filters, segmentation, classification etc./. The particular image processing task can be solved by combining these means to create a complex and complete picture processing program. In most cases the application domain oriented user is not able to do that because of the absence of experiences and knowldge in the field of computer-aided picture processing.

Expert systems can be used to overcome this difficulty in a very flexible manner by knowledge controlled planning of picture processing algorithms. The dialogue between the domain-specific user and the expert system takes place within

the problem space of the user and supports together with parameters of the input picture the specification of the program
to be planned. The paper describes the expert system XAMBA for
computer-aided planning of picture processing programs applied
to the automated analysis of a special type of pictures within
the picture processing system A6471-AMBA/R from Kombinat Robotron (Voss,1985).

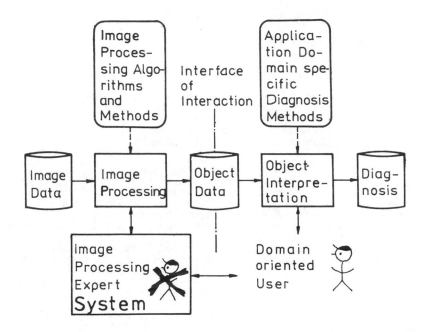

Fig. 1   Picture processing as layered process

COMPUTER-AIDED PROGRAM PLANNING

     The specification of computer programs can be considered
as hierarchical process of planning actions to achieve a
complex goal. The applied principle of stepwise refinement
leads to the following method:
  (1) top-down refinement of the programming task by stepwise
      decomposition of complex goals into sets of functional
      independent but organisational and informational connec-
      ted subgoals with less complexity until there will be
      only terminal subgoals which can be achieved by elemen-
      tary actions (predefined program fragments);
  (2) buttom-up construction of the program according to the
      goal decomposition of step (1) by arranging predefined
      program fragments (actions to achieve terminal subgoals)
      which are connected by predefined data structures to
      ensure the necessary interaction.
The algorithms of goal specification and program construction
are generated by combination of the components of a knowledge

106

base and specified by using information of a data base obtained from the user communication or the input data of the program to be planned. The refinement of one subgoal is independent of other subgoals and considers only local circumstances. The interaction of program actions takes place via global data according to the interaction of the corresponding subgoals.

REPRESENTATION OF PROGRAM PLANNING STATES

A frame concept was developed to represent the actual refinement stage of the program to be planned. Each potential program action is described as special data structure with the following slots:
(1) a verbal description of the goal to be achieved,
(2) functional complexity (terminal goal or not),
(3) successor(s) of the action during program processing,
(4) a pattern for addressing the starting rule of the refinement algorithm,
(5) father and children frame names according to the position with in the refinement hierarchy,
(6) a fragment of the program's source code which represents the action to achieve the goal,
(7) arguments which are data structures and connect the action with others.
The whole program stage is represented by the full list of the frames generated during the past refinement steps. This list represents a so-called program net being a set of related program graphs. Fig. 2 shows a cut-out of an example program net in form of two connected program graphs.

KNOWLEDGE-BASED REFINEMENT

The basic algorithm of goal decomposition and action generation is as follows:
(1) characterization of the goal to be decomposed during user-system-interaction and storing the obtained predicate values in the data base,
(2) classification of the goal applying production rules of the knowledge base which match the given goal characteristics of (1),
(3) allocation of a predefined subgoal structure belonging to the infered goal class (children of the goal),
(4) specification of the subgoal structure by data structures in order to fit it according to the real conditions of the goal to be refined.
This algorithm is applied repeatedly until there are only terminal subgoals within the frame list. Then a program generator will produce the source code of the picture processing program. Therefore the knowledge base contains three different kinds of knowledge units: production rules, predefined subgoal structures and predefined data structures.

To prepare the expert system for a special application domain the knowledge must be formulated with the help of a comfortable graphic-supported knowledge aquisition module. The separate program generator allows easily to adapt the system to different picture processing languages. Fig. 3 shows the general structure of the overall system.

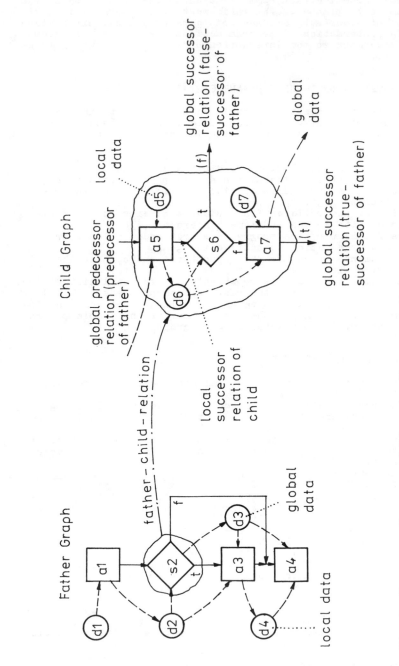

Fig. 2 Program Representation by Program-Net

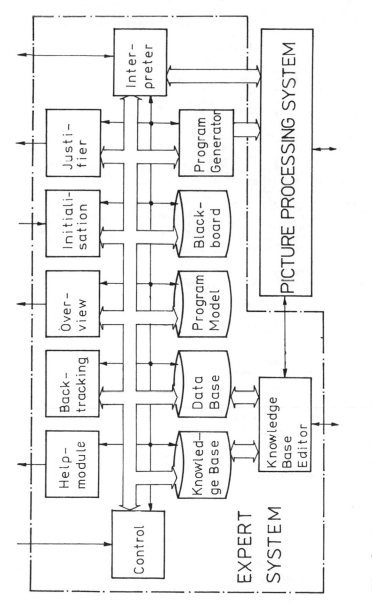

Fig. 3 General System Overview

APPLICATION TO ISOLATED OBJECT ANALYSIS

The system was implemented within the picture processing system A6471-AMBA/R from Kombinat Robotron applied to analysis of isolated objects in more or less homogeneous background.

The knowledge base contains

(1) basic program structures for
  - repeatedly processing of series of pictures
  - processing of single pictures

(2) substructures for
  - picture input
  - picture preprocessing
  - object search and object isolation
  - object classification
  - object class evaluation
  - result output

(3) data structures for
  - class membership criterions
  - class features
  - output formats
  - parameters of functions.

The generated picture processing programs are written in the language LAMBA (Robotron,1985).

During the test phase the following results could be stated:

  - The time-out of the system between two user requests is some seconds. The development of a typical program takes about half an hour.   •
  - The generated programs are well structured and commented.
  - The efficiency of the automatically generated programs doesn't sicnificantly differ from that of hand-made programs.
  - Additionally inserted verbal or graphic help and overview functions make the handling of the system very comfortable to the domain-oriented user.

Up to now the system was tested in analysis of microscopical slides in pathology. Further applications in related fields will be prepared.

REFERENCES

Cohen, P. R., Feigenbaum, E. A., 1982, The Handbook of Artificial Intelligence, Heuristech Press, Stanford, C. A.
Robotron Vertrieb Berlin (VEB), 1985, Dialog-und-Programmier-System AMBA/R, Berlin
Sacerdoti, E. D., 1977, A Structure for Plans and Behavior, Artificial Intelligence Series, Elsevier North-Holland Inc., New York
Voss, K., Hufnagl, P., Klette, R., 1985, Interactive Software Systems for Computer Vision, Progress in Pattern Recognition 2 (Kanal, L. N., Rosenfeld, A., eds.), North Holland, 57:78

INTELLIGENT TEST STANDS

FOR MULTIFUNCTIONAL USE

Jaromír Trenčina

Applied Physic Department
Research Institute of Civil Engineering
815 37 Bratislava, ČSSR

INTRODUCTION

The industry automation brought about at present by scientific technical progress requires for the monitoring and control of processes the utilization of up-to-date intricate, fast
and intelligent measuring information systems. For a longer time has been valid the principle that the aim of the experimental activity shall be the creation of a mathematical model of
object or process monitored. In addition at present in many
cases it is required to obtain results of experimental activity in real time so that they may be used to influence the process or its optimalization. To meet these requirements, which
have general validity in all branches of national economy even
in nonproductive sphere /health service, education/ at present
are being developed intelligent information measuring systems
of universal character whose basic configuration comprises
sensors, a multichannel data logger and a computing system.
The single information measuring systems differ from one another only by the number of measuring channels, type and intrica-
cy of the process followed and the dose intelligence stored in
all 3 parts of the basic system: sensors - data logger - computer. The intelligence distribution to single chain links has
been enabled by rush development of microprocessor technique.
Further are described 2 information measuring systems with
distributed intelligence we use in the building testing. The
first one is aimed at monitoring the static and quasistatic
experiments and the second one to analyse the dynamic processes.
The utilisation of such apparatuses grants at present in production, development and research large economic effects and
substantially increases the qualitative parameters of materials
and products and often leads to a rational design of new products. All this work with measuring information systems conducting to the determination of qualitative parameters in production are called computer diagnostic.

INFORMATION MEASURING SYSTEM FOR STATIC TESTS

Its block diagram is in Fig. 1. The system comprises sensors, a data logger and a computing system. The intelligence distribution as well as the technical parameters are described in single parts.

Sensors

Different sensors of nonelectrical quantities are used for the observation of the mechanical state of structures made from metal, concrete or plastic materials. Measurement of strain, deformation, strength, pressure, temperature, inclination, moisture content etc. is based on various principles. For the time being the built-in intelligence in sensor is relatively small and deals with the linearisation of the output signal, with the compensation of temperature effects and sometimes with the conversion of analog signal value into numerical value. In the future we may expect implementation of the microprocessor circuits into the sensor, increasing herewith substantially their accuracy, reliability and resistance to exterior disturbing influences. The sensors will be independent upon the supply voltage flutter, the length of leads and will have embedded a software from the sphere of mathematical statistics /standard deviation, mean varelue etc./. The data processed in this way will enter into the second part of the system.

Data logger

Its characteristic technical parameters are as follows:
1. The possibility of measuring and processing data from a large number of points measured /1-500/.
2. The fast automatic scanning of measured points /100-1000 channels/sec/.
3. The processing of analog DC and AC signals.
4. The processing of discrete signals.
5. The possibility of communication and further processing of data in real-time by means of data transfer over the RS232, RS422 and IEEE 488 /IMS-2/ communication interfaces.
6. The storing of measured data in a built-in casette tape or disk unit.
7. The possibility of battery feeding for the purposes of in--situ measurement.

In the sphere of data processing and experiment control there is implemented in the apparatus the following intelligence:
8. The conversion of eletrical signal measured quoted under points 3 and 4 into physical quantities.
9. The selection of a suitable measuring method for every programmed channel and the utilization of due algorithm for evaluation.
10. The evaluation of the course of the quantity observed in form of tables or plots.
11. The signalling of exceeding the quoted values defined in advance.
12. The correction of the values measured owing to temperature effects and the supply voltage flutter.
13. The generation of control signals for actuators of experimental devices on the basis of task and due to evaluated measured data.

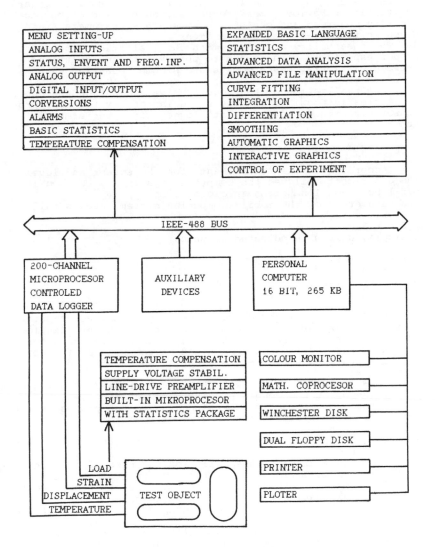

MENU SETTING-UP		EXPANDED BASIC LANGUAGE
ANALOG INPUTS		STATISTICS
STATUS, ENVENT AND FREQ. INP.		ADVANCED DATA ANALYSIS
ANALOG OUTPUT		ADVANCED FILE MANIPULATION
DIGITAL INPUT/OUTPUT		CURVE FITTING
CORVERSIONS		INTEGRATION
ALARMS		DIFFERENTIATION
BASIC STATISTICS		SMOOTHING
TEMPERATURE COMPENSATION		AUTOMATIC GRAPHICS
		INTERACTIVE GRAPHICS
		CONTROL OF EXPERIMENT

IEEE-488 BUS

200-CHANNEL MICROPROCESOR CONTROLED DATA LOGGER

AUXILIARY DEVICES

PERSONAL COMPUTER 16 BIT, 265 KB

TEMPERATURE COMPENSATION
SUPPLY VOLTAGE STABIL.
LINE-DRIVE PREAMPLIFIER
BUILT-IN MIKROPROCESOR
WITH STATISTICS PACKAGE

COLOUR MONITOR

MATH. COPROCESOR

WINCHESTER DISK

DUAL FLOPPY DISK

PRINTER

PLOTER

LOAD
STRAIN
DISPLACEMENT
TEMPERATURE

TEST OBJECT

FIG. 1

113

14. The use of programs of mathematical statistics with the view of the determination of errors and deviations and theapplication of regression analysis methods.
15. The setting-up of the whole measuring process with regard to its time behaviour, measurement sequence, kind of physical quantity observed and evaluation form. These parameters are given in form of the so called "menu".
16. The possibility of programming the evaluation procedures proper especially with regard to observe, the correlations between particular measurements and single measuring channels.

The realization of the so extensive evaluation and control procedures is possible only by embedding of several microprocessor circuits directly in data logger with programmes in ROM memories.

Computing

According to the requirements for the extent and level of the experiment followed the computing system may be used in off-line or on-line often in closed loop.
The parameters of the mobile computing system, are as follows:
- 16 bit word
- 256 k RAM
- display unit with coloured graphics
- printer
- X-Y plotter
- fixed disk
- 2 floppy disk units
- mathematical coprocessor

As it is visible from the block diagram the PC computer has a standard hardware. The decisive factor for the utilization of the PC is its software i.e. its intelligence. The characteristical parameters of the software aimed at the sphere of static and quasistatic observation, evaluation and control of the experiment are:
- The DOS operating system with BASIC 2.0 or BASIC 3.0 language.
- Floating point and double precision computations.
- Coloured graphic records corresponding to the proposals for the accuracy of evaluation according to specific conditions of the experiment.
- Drawing up protocols and tables documenting the whole experiment course.
-Computer aided drawing especially for the need of creating detailed and axonometric views of the experimental object model.
- Finite elements method for purposes of the structure mechanical state calculation in case of load change.
- Mathematical statistics for more complex utilisation of results measured and for assuring correlation analyses, creation of hypotheses, distribution of probability, smoothing etc.
- Possibility of programme control of the whole information system most often through the standard IEEE 488 bus.
- Possibility of creation of user subroutines and programmes according to the specific requirements of the experiment.
- Monitoring and control of the whole experiment with regard to its time behaviour and assuring protection against power failure and other undesirable influences.
The programm assuring all activities referred lays great claims to memory capacities of the computer /2 MB/ but determines the general level of the inforamtion measuring system with distributed intelligence.

114

THE INFORMATION MEASURING SYSTEM FOR DYNAMIC TESTS

Its block diagram is in Fig. 2. The system is composed of sensors of dynamic parameters of an object, a signal analyzer and a computing system.

Sensors

For measuring dynamic parameters of the object the sensors of amplitude, speed and acceleration are used. The due conversion among single physical quantities requiring the derivation or integration of measured values in such systems carried out means of the computer. Even for these types of sensors may be expected the intelligence increase after completing them with appertaining microlectronic circuits. These will be able in the future in many cases to change arbitrarily their frequency characteristics, to assure automatic calibration to correct exterior effects etc.

Signal analysis

Its characteristic parameters are:
- Number of channels:2
- Frequency range 0-100 kHz
- A/D convertor 12 bit, 2,4 us
- Input sensitivity 1mV-100V /full deviation/
- Digital frequency zoom lens
- FFT analysis of courses in real time up to 2 kHz /800 spectral lines/
- Data recording on floppy disk
- Number of samples 512/1024/2048

Remarkable, however, is the analyzer software written into the ROM memory and performed by several microprocessors. To the properties determining the intelligence degree of the device belongs particularly, programme package of wave analysis permitting the intricate evaluation of dynamic courses measured either directly or following the data record on the floppy disk unit. It bears especially on evaluation in the following domains:
- Time domain - averaging, autocorrelation, crosscorrelation, analysis of transfer effects.
- Frequency domain - FFT, complex spectrum, linear spectrum, cross spectrum, transfer characteristic, coherence, Nyquist's characteristic etc. convelution.
- Amplitude domain - probability of distribution, distribution function.
- Other domains: Cepstrum
- Data analysis is possible from the display, floppy disk or externally through the IEEE 488 bus.

Computing system

It is utilized to processing a large quantity of information obtained from the analyzer with the view of creating a mathematical model of the object observed. Its hardware configuration is identical to the computation system for static tests. In the sphere of software its intelligence is often completed with the following so called engineering tools /CAE/:
- CA Dynamic method of finite elements /CAFE/.
- Modal analysis of structures as far as to the animation of the observed model of structure on the display.
- Simulator of influence of changed dynamic parameters on the structure.

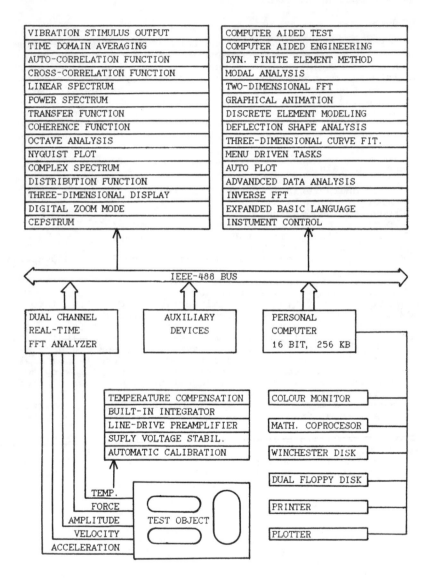

VIBRATION STIMULUS OUTPUT
TIME DOMAIN AVERAGING
AUTO-CORRELATION FUNCTION
CROSS-CORRELATION FUNCTION
LINEAR SPECTRUM
POWER SPECTRUM
TRANSFER FUNCTION
COHERENCE FUNCTION
OCTAVE ANALYSIS
NYQUIST PLOT
COMPLEX SPECTRUM
DISTRIBUTION FUNCTION
THREE-DIMENSIONAL DISPLAY
DIGITAL ZOOM MODE
CEPSTRUM

COMPUTER AIDED TEST
COMPUTER AIDED ENGINEERING
DYN. FINITE ELEMENT METHOD
MODAL ANALYSIS
TWO-DIMENSIONAL FFT
GRAPHICAL ANIMATION
DISCRETE ELEMENT MODELING
DEFLECTION SHAPE ANALYSIS
THREE-DIMENSIONAL CURVE FIT.
MENU DRIVEN TASKS
AUTO PLOT
ADVANDCED DATA ANALYSIS
INVERSE FFT
EXPANDED BASIC LANGUAGE
INSTUMENT CONTROL

IEEE-488 BUS

DUAL CHANNEL REAL-TIME FFT ANALYZER

AUXILIARY DEVICES

PERSONAL COMPUTER 16 BIT, 256 KB

TEMPERATURE COMPENSATION
BUILT-IN INTEGRATOR
LINE-DRIVE PREAMPLIFIER
SUPLY VOLTAGE STABIL.
AUTOMATIC CALIBRATION

COLOUR MONITOR

MATH. COPROCESOR

WINCHESTER DISK

DUAL FLOPPY DISK

PRINTER

PLOTTER

TEMP.
FORCE
AMPLITUDE
VELOCITY
ACCELERATION

TEST OBJECT

FIG. 2

- Simulator of force effects on the structure.
- Set of special programmes for the behaviour analysis of the structure in dynamic mode.
The extent of this software attains several MBytes in ROM memory or on the fixed disk of the computer.

ARTIFICIAL INTELLIGENCE

Artificial intelligence can find a large application in particular parts of systems where it will gradually replace the deciding activity of the user on which measuring and evaluation method is optimal for the given problem.
The artificial intelligence application will assure first of all the optimal distribution of diagnostic process into single parts of the whole measuring and evaluation chain.

CONCLUSION

The described systems together with 30 measuring apparatuses connectable to the IEEE-488 /IMS-2/ bus utilized at our workplace create conditions for rational and efficient diagnostic of the qualitative parameters of building materials, precast elements, machines and structures and represent very expensive objects.

THE UNIVERSAL WAVEFORM ANALYZER AS INTELLIGENT MEASUREMENT APPARATUS

P.DAPONTE (*) C.LANDI (**)

(*) Department of Electrical Engineering
     University of Cosenza
(**) Department of Electrical Engineering
     University of Naples   Italy

1. INTRODUCTION

Today the market offers a highly sophisticated measurement apparatus, called "Universal Waveform Analyzer", which allows to analyze signals with (i) nominal accuracy of up to 14 bit and 8 bit with 100 kHz and 100 MHz acquisition rate respectively, (ii) a very fast execution time of simple arithmetic operations (averaging, squaring, etc.) and more sophisticated mathematical functions (Fast Fourier Transform, convolution, correlation, etc.) by means of programming keys, (iii) a repetitive high speed recording, (iv) a very interesting price/performance ratio, (v) the possibility of linking other digital devices via a standard interface IEEE-488 at a trasmission rate of up to 1 Mbyte/s.

The widespread diffusion of low cost microcomputer systems with wide mass storage memory and good graphic performances, has prompted the Authors to improve the performances of a Universal Waveform Analyzer by interfacing it with a suitable Personal Computer in order to achieve the following advantages: (i) automatic choice of the most convenient acquisition and conversion parameters, (ii) appropriate presentation and storage of the acquired data, (iii) higher speed during the execution of complex measurement procedures, (iv) reduction of operator errors, (v) more programming flexibility.

In the present paper the Authors describe an Intelligent Measurement Apparatus built by coupling a DATA 6000 Universal Waveform Analyzer to an IBM AT Personal Computer with 30 Mbyte hard disk, Data Acquisition Control Adapter and Professional Graphic Adapter programmed by means of the Graphic Kernel System.

The most interesting applications carried out at the Electrical Department-University of Naples and at the Institute of Ophthalmology-II Medical Faculty of the same University are summarized hereunder.

2. THE UNIVERSAL WAVEFORM ANALYZER

The DATA 6000 provides human-engineered, key-programmed control of precision signal acquisition, conditioning, digitizing, and storing as well as computing, displaying, and transferring functions. Modularly designed, this instrument offers reconfigurability of both hardware and software to meet a broad range of signal processing problems in research, development, and manufacturing. The instrument mainframe accepts plug-in front-end modules which contain their own on-board ROM to communicate their protocols, capabilities, and configurable characteristics to the DATA 6000's managing software. The internal storage capability is 56 kwords and may be expanded with two 360 kbytes floppy disk drives. In addition, it uses software programmed groups of simple direct-performing push-buttons in a multilevel access mode for setup and control. For the analysis of repetitive waveforms, several signal processing features have been included in the instrument. Sweep to sweep weighted averages of the input signal may be generated. The user may select the number of sweeps to be averaged. A min/max envelope function allows simultaneous display of present values of the envelope over any time period. Four overlaid traces can display complete signal excursion along with the present value. Signal comparison may be made in a number of ways. A live signal may be compared to one displayed from the internal memory or from a floppy disk. All functions, including integration and differentiation, auto and cross correlation, are executed by a single keystroke, and can be automatically re-executed for each new data acquisition. Fast Fourier Transform is performed with a single keystroke from one real or two (real and imaginary) inputs.

The price/performance ratio of the DATA 6000 is reasonably low but, as a direct consequence of this, the apparatus presents some limits, the most important of which are: the keyboard is awkward to use, the graphic display is oscilloscope oriented, programming flexibility is scanty (essentially due to the high level of the macro functions implemented). In order to solve these problems, an appropriate software interface with an IBM AT Personal Computer was implemented by the Authors.

3. THE INTELLIGENT MEASUREMENT APPARATUS AND ITS APPLICATIONS

In order to make possible an intelligent use of the PC, analog and digital input controls together with digital and analog output signals are necessary. This is accomplished by means of the Data Acquisition Control Adapter board which includes (i) four analog channels connected to a 12 bit A/D converter through a 4/1 multiplexer (ii) two analog output signals coming from two separate 12 bit D/A converters, (iii) a 16 bit binary input port, (iv) a 16 bit binary output port and (v) one channel 16 bit event counter (cfr Fig.1).

Many applications which put in evidence the high power and the great flexibility of the implemented Intelligent Measurement Apparatus shown in Fig.1 have been carried out; some of them are illustrated in the following paragraph.

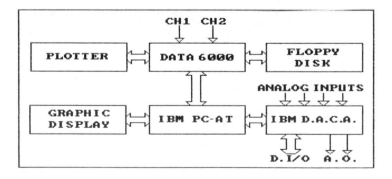

Fig.1 Intelligent Measurement Apparatus configuration

3.1 Passive components characterization

As reported in /1/ and in subsequent papers, the a.c. impedance measurements of a steel/concrete specimen allow to determine its kinetic parameters and to hypothize an equivalent circuit, which may vary with supply frequency. The impedance measurements were carried out by utilizing the well known transfer function analysis:

$$H(j\omega) = \frac{FV(t)}{FI(t)} = \frac{V(j\omega)}{I(j\omega)} = Z(j\omega)$$

where I(t) and V(t) are respectively the current excitation and the voltage response of the steel/concrete electrode hypothized as linear electrical system and the FV(t) and FI(t) quantities are the Fourier Transform Signals of V(t) and I(t).

The main advantages in the use of the Intelligent Measurement Apparatus in such an application are the great resolution (14 bit) of the acquired data and the fast computation speed in the on-line evaluation of the two Fourier Transforms. The PC-AT is used not only as measurement controller but also (i) to supply directly the specimen with a digital waveform generator according to the measured impedance values , (ii) to display graphically the complex impedance versus the frequency on the monitor during the test, and (iii) to change the equivalent circuit according to the frequency supply.

3.2 Electrical equivalent circuit identification

A typical approach to the study of a lead-acid battery behaviour is the identification of equivalent models whose parameters can be determined through electric measurements on the terminals of the cell /2/. These parameters are dependent on many influencing factors (stress, temperature, electrolyte density, aging, etc.). For the identification of the model a fitting algorithm was used: it consists in comparing the difference between the evolution of real and simulated voltage versus time, after a current step, and to minimize the differences by varying the model parameters. The PC-AT is used to optimize the fitting algorithm by adapting the sampling rate to the slope of V(t) diagram and to change battery working conditions, with the aim of creating a data base; the latter is the first step towards the implementation of an expert system for the Electric Vehicle battery handling, assessment, and diagnostics.

3.3 Echographic signal digital analysis

At present, the most significant application of the Intelligent Measurement Apparatus is tissue characterization in ophtalmic echography. It consists in an attempt to extract from ultrasonic signals those quantitative parameters which are intrinsic in the tissue itself and useful for medical purposes.

To this aim A/D conversion techniques seem to be very promising in ophtalmic echography tissue characterization /3/. They present, with respect to the consolidated techniques based on the storing of the echography result on Polaroid films, the following advantages: (i) more accurate data acquisition and presentation, (ii) signal digital storage, hence infinite persistence of the echograms on the monitor, (iii) accurate time measurement, hence distances and thickness may be evaluated by means of a suitable cursor,(iv) large possibilities of data processing, (v) creation of large data bases with fast and easy access. Both radio frequency and demodulated signals coming from the OPHTALMOSCAN 200 echograph were processed.

These operations are carried out with the PC-AT acting as intelligent measurement system controller. The PC-AT chooses the most appropriate

sampling rate, the gain of the input amplifier the trigger conditions for each selected signal frame and controls the information retrieval of echograms stored on the hard disk.
The major advantage of the Intelligent Measurement Apparatus has been found in digital average filtering and, more generally, in signal conditioning by reducing the influence of noise and disturbing signals. Moreover signal displays in axonometric form permits to more easily detect the presence of mobile parts (such as detached retina, melanoma, etc.) in the ocular bulbs.

4.CONCLUSION

The linking of a waveform analyzer to the IBM PC-AT allows to exalt the functions of the former,assigning to the latter the task of intelligent management of the acquired data. The experimental results obtained up to now, show that the Intelligent Measurement Apparatus built by the Authors may be considered as a new research tool whose usefulness was found very interesting for ophthalmic ultrasonic echography.

5. ACKNOWLEDGEMENTS

The Authors whish to express their gratitude to Proff. G.Savastano and F. Cennamo for the many stimulating discussions which gave a significant contribution to the development of this paper.

5. REFERENCES

/1/ Arpaia M.,Trisciuoglio G.: Identification of the Electrical Equivalent Circuit for Steel/Concrete Interface.IMEKO Symp.on Meas. and Estimation,Bressanone,1984.
/2/ Bernieri A., Noviello E.I.: Identification of Lead-Acid Battery Equivalent Models: Internal Parameter Measurement. IASTED Conf. MIC'85, Grindelwald,1985.
/3/ Cennamo F., Luciano A.M., Savastano M.: Real Time Analysis of Echographic Signals. 11th IMACS World Congress, Oslo, August, 1985.

DYNAMIC RANGE OF DIGITAL SPECTRUM ANALYZERS

István Kollár

Technical University of Budapest
Department of Measurement and Instrument Eng.
H-1521 Budapest, Hungary

INTRODUCTION

Digital spectrum analyzers belong to the most difficult-to-understand instruments. Their readouts and especially the possible errors present in their readouts cause a lot of headache for most users. The situation may be even worse, if an appropriate type is to be chosen on the basis of the available data sheets. Neither is it easy to calculate the performance of the usually available computer programs for spectral analysis: the non-specialist user is often helpless even with the best data processing package.

This paper deals with one of the key features of spectrum analyzers - their dynamic range. Even its definition is very often not quite clear - not to speak of the nature of the phenomena, which limit its value. In the paper these phenomena are briefly summarized, and expressions are derived to estimate the effect of two of them: input quantization and the often used block-float FFT.

THE MEANING OF THE DYNAMIC RANGE

The most frequently used definition is as follows: it is the ratio of the largest signal to the smallest signal that can be displayed simultaneously with no analyzer distortion products /Hewlett-Packard, 1974/, that is, it is possible to detect both signals unambiguously, with sufficiently exact readout. The auxiliary verb "can" means that this must hold true only if the signal levels are adjusted appropriately - the greater one is generally to be adjusted to the full-scale level. However, the dynamic range exhibits no direct relationship with the input range, which usually can be much greater.

In the resulting spectrum two different types of components can be distinguished, which put a limit to the dynamic range. First, there are spectral components depending on the greater signal itself, as e.g. the results of spectral leakage, or of nonlinear distortions. They are of deterministic nature, and thus they cannot be modified by averaging. Second, there are "spurious responses", consisting of the spectrum estimates of

the noises produced by the analyzer itself. Since the periodogram of such noises has great /≈100%/ variance, averaging can
help in decreasing their effect.

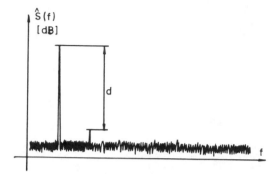

Fig. 1   Definition of the dynamic range

ERROR SOURCES IN THE MEASUREMENT OF SPECTRA

        Considering the architecture and usual operating principles
of Fourier analyzers, the effects of spectral error sources,
affecting the dynamic range, are as follows:
a./ Consequence of digital computations: bias and variance due
    to
      - the uniform input quantizer;
      - the window function coefficients, which are stored with
        limited accuracy;
      - the finite wordlength FFT;
      - averaging with finite wordlength;
      - the finite wordlength complex demodulator /if there is
        any/.
b./ Consequence of the use of the /modified/ periodogram:
    spectral leakage.
c./ The effect of non-ideal components in the analyzer:
      - nonlinear distortion of the input amplifier and the anti-
        aliasing filter;
      - finite rejection of the anti-aliasing filter;
      - time jitter of sampling, resulting in slight modulation;
      - differential and integral nonlinearities of the A/D con-
        verter;
      - statistical error of quantum levels /internal noise of
        the ADC/;
      - finite rejection of the low-pass filter in the complex
        demodulator.
The effect of some of the above error sources is rather obvious,
thus the errors can be effectively reduced by careful design,
e.g. a window can be chosen with very good sidelobe behaviour
/see Harris, 1978; Nuttall, 1981; Cox, 1978/, reducing spectral
leakage, etc. However, some others have not so trivial an
effect. In the following lines we are going to deal with two
error sources of the latter type: the input quantizer and the
finite wordlength FFT, using the commonly used block-float
number representation.

124

THE EFFECT OF THE INPUT QUANTIZER

Nonlinearity

The most awkward feature of A/D converters is their /differ-
ential and integral/ nonlinearity. A nonlinearity of ± 0.5 LSB
may result in an additive harmonious distortion component of
amplitude 0.64 LSB /see Fig.2./, which means,   using a

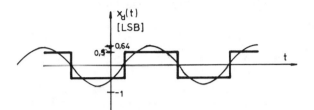

Fig.2   Harmonious distortion component
caused by nonlinearity

B-bit A/D converter and a $(2^{B-1}-1)$ LSB amplitude sine wave as
an input signal   that the dynamic range is not more than

$$d = 20 \cdot \lg\left(\frac{2^{B-1}-1}{0.64}\right) \approx (20 \cdot \lg 2)\,B - 20 \cdot \lg 1.3 \approx 6B - 2.1 \ , \qquad /4.1/$$

that is, 70.1 dB for B=12, and 58.1 dB for B=10. Thus, very
linear ADC-s must be used, and a dither signal with amplitude
of several quantum sizes /LSB-s/ is to be applied. This dither,
"moving" the measured signal along the ADC characteristic,
"averages" differential nonlinearities.

Quantization Noise

In addition to the above distortions, quantization noise
is always present in the processed signal. This is the quantization error of an ideal uniform quantizer, and it usually can be
modelled by means of a uniformly distributed, independent, additive white noise /Widrow, 1961/. This noise results at every
estimated point of the spectrum a component of second-order
chi-square /that is, exponential/ distribution:

$$\hat{S}_{n_q}\left(-k\Delta f\right) = \hat{S}_{n_q}\left(k\Delta f\right) = \frac{q^2}{12}\,\frac{1}{2}\,\chi_2^2, \quad k=1,2\ldots\frac{N}{2}, \qquad /4.2/$$

assumed that the length of the sampling interval, Δt, is chosen
as time unit, and the spectral estimator is

$$\hat{S}\left(k\Delta f\right) = \frac{1}{N_e}\left|X_w\left(k\Delta f\right)\right|^2, \qquad /4.3/$$

where

$$N_e = \sum_{i=0}^{N-1} w^2(i\Delta t) = \frac{1}{N}\sum_{k=0}^{N-1} w^2(k\Delta f), \qquad /4.4/$$

125

w(t) is the linear window function, and

$$X_w(k\Delta f) = \sum_{i=0}^{N-1} w(i\Delta t)x(i\Delta t)e^{-j2\pi\frac{ki}{N}}, \qquad /4.5/$$

that is, for smooth spectra $\hat{S}(k\Delta f)$ is approximately unbiased.

The components shown in /4.2/ are independent of each other. We assume a peak in the spectrum to be detectable, if it is significantly /with 95% confidence/ larger than the spurious peaks of the spectrum estimate of the quantization noise. This means in an N-point spectrum (N/2 independent spectral components) that

$$\frac{X^2}{4N_e}w^2(0) > S_{n_q}\left[-\ln\left(1-\sqrt[N/2]{0.95}\right)\right] \approx S_{n_q}\ln\left(\frac{N}{2\ 0.05}\right) = \frac{q^2}{12}\ln\left(\frac{N}{0.1}\right),$$

$$/4.6/$$

where X is the amplitude of the sine wave, and the picket fence effect is not taken into account. To consider the effect of this phenomenon, on the left

$$\frac{X^2}{4N_e}w^2\left(\frac{\Delta f}{2}\right)$$

should be written.

From /4.6/, using $X_{max} \approx 2^{B-1}LSB = 2^{B-1}q$, we obtain for the dynamic range:

$$d_1 = 20\ \lg\ \frac{2^B}{2\sqrt{\frac{N_e}{3w^2(0)}\ \ln\left(\frac{N}{0.1}\right)}} = 20\ \lg\ \frac{2^B}{\sqrt{\frac{4K_w}{3N}\ \ln\left(\frac{N}{0.1}\right)}} \approx$$

$$\approx 6B - 10\ \lg\left[\frac{4K_w}{3N}\ln\left(\frac{N}{0.1}\right)\right], \qquad /4.7/$$

where K_w depends on the window shape only:

$$K_w = \frac{N_e}{w^2(0)}\ N = \frac{\sum_{k=0}^{N-1} w^2(k\Delta f)}{w^2(0)}. \qquad /4.8/$$

If the spectra are averaged, because of the central limit theorem normal distribution may be assumed instead of χ_2^2:

$$d_m \approx 20\ \lg\ \frac{2^B}{\sqrt{\frac{4K_w}{3N}\ 1.65\ \frac{1}{\sqrt{m}}}} \approx 6B-10\lg\left(\frac{4K_w}{3N}\ 1.65\ \frac{1}{\sqrt{m}}\right),$$

$$m \gg 1. \qquad /4.9/$$

126

In the case of $N=256$, $K_w=3.6$ /Flat Top Window, see Cox, 1978; Kollár-Nagy, 1982/:

	B=10	B=12
d_1	68.5dB	80.6dB
d_{16}	81.3dB	93.4dB
d_{64}	83.3dB	96.4dB

THE EFFECT OF THE BLOCK-FLOAT FFT

In his classical paper Welch /1969/ suggests the following formula:

$$\frac{rms\{error\}}{rms\{result\}} < \frac{C \cdot N \cdot 2^{B_{FFT}}}{rms\{initial\ sequence\}},\qquad /5.1/$$

where B_{FFT} is the number of bits /<u>without</u> sign bit/: the number representation is B_{FFT} bits plus a sign; the initial sequence is adjusted in such a way that its maximal value is approximately 1, and C is a slightly waveform-dependent constant, which may be chosen in our case for $C=1.1$. Using a similar noise model as in the previous section /generally no better model can be suggested/,

$$d_1 = 20\ \lg \frac{2^{B_{FFT}}}{\sqrt{5 \cdot \ln \frac{N}{0.1}}} \approx 6\ B_{FFT} - 10\ \lg\left(5 \cdot \ln \frac{N}{0.1}\right),\qquad /5.2/$$

and in the case of averaging

$$d_m = 20\ \lg \frac{2^{B_{FFT}}}{\sqrt{8.25 \frac{1}{\sqrt{m}}}} \approx 6\ B_{FFT} - 10\ \lg \frac{8.25}{\sqrt{m}}\ ,\quad m \gg 1 \qquad /5.3/$$

are obtained. With $B_{FFT}=15$ and $N=256$ we get

d_1	74.4dB
d_{16}	87.2dB
d_{64}	90.2dB

COMBINATION OF THE ABOVE LIMITS OF THE DYNAMIC RANGE

Since the input quantization noise and the FFT roundoff noise are independent and approximately normal, in the result of the FFT their combination appears again as an additive normal noise with a variance

$$VAR_c = VAR_q + VAR_{FFT}.$$ /6.1/

Thus, an additive noise can be taken into account with the above variance. Concerning the values of the dynamic range, they should be transformed back to variances, and then summed:

$$d_{mc} = -10 \lg \left(10^{-\frac{d_{mq}}{10}} + 10^{-\frac{d_{mFFT}}{10}} \right).$$ /6.2/

E.g. from d_1=80.6dB and d_1=74.4dB the result is d_1=73.5dB.

On the basis of the above results it is understandable, why in modern Fourier analyzers usually 12-bit ADC-s and 16-bit block-float FFT processors are used, and why the dynamic range is commonly given as being 70-75dB. Also, the performance of a spectral analysis program can be easily checked using the above formulae.

REFERENCES

Cox,R.G., 1978, Window Functions for Spectrum Analysis, Hewlett-Packard Journal, 29/13:10-11.
Harris,F.J., 1978, On the Use of Windows for Harmonic Analysis with the Discrete Fourier Transform, Proc. IEEE, 66/1:51-83.
Hewlett-Packard Company, 1974, Spectrum Analysis ... Spectrum Analyzer Basics, Application Note 150.
Kollár,I., 1984, Fourier Analyzers from the Point of View of Measurement Theory, Cand. Sci. Diss., Hung. Acad. of Sci. /In Hungarian/.
Kollár,I., 1986, The Noise Model of Quantization, 1st IMEKO TC4 Int. Symp., Noise in Electrical Measurements, Como, Italy, June 19-21.
Kollár,I., Nagy,F., 1982, On the Design and Use of FFT-Based Spectrum Analyzers, Periodica Polytechnica - El. Eng., 26/3-4:295-315
Nuttall,A.H., 1981, Some Windows with Very Good Sidelobe Behavior, IEEE Trans. ASSP, 29/1:84-89.
Welch,P.D., 1969, A Fixed-Point Fast Fourier Transform Error Analysis, IEEE Trans. AU, 17/2:151-157.
Widrow,B., 1961, Statistical Analysis of Amplitude-Quantized Sampled-Data Systems, Trans. AIEE, II., Appl. and Ind., 79/52:555-568.

COMPUTER BASED REAL-TIME ANALYSIS OF SIGNAL PATTERNS IN A FREQUENCY RANGE OF ABOUT 5 DECADES

Klaus Faulstich, Frank Baldeweg, Joe Klebau

Akademie der Wissenschaften der DDR
Zentralinstitut für Kernforschung Rossendorf
DDR - 8051 Dresden, P.O.B. 19, GDR

INTRODUCTION

In many fields of nature and technique analog measuring variables occur in a dynamic range of several decades. An on-line analysis of this kind of signal patterns is very difficult in the case of using traditional measuring devices. Essential problems are:

- a highly complicated measuring equipment for analysis in the whole dynamic range, respectively

- a restriction of analysis of signal patterns in a subrange of dynamics due to hardware.

The application of computer based intelligent measurement devices offers a solution of these problems. The measuring equipment in this case can be very simplified and the analysis in the whole dynamic range can be carried out with an accuracy depending on the implemented software. By variation of software it is possible to solve various kinds of problems. Possible applications of this kind of analysing systems are:

- process monitoring in a given frequency range

- monitoring of seismic disturbances

- testing of analog signal patterns, produced by protection systems, using infrared or microwave technique /1/.

HARDWARE

For a real-time analysis system, which is able to gather and test signal patterns in a frequency range from 0.0009 to 25 c/s, a single-board computer based an the U880-CPU was developed. This system can be installed near by the place of measuring. Using a serial interface (20 mA current loop) a special service unit allows to display gathered data and results of the analysis, as well as to set parameters of the analysis. The block diagram is shown in Fig. 1.

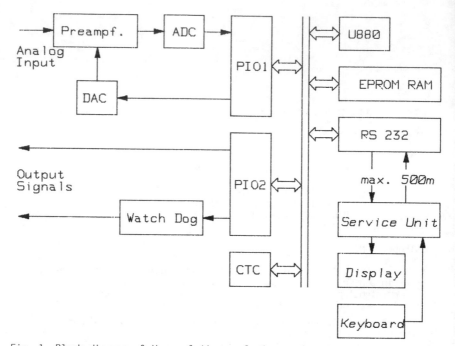

Fig. 1 Block diagram of the real-time analysing system

SOFTWARE

A real-time analysis of analog signal patterns in the above mentioned
frequency range requires:

- a subdivision of the whole range in separate subranges
- an optimizing of the software regarding computing accuracy and computing
  speed.

In the given application a data format of 8 bit was used. For converting
an ADC with a word length of 10 bit is provided. By a four stage cascade
of digital filtering and compressing algorithms, the signal pattern is
manipulated and then temporally scaled, stored in ring buffers (128 byte
long). For a scaling factor of 8:1 the analysis may be prepared in four
ranges of time:

- 2.5 sec with 0.02 sec resolution (high dynamic)
- 20.0 sec with 0.16 sec resolution (medium dynamic)
- 2.7 min with 1.26 sec resolution (little dynamic)
- 22 min with 20.5 sec resolution (very little dynamic)

The analysis of measured data concerns various parameters sequentielly
executed in the separate ranges of time. The upper limit for the execution
time is less then 1 ms. Extremely slow alterations in the signal patterns
are investigated by an indirect analysis of the preamplifier's control
voltage. In this case the preamplifier is assumed to be controlled by the

130

DAC after a restart as long as the value of the undisturbed signal patterns, which is gathered and stored in the computer, has not amounted to a medium value of the analysis region, that means about 256/2. The control of preamplifier takes place at discrete time points in dependence on the difference of the actually gathered signal pattern and the value of the undisturbed signal. Only one control step is activated each at time to compensate for the difference. By providing a threshold value and converting the actual deviation in control steps of the DAC (dependend on the operating point of the preamplifier), an analysis of extremly slow changing analog values is possible. In that case the ringbuffer containing the filtered and compressed data of the DAC will be analysed too /2/. By applying battery back up for the RAM, the corresponding ring buffer-regions containing date, time and features of analysis may be stored, if certain criteria are fulfilled. Thus, later analysis is possible even when power supply fails. The capability of self monitoring increases the availability of this analysing system.

TEST OF THE ANALYSING SYSTEM

To test analysing systems with the above mentioned dynamic ranges is often problematic because reproducible signal patterns of defined types of disturbances demand sophisticated storage technology. In this case it is advantageous to use a microcomputer in whose memory digitalised signal patterns for each disturbance type is stored /3/. These signal patterns may be theoretically derived or get from the real process. The stored signal pattern can be put out at discrete time intervals via a DAC with a eligible number of intermediate data got by interpolation. The eligible stretching of the signal pattern that was put out the DAC enables to test the analysis system in each subrange of time. In the case that the stored signal pattern was not put out after a call, the DAC provides the value of the undisturbed signal pattern.

RESULTS

A real-time analyser of this kind was installed in July 1982 for a protection system in the ZfK Rossendorf. The selected algorithms proved to meet all requirements, their resolution (8 bit) and computing speed. The analysis in the certain ranges of time for given features and criteria shows in the case of disturbances well defined statement concerning the signal patterns in frequency range from 0.0009 to 25 c/s. The system is working with high reliability in a temperatur range from -20 °C to +50 °C. The working time corresponds to a MTBF of about 30 000 hours.

REFERENCES

1. Bombrini, A., Valutorre o delegare, antifurto ago 84(40-42)

2. Faulstich, K., Verfahren zur Bewertung von Signalen in Überwachungs-anlagen, Patent DD 224982A1

3. Faulstich, K.; Bäumert, K.-H.; Voitel, D., Verfahren zur Simulation von Analogmeßstellen, Patent DD 228375A1

A NEW METHOD FOR MICROPROCESSOR-BASED REAL-TIME MEASUREMENT

OF PITCH DETECTION ON SPEECH SIGNALS

Wolfgang Lüdge, Peter Gips

Centre for Scientific Instruments
Academy of Sciences of GDR
DDR-1199 Berlin, German Democratic Republic

INTRODUCTION

The exact measurement of the fundamental frequency of speech signals is an important supposition for the diagnosis of neural disabled
probands by acoustic signals. Similarily, changes of the fundamental
frequency can yield statements about unbalances in the technical diagnoses of rotating parts.

Many methods are known to determine the fundamental frequency from
recorded signals /1/. But the number of methods is decreasing, when it
is necessary to determine the fundamental frequency in real-time. There
is a further limitation of the method number, if the choiced procedure
must be realized in a mobile device.

That's the point that we developed a method realizing the real-time
pitch detection with a relatively small effort using the commercial device "Modular Fourier Analyzer MFA 104". The device MFA 104 includes an
internal microcomputer, a 10-bit data acquisition up to 30 kHz sampling
frequency, an alphanumeric and grafic display and output possibilities
for plotter and printer. The chosen real-time algorithm is based on the
calculation of the average magnitude difference function (AMDF). Contrary to the known realizations (/1/, /2/, /3/) we used a time window
length being controlled by the algorithm itsself. The time-critical
operations are realized by a specially developed signal processor. For
the adaptive control of the time window length the "intelligence" of
the microprocessor is used.

COMPARISON FFT - AMDF FOR PITCH DETERMINATION

In the last years, many portable FFT analyzers were produced internationally. It follows the question, why it is not possible to use these
devices directly for pitch determination and why it is necessary to
develope more and more new methods and devices for that problem.

An FFT-analysis needs a certain time window lenght T during which
a number of N values is collected and transformed. The frequency-domain
resolution is the reciprocal time window lenght. Therefore, short time
window lengths being in the range of the pitch distances result in a bad

frequency-domain resolution. Longer time window lengths yield a better frequency resolution but the time resolution of the pitch fluctuation is bad, resulting in sidebands around the fundamental frequency. Additional problems arise from the leakage effect.

Pitch determinations yield better accuracy computing the Cepstrum. For this, a twice done Fourier transform, a squaring and the logarithm computation are necessary. That needs much computing time. For a time resolution of 10 ms, a real-time processing is not possible with commercial FFT analyzers. Last but not least, the significant peak of the AMDF-function is sharper than the corresponding peak of the cepstrum and the computation of the Average Magnitude Difference Function AMDF requires much less computing time. This function is defined as follows:

$$\text{AMDF}(p) = \frac{1}{M} \sum_{v=q}^{q+M-1} |x(v)-x(v+p)|$$

The AMDF performs a comparison for similarity of adjacent signal parts. Contrary to the correlation function, the AMDF has a minimum if the signal parts have the best match. This minimum is sharper than the maximum of the correlation function. The computation of the AMDF does not need multiplications what leads to a considerable computing time advantage.

METHOD

A direct running of the definition equation requires for the value range p M * M computing operations. One computing operation consists of the suboperations difference forming, absolute value computation and summation addition. A resolution of one per cent requires 10 000 operations during the time frame of 10 ms, that means to perform these three suboperations in 1 μs. Moreover, the minimum test and the output respectively the display of the computed value p in Hz are necessary.

Therefore the AMDF is computed over the full value range from 1 to M only at the beginning of the measurement and after a speech pause of 2.5 s. The processing of the next time window begins with a change of the shift p by the microcomputer being +30 % of the last computed average of two periods. The speech part length under test of 25 ms and the interval of 10 ms between the initial addresses of the time window remain constant for the next pitch determination.

In spite of an adaptive control of the value range of the AMDF, a hardware support of the time critical operations is necessary. That's why a special signal processor was developed, which computes the so called City Block Distance as a part of the AMDF for an initial address v and a number N. This stupid but very quick modul makes these three mentioned suboperations - difference and absolute value computation and summation addition - during 1.2 μs. The intelligence of the microcomputer needs more computing time and is used for adaptive control of the value range of p, for determination of the minimum, for curve display and for on-line computation of statistical parameters.

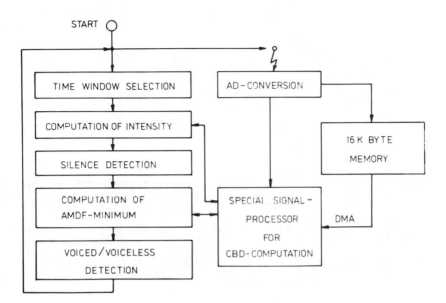

Fig 1   Flowchart of Real-Time Pitch Detection

With the beginning of the pitch determination program, the analog-digital-converter continously stores the data in a buffer memory. The 8 bit microprocessor U880 of the instrument's computer only interrogates the highest 8 bit from the 10 bit of the analog-digital converter with a sampling frequency of 6400 Hz. The buffer memory has 16 kbyte. After reaching the final address, the previous values will be overwritten continously. The program begins with a value range of shift p of 14.3 ms (e. g. 70 Hz). This corresponds to a number of 91 measuring values.

The computation of the pitch period is done within the time interval between the data acquisition and begins with the setting of the time window length T and the computation of the normalized intensity I within the time window length according to the following equation.

$$I = \frac{1}{N} \sum_{v=1}^{N} |x_v|$$

The intensity value I is a measure for the silence times. If I exceeds a silence threshold, the City-Block-Distance CBD is computated for the first value of the shift p as follows:

$$CBD = \sum_{v=1}^{n} A(v){-}B(v)$$

The acceptance of the number M of shifts and of the initial addresses v and v + p by the signal processor is realized as a port output of the microprocessor. After this the signal processor performs an autonomous Direct Memory access to the buffer memory and computes the City Block Distance. By a port-input, the City Block Distance is ac-

cepted by the microprocessor. Afterwards, the microprocessor tests the
accepted value, whether it is a minimum within the value range of shift
p.

By computing the ratio between the CBD value and the intensity, a
following test decides, whether the speech part to be tested is valued
to be voiced or voiceless. After that, the addresses incremented by 1, v
and v + p are output and the algorithm is repeated till the final value
of the shift is reached. The shift p having a minimum of the CBD values
is transformed into a value of the fundamental frequency and displayed
as a curve point. After 10 ms, the next pitch determination is performed. This distance of 10 ms as well as the sampling frequency and the
monitoring of speech pauses after 2.5 seconds are determined by a CTC
circuit and by interrupts.

In this manner, the curve of the fundamental frequency is displayed
for a duration of 8 s and overwritten continously. The microprocessor
parallely stores the function of the fundamental frequency from 40 s
speech and the time behaviour of the actual speech part of 2.56 s.

Figure 2 shows the result of a pitch period analysis as a hardcopy
from the display. The output to the display is carried out synchronously
in steps of 10 ms. The remaining computing time between the data acquisitions is used for the real-time computation of the frequency distribution of the pitch periods. After the end of the measurement the output
of the pitch period histogram and the statistic table is possible.

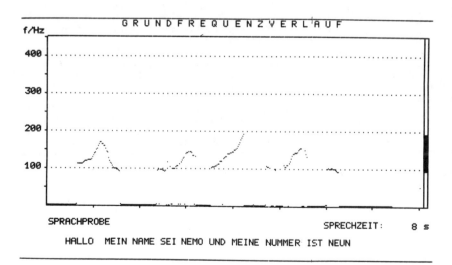

Fig. 2   Hardcopy of the fundamental frequency curve of a male speaker

136

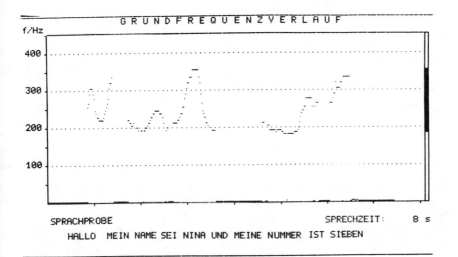

Fig. 3   Hardcopy of the fundamental frequency curve of a female speaker

REFERENCES

1. Hess, W.: Pitch Determination of Speech Signals.
    Springer-Verlag. Berlin, Heidelberg, New York 1983

2. Ross, M. J.; Shaffer, H. L.; Cohen, A.; Freudberg, R.;
    Manley, J. H.: Average Magnitude Difference Function Pitch
    Extractor. IEEE Transaction on Acoustics, Speech and Signal,
    Processing. Vol ASSP-22 (1974) No 5, pp. 353 - 362

3. Ney, H.: Bestimmung der Zeitverläufe von Intensität und Grund-
    periode der Sprache für die automatische Sprechererkennung.
    Frequenz 35 (1981) 10, pp. 265 - 270

4. Lüdge, W.; Schneider, W.; Bomm, H.; Modularer Fourieranalysator
    MFA 100. radio fernsehen elektronik, Vol. 34 (1985) No 9,
    pp. 583 - 587

ON RECONSTRUCTION OF MEASUREMENT SIGNALS WITH DISCONTINUITIES

Roman Z. Morawski

Institute of Radioelectronics
Warsaw University of Technology
PL-00665 Warsaw, Poland

INTRODUCTION

The problem of reconstruction of not-directly-measurable signals on the basis of other measured signals appears in many different areas of measurement-signal processing: in seismography[1], in calorimetry[2], in acoustics[3], in spectrometry[4,5], in defectoscopy[6], in pattern recognition[7]. It is of vital importance for all dynamical measurements since its particular case arises when errors introduced by the inertia of a measurement channel are to be corrected [8,9,10,11]. In all above-mentioned situations, an ordinary differential equation (ODE)

$$f_Y(t,y(t),\dot{y}(t),\ldots)=f_X(t,x(t),\dot{x}(t),\ldots) \tag{1}$$

is used most willingly as a model of the relation between a measured signal (y) and reconstructed one (x). It is usually assumed that the structure and parameters of the model are known and discrete values of y(t) obtained from measurements, inevitably subject to random errors, are given. The signal x(t) is to be estimated.

Numerical treatment of this logically simple problem meets many practical obstacles, caused by its numerical ill-conditioning[12]. It means that even relatively small errors of measurement data may cause unacceptably great errors of reconstruction. Those last ones have a tendency to grow unlimitedly when a sampling interval is reduced. Thus, in order to avoid excessive errors of signal processing, one must keep it about an optimal value, guaranteeing a minimum of the global error of reconstruction. Nevertheless, additional difficulties appear when the rising time of the signal x(t) becomes comparable with the sampling interval. The signal should be treated then as discontinuous one and reconstructed as a sequence of slowly-changing continuous signals. It would be easy if the right-hand conditions for each interval of continuity were known; but it is usually not a case since the reconstructed signal is not directly measurable.

The paper shows a simple method of overcoming that diffi-

culty. The method is developed for direct procedures of reconstruction, but may be useful also when other procedures are
applied.

NOTATION AND ASSUMPTIONS

    The following system of notation is used along the paper:
t – scalar real variable modelling time;
T – point of discontinuity;
h – sampling interval;
$f(t)$ – piecewise-continuous function modelling a physical signal
        (reconstructed one if $f=x$, measured one if $f=y$, etc.);
$\overset{(k)}{f}(t)$ – kth derivative of $f(t)$;
$\{\hat{f}_n\}$ – sequence of samples of $f(t)$ subject to measurement-or
        -computational errors;
$Var(\hat{f}_n)$ – variance of random variable $\hat{f}_n$;
s – complex frequency (differentiation operator);
$F(s)$ – Laplace transform of $f(t)$;
$G(s)$ – transfer function of the model;
$D(s)=d_0+d_1s+\ldots+d_{LD}s^{LD}$ – denominator of $G(s)$;
$N(s)=n_0+n_1s+\ldots+n_{LN}s^{LN}$ – numerator of $G(s)$;
$Q(s)=q_0+q_1s+\ldots+q_{LD-LN}s^{LD-LN}$ – quotient $\left.\rule{0pt}{20pt}\right\}$ resulting from
$R(s)=r_0+r_1s+\ldots+r_{LN-1}s^{LN-1}$ – remainder $\quad$ division $D(s)/N(s)$

The relation between x and y is modelled by a linear stationary
dynamical system

$$d_{LD}\overset{(LD)}{y}+\ldots+d_1\dot{y}+d_0y=n_{LN}\overset{(LN)}{x}+\ldots+n_1\dot{x}+n_0x \qquad 0<LN<LD \qquad (2)$$

which is assumed to be stable and minimal-phase. All the sig-
nals characterizing the system satisfy zero initial conditions.
Its response to any piecewise-continuous $x(t)$ is continuous
(LN < LD).

IDENTIFICATION OF THE PROBLEM

    The task of reconstruction may be logically decomposed
into two simpler ones:
– estimation of the left-hand side of (1) on the basis of measurement data, which results in $\hat{f}_y(t)$,
– subsequent solution of the ODE

$$f_x(t,x(t),\dot{x}(t),\ldots)=\hat{f}_y(t) \qquad (3)$$

with respect to the reconstructed signal, which results in $\hat{x}(t)$.
Direct methods of reconstruction consist in numerical differentiation of the data and numerical solution of (3). A corresponding scheme for differentiation should ensure limited amplification of measurement errors while a scheme for solving (3) must
be stable[13].

    Let us consider reconstruction of $x(t)$ which is disconti-
nuous in one point only, viz. for $t=T$. Since any direct-method
-based procedure of reconstruction is of recursive type, the
values of $\hat{x}(t)$ for $t > T$ are computed on the basis of earlier

estimated values of $\hat{x}(t)$ for t<T. It means that the effect of discontinuity at t=T is equivalent to the false assumption on the initial conditions for the second interval of reconstruction, i.e. for t>T.

Example 1. The system

$$\ddot{y}+3\dot{y}+2y=\dot{x}+3x \qquad (4)$$

excited by $x(t)=1(t-T)$ responds with

$$y(t)=1(t-T)\cdot[1.25-2e^{-(t-T)}+0.5e^{-2(t-T)}] \qquad (5)$$

Reconstruction of $x(t)$ by means of direct methods leads to the ODE

$$\dot{x}+3x=3\cdot1(t-T)+\delta(t-T)$$

Its solution for t>T

$$\hat{x}(t)=1+[\hat{x}(T+)-1]e^{-3(t-T)} \qquad (6)$$

depends on the estimate of x(T+). The error of reconstruction becomes zero if and only if $\hat{x}(T+)=x(T+)=1$ ●

DEVELOPMENT OF THE ALGORITHM

A very simple solution of the problem follows from its analysis in the domain of Laplace transforms where reconstruction may be described by the formula

$$X(s)=\frac{1}{G(s)}\cdot Y(s)=[Q(s)+\frac{R(s)}{N(s)}]\cdot Y(s) \qquad (7)$$

This natural decomposition of the inverse transfer function $1/G(s)$ means that the inverse filter is equivalent to the parallel connection of two filters whose transfer functions are $Q(s)$ and $R(s)/N(s)$. The first one is a differentiating filter of order LD−LN whilst the second one performs so called "real" integration.

The decomposition (7) implies corresponding decomposition of $x(t)$

$$x(t)=x_D(t)+x_C(t) \qquad (8)$$

where $x_C(t)$ − continuous component of the reconstructed signal, obtained as the solution of the ODE

$$n_{LN}\,x_C^{(LN)}+\ldots n_1\dot{x}_C+n_0x_C=r_{LN-1}\,y^{(LN-1)}+\ldots+r_1\dot{y}+r_0y \qquad (9)$$

and $x_D(t)$ − component with discontinuities of the first kind, computed as a linear combination of the derivatives of y(t), viz.

$$x_D=q_{LD-LN}\,y^{(LD-LN)}+\ldots+q_1\dot{y}+q_0y \qquad (10)$$

Since the system (2) is assumed to be of minimal phase, all the zeros of N(s) must have negative real parts. As a consequence, the solution of (9) with respect to $x_C(t)$ must be stable. Thus, the problem of reconstructing $x_C(t)$ on the basis of y(t) is numerically well-conditioned since $x_C(t)$ may be expressed in terms of the convolution of $y(t)$ and $\mathcal{L}^{-1}\{R(s)/N(s)\}$ which is a linear combination of functions of the type

$$t^k \exp(-|\alpha|t)\cos(\omega t+\varphi)$$

dwindling in time.

Example 2. The transfer function of the system (3) is

$$G(s)=-\frac{s+3}{s^2+3s+2} \tag{11}$$

Its inversion may be decomposed according to (7)

$$\frac{1}{G(s)} = \frac{s^2+3s+2}{s+3} = s + \frac{2}{s+3} \tag{12}$$

This means that the components of the reconstructed signal, $x_C(t)$ and $x_D(t)$, satisfy the following ODE's

$$\left.\begin{array}{r}\dot{x}_C(t)+3x_C(t)=2y(t)\\ x_D(t)= \dot{y}(t)\end{array}\right\} \tag{13}$$

where $y(t)$ is given by (4). Under zero initial conditions

$$\left.\begin{array}{l}\hat{x}_C(t)=1(t-T)\cdot[1-2e^{-(t-T)}+e^{-2(t-T)}]\\ \hat{x}_D(t)=1(t-T)\cdot[2e^{-(t-T)}-e^{-2(t-T)}]\end{array}\right\} \tag{14}$$

Thus, the reconstruction will be correct: $\hat{x}(t)=\hat{x}_C(t)+\hat{x}_D(t)=x(t)$●

SOME EXPERIMENTAL RESULTS

There have been compared numerical properties of two algorithms of reconstruction: A1 — without decomposition, and A2 — with decomposition given by (7). For approximation of the derivatives of $y(t)$, the symmetrical Nyström scheme of numerical differentiation

$$\dot{y}(t) \cong \frac{y(t+h)-y(t-h)}{2h} \tag{15}$$

has been applied, while for approximation of the derivatives of $x(t)$ — the backward Euler scheme

$$\dot{x}(t) \cong \frac{x(t)-x(t-h)}{h} \tag{16}$$

The solution of difference equation resulting from discretization has the form

$$\hat{x}_{in} = \sum_{\nu=-Ni}^{n} c_{i\nu}\hat{y}_{n-\nu} \qquad \text{for Ai, i=1,2} \tag{17}$$

where Ni — small, positive, integer numbers. Under the assumption that samples $\hat{y}_n$ are subject to random uncorrelated errors only, one may gather that

$$\text{Var}(\hat{x}_{in}) = \sum_{\nu=-Ni}^{n} c_{i\nu}^2 \text{Var}(\hat{y}_{n-\nu}) \qquad \text{for i=1,2} \tag{18}$$

and, if $\text{Var}(\hat{y}_n)=\sigma_Y^2$ for $n=0,1,\dots$, then

$$\text{Var}(\hat{x}_{in}) = K_{in} \sigma_Y^2 \quad \text{where} \quad K_{in} = \sum_{\nu=-Ni}^{n} c_{i\nu}^2 \quad \text{for } i=1,2 \qquad (19)$$

The coefficients K_{in} characterize algorithms sensitivity to random errors of data, independently of the signal to be reconstructed. Therefore, they may be used for comparison of the algorithms A1 and A2. If long-term signals are considered, then the limit value

$$K_i = \lim_{n \to \infty} K_{in} \qquad (20)$$

is of particular usefulness since it has the sense of "steady-state" amplification of random errors of data.

Numerical experiments, carried out for several transfer functions $G(s)$ and different values of h, have revealed some regularities: for those values of h which are of practical importance, the values of K_2 are significantly less than corresponding values of K_1.

Example 3. For the transfer function (11), the following values of K_i have been computed:

h	1	0.1	0.01	0.001	0.0001
K_1	0.77	2.1	14	80	450
K_2	0.89	2.6	8.4	27	84	●

The algorithms A1 and A2 have been also compared with respect to discretization errors in the frequency domain, viz. error functions

$$E_i(\omega) = \left| C_i(\exp(j\omega h)) - G^{-1}(j\omega) \right| \quad \text{for } i=1,2 \qquad (21)$$

have been computed for several transfer functions $G(s)$ and different values of h. As a rule, the inequality

$$E_2(\omega) < E_1(\omega)$$

has been observed for $\omega < \pi/h$.

Example 4. For the transfer function (11), the following values of $E_i(30)$ have been obtained:

h	1	0.1	0.01	0.001	0.0001
$E_1(30)$	30.5	29.7	4.45	0.446	0.0446
$E_2(30)$	30.8	28.5	0.445	0.0044	0.0001	●

Thus, the decomposition given by (7), designed for reconstruction of measurement signal with discontinuities, has turned to be a useful tool when applied to any other signal.

CONCLUSION

The above-described procedure was successfully applied in a calorimetric system for reconstruction of thermokinetics, i.e. of the course of heat power, on the basis of the course of temperature measured in the reaction vessel. It may be linked up with all direct methods of reconstruction, based on the model (2). The underlying it decomposition of the inverse transfer function (7) leads to a very useful decomposition of the task into two parts: pure differentiation of the measurement data (operator $Q(s)$), and elementary convolution (operator $R(s)/N(s)$). Since only the first one is numerically ill-conditioned, all regularization endeavours[12] may be concentrated on it. Thus, the algorithm seems to be of significance for other than direct methods of reconstruction.

REFERENCES

1. V. K. Arya, and H. D. Holden, Deconvolution of Seismic Data – An Overview, IEEE Trans. Geosci. Electron , vol. GE –16, No 2, 95:98 (1978).
2. E. Cesari, P. C. Gravelle, J. Gutenbaum, J. Hatt, J. Navarro, J. L. Petit, R. Point, V. Torra, E. Utzig, and W. Zielenkiewicz, Recent Progress in Numerical Methods for the Determination of Thermokinetics, J. Thermal Analysis , vol. 20, 47:59 (1981).
3. J. Mourjopoulos, P. M. Clarkson, and J. K. Hammond, Dereverberation of Speach Using Optimum Control, Proc. Conf. Digital Signal Processing – 84, Elsevier Sci. Pub., 415: 419 (1984).
4. A. Miękina, and R. Morawski, Noise Reduction in Spectrophotometric Data Processing Using Tikhonov Regularization, Proc. 1st IMEKO–TC4 Symp. Noise in Electr. Meas., 217:222 (1986).
5. V. V. Raznikov, and M. O. Raznikova, Deconvolution of Overlapping Mass Spectral Peaks Following Ion-Counting Data Acquisition, Int. J. Mass Spectrom. and Ion Process. No 2/3, 157:186 (1985).
6. L. D. Sabbagh, and H. A. Sabbagh, Some Numerical Techniques for Inverse Problems. Rev. Progr. Quant, Nondestruct. Eval., Proc. 10th Annu. Rev., vol. 3B, 899:906 (1984).
7. V. R. Martirosjan, Iteracjonnyj algoritm vosstanovlenija izobraženij ..., Avtometrija , No 3, 87:92 (1984).
8. P. M. Clarkson, The Application of Optimal Control Methods to the Deconvolution of Velocity Meter Signals, ISVR Tech. Rep. No 128, Univ.Southampton (1985).
9. K. Hejn, and A. Leśniewski, Digital Correction of the Dynamic Error Arising in the Input Circuits of the Measuring Instruments, Proc. 4th IMEKO–TC7 Symp. Meas. and Est., 51:60 (1984).
10. J. Jakubiec, Real-Time Numerical Correction of Transient Measurements, Proc. 3rd IMEKO–TC7 Symp. Comput. Meas., 97:100 (1981).
11. O. N. Tikhonov, Korrekcija rezultatov izmerenija ..., Izmier. Tekhnika , No 9, 11:14 (1967).
12. M. P. Ekstrom, A Spectral Characterization for the Ill-Conditioning in Numerical Deconvolution, IEEE Trans. Audio Electroac. , vol. AU–21, No 4, 344:348 (1973).
13. G. Hall, and J. M. Watt, (eds), "Modern Numerical Methods for Ordinary Differential Equations", Clarendon Press, Oxford (1976).

NUCLEAR BELT WEIGHER MODIFIED BY SCANNING METHOD

Johanngeorg Otto

Institut für Prozeßmeßtechnik und Prozeßleittechnik
Universität (TH) Karlsruhe
D-7500 Karlsruhe

1. INTRODUCTION

Belt weighers determine the flow of mass on continuous conveyor systems by measuring belt velocity and belt loading. Electro-mechanical belt weighers use the gravimetric force for metering the belt loading, nuclear belt weighers employ the effect of absorption of gamma radiation in matter with the advantage of contactless measurement. For this purpose a rod source, which can also consist in a chain of many point sources (Co^{60}, Cs^{137} or Am^{241}), is fixed below the conveyor belt. A cylindric szintillation counter above the belt detects the whole radiation arriving from the region of the conveyor belt.

The attenuation of gamma radiation by passing through matter depends on the density ρ, the mass coefficient of absorption μ and the thickness of the material h. If density, mass coefficient of absorption, incoming quantum rate K_0 and outgoing quantum rate K are known, in principle the thickness of the absorber can be computed. If the experimental setup is "ideal", i.e. consisting of point source, collimator, divided absorber, aperture stop and gamma detector, one can apply the exponential law of absorption according to Lambert-Beer, which is valid for "ideal" gamma rays only:

$$K = K_0 \exp(-\mu\rho h) = K_0 \exp(-\mu F), \qquad (1)$$

where $F = \rho h$ denotes the mass per unit area.

The real setup (see fig. 1 and 3) contains neither collimator nor aperture stop nor a devided absorber. The gamma radiation is not collimated at all in y-direction, that means transverse to transport direction. Additional to the radiation according to Lambert-Beers law of absorption there exists

145

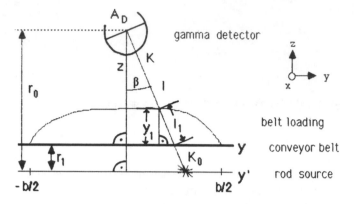

Fig. 1   geometrical setup and notations

scattered radiation, produced by the Compton effect, which reaches the gamma detector. Therefore the exponential law of absorption has to be modified with an additional factor, the buildt-up factor, which depends on the height of the absorber, the kind of material on the belt, the energy of the gamma quantums, the sensitivity of the detector and the geometrical arrangement of the belt weigher. The buildt-up factor must be determined empirically, as described in ref. /1/. For the following considerations we assume the law of Lambert-Beer already modified.

2. THE PROFILE ERROR

The total quantum rate K the szintillation detector counts is an integral of the quantum rate over the whole breadth of the conveyor belt:

$$K = \int_{-b/2}^{b/2} K_0(y) \exp(-\mu F(y)) \, dy \quad , \tag{2}$$

where $K_0 = K_0'(y'(y)) \cdot (A_D \cos\beta(y) \exp(-\mu \, \delta F(y))/4\pi r_0^2) \cdot f(y)$. $\tag{3}$

$K_0'(y'(y))$ denotes the activity per unit length of the rod source, the second factor of the right side of equation (3) is a geometric factor, which considers the angle β of the gamma rays, the distance between source and detector and the size A_D of the detector. The last factor $f(y)$ indicates the attenuation caused by the housing of the rod source and the belt.

The total quantum rate K is desired to be a one-to-one correlation to the belt loading resp. to the mean mass per unit area $\overline{F}$. Let us consider the

146

change of the total quantum rate ΔK when changing a standard material distribution $F_N(y)$ by a small deviation $\Delta F(y)$ to

$$F(y) = F_N(y) + \Delta F(y) \quad , \tag{4}$$

where $\Delta F(y) \ll F_N(y)$, and $\mu \Delta F(y) \ll 1$. This leads to

$$\Delta K = -\mu \int_{-b/2}^{b/2} K_0(y) \exp(-\mu F_N(y)) \, \Delta F(y) \, dy \quad , \tag{5}$$

which is proportional to the change of the mean mass per unit area $\overline{\Delta F}$ only if in case of the standard profil $K_0(y) \exp(-\mu F_N(y)) = \text{constant} = K_N$, and we have

$$\Delta K = -\mu K_N \overline{\Delta F} \quad \text{resp.} \quad K = K_N \int_{-b/2}^{b/2} \exp(-\mu \Delta F(y)) \, dy \quad . \tag{6}$$

So the rod source must be activated in such a manner, that all gamma rays penetrating the standard material distribution contribute the same fraction to the total quantum rate. Under these circumstances small changes of the belt loading can be detected correctly.

Large deviations from the standard profile cause errors in determining the belt loading, which up to now restrict the use of nuclear belt weighers. These so called "profile errors" can only be excluded in case of fine grained materials, whose profils can be formed by some device according to an expected standard profile. Let us assume a large but constant deviation $\Delta F(y) = \text{constant} = \overline{\Delta F}$. The conventional nuclear belt weigher estimates

$$\widehat{\overline{\Delta F}} = -\frac{1}{\mu} \ln \frac{K}{K_0} = -\frac{1}{\mu} \ln\{ K_N \int_{-b/2}^{b/2} \exp(-\mu \overline{\Delta F}) \, dy \, / \, K_N b\} = \overline{\Delta F} \quad , \tag{7}$$

which is the correct value. So constant deviations from the standard profile make no profile error.

Figure 2 shows a standard profile $F_N(y) = F$ constant over the whole breadth of the belt and two charakteristic deviations $\Delta F(y)/F$, which represent slowly variable profils and discontinuous profils:

fig. 2a): $\Delta F(y)/F = 2cy/b$ (8a)

fig. 2b): $\Delta F(y)/F = +c$ (8b)

Parameter c denotes the maximum deviation. Both profiles have $\overline{\Delta F} = 0$. The relative profile errors follow from equations (6), (7) and (8):

147

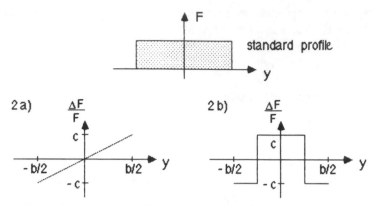

Fig. 2a/b standard profile and characteristic deviations

$$\widehat{\Delta F}(y)/F = - \ln((1/c) \cdot \sinh c) \qquad (9a)$$
$$\widehat{\Delta F}(y)/F = - \ln(\cosh c) \qquad (9b)$$

In case a) the maximum value $c = 1$ means that a fourth of the belt loading is shifted from the left to the right half of the conveyor belt, resulting in a profile error of -16%, in case b) $c = 1$ describes the situation that the material is concentrated on half of the breadth of the belt, and the conventional nuclear belt weigher shows an error of -43%.

3. THE SCANNING METHOD

The profile error exists, because the total quantum rate and the belt loading are not one-to-one correlated, if the profile form changes. The only way to resolve ambiguities is to analyse each gamma ray separately.

Figure 3 shows a nuclear belt weigher with conveyor belt (1), belt loading (2), rod source consisting of 16 point sources (3) and detector (4). There is a modification compared to the conventional nuclear belt weigher: a lead cylinder (5), which is placed between rod source and belt. The lead cylinder has N (here N = 3) pairs of slots helical arranged on a srew line. Driven by a motor (7) the cyclinder opens the way of the gamma rays from rod source to detector part by part, the material distribution is scanned within N scanning intervals. The partial quantum rate measured in each interval is used to evaluate the belt loading of the intervals separately. The scanning algorithm computes the mean value $\widehat{\Delta F}_s$ of the loadings $\widehat{\Delta F}_i$ of the intervals:

$$\widehat{\Delta F}_i = \frac{1}{N} \sum_{i=1}^{N} \widehat{\Delta F}_i \qquad (10)$$

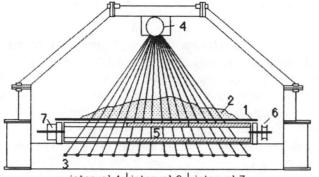

interval 1 | interval 2 | interval 3

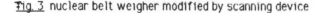

Fig. 3 nuclear belt weigher modified by scanning device

If there is an adequate high number of scanning intervals, the profile variations within the intervals can be neglected and the profile error disappears. The ratio of the error of the scanning methode with N intervals compared to the profile error of the conventional nuclear belt weigher is /1/:

$$\widehat{\Delta F_s}/\widehat{\Delta F} = 1 / N^2 \quad \text{(profile 2a)}, \quad (11a)$$

$$\widehat{\Delta F_s}/\widehat{\Delta F} = N_1 / N \quad \text{(profile 2b)}, \quad (11b)$$

N_1 is the number of discontinuities.

The laws of decay of radionuclides and of interaction of radiation with matter are statistical laws. According to the Poisson statistics the number of gamma quants detected per unit time fluctuates, even if the belt loading and the profile form are constant in time. The variance of the measured quantum rates leads because of the nonlinearity of Lambert-Beers law to a bias in calculating the belt loading, whose expectation value is /1/:

$$E\{(\widehat{\bar{F}}-\bar{F})/\bar{F}\} = K / (2K_N^2 T\mu\bar{F}) \cong 1 / (2K_N T) . \quad (12)$$

If the product of of measuring time T and quantum rate is K_N small, the scanning algorithmus has to be corrected by (12).

Scanning in connection with the motion of the belt causes loss of information, i.e. only 1/N of the area of the belt is irradiated. The scanning cycles must be so fast, that there are no important changes of the belt loading within one cycle. The variance of the measured belt loading increases due to radiation noise and scanning in comparison to the conventional nuclear belt weigher at least by the factor N.

149

4. EXPERIMENTAL RESULTS

The experimental results confirm the theoretical considerations. Figure 4 shows the measured belt loading (B_g) versus real loading (B) for scanning (B_s) and conventional (B_k) algorithm. Scanning gives the exact values of the belt loading, while the conventional method computes wrong values. As theoretical expected (see calculated continuous line), the conventional algorithm gives too small values, not only for this but for all tested profil forms. So the scanning method extends the applicability of nuclear belt weighers to totally arbitrary material distributions.

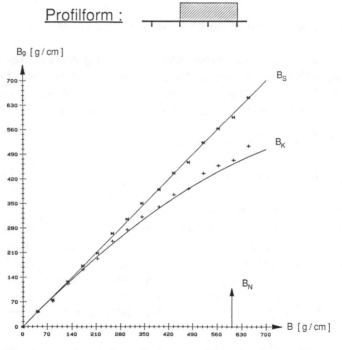

Fig. 4   experimental results in case of non standard profiles

5. REFERENCE

1.   Otto, Johanngeorg :
     Profilunabhängige radiometrische Messung bei Bandwaagen
     Dissertation am Institut für Prozeßmeßtechnik und Prozeßleittechnik
     der Universität Karlsruhe 1986
     Fortschritt-Bericht Reihe 8, Nr.111, VDI-Verlag Düsseldorf 1986

DEVELOPMENT AND APPLICATION OF A MULTIAXIAL CLASSIFICATOR ON PSYCHIATRIC DECISION SUPPORT

Gerd Koselowski, Hans-D. Hempel, Gert-E. Kühne

Department of Psychiatry, Clinic of Psychiatry and Neurology "Hans Berger", Medical School of Friedrich-Schiller-University of Jena, G.D.R.

INTRODUCTION

The support of multivariate statistics and clinical judgement simultaneously as a way for classifying psychiatric patients will be illustrated. By and large the traditional diagnostic groupings were usually based on an intuitively evaluation of a large number of complex variables. That means for the clinician difficulties and impossibilities of grasp these complex patterns of multifactorial determined conditions of the mental disorders. It has been difficult to isolate clear-cut diseases with unified aetiology, psychopathology, clinical manifestation and treatment (Paykal, 1981). In fact, multivariate statistical methods can assist in this process and are capable of defining and classifying large sets of data in specific clinical syndrome (Bartko et al, 1981).

In this paper we pursue the aim to describe a methodic for classifying psychiatric disorders by considering psychopathological symptoms and other components of the disease, e.g. age, sex, ICD-9 concept, course and psychopharmacotherapy. Nevertheless increasing of neurobiological and psychosocial connections psychopathological criteria have further a broad application and the operational criteria in psychiatric diagnosis permit an objective evidence of diagnostic concepts.

Using prognostically course data, premorbid personality and psycho-social constellations the Vienna Scientific Criteria were applied by Berner (1982) to a multiaxial method. Kühne et al (1984) used a polydiagnostic concept to estimate its for characterizing psychopathological basis syndroms. The analysis proves the importance of these multiaxial methods in order to receive results on the psychosis structure and therapeutic concepts.

METHOD

A combined application of multivariate techniques and
clinical judgements for developing a multiaxial classificator on psychiatric disorders will be demonstrated. The
psychiatric patients were examined with the Structured Psychopathological Assessment Scales (SPES) by Kühne and Grünes
(1983). This multiaxial system presented allows to structure
furthermore a large set of data.

At first the results of typical psychopathological syndro-
mes were derived by some steps of cluster analyses on the
levels of affectivity, perception disturbance, thought
disturbance vegetative symptoms and drive by explorative and
non-explorative assessments. In this way characteristic
types of psychopathological constellations were generated by
this multiaxiale classification.

After that the best cluster solution was selected by
statistical and clinical judgement and by cluster analysis
new calculated. Other statistical methods were applied in
addition to validating the results.

The validation was placed by multiple techniques
- Methods of discrimination
- Cross validation by $\mathcal{T}$-method
- Validation by outside criteria (ICD-9, sex, age, treatment)

At last quantitative methods and the configuration
frequency analysis were realized.

RESULTS

44 psychopathological types including precise psychopathological patterns were founded. These psychopathological
types represent clinical reliable and statistically valide
samples based on highly discriminating symptoms. The psychopathological types were determined in the procedure of
diagnostic judgement not only by the presence or absence of
the items but regarding the value of the components of the
middle vector appoint on a specific hierarchical rank. Within the types items with a lower component are not obligatory
on the classification of patients (Kendell, 1978), rather
these symptoms assist analogously as accessory items.

sample 1:
Type of Anxiety

Item	components of middle vectors
Restlessness	1,1
Restricted body feeling	1,07
Ambivalence	0,96
Feeling of tension	0,95
Restricted life feeling	0,95
Anxiety with fear of death	0,94
Feeling of inferiority	0,73
Feeling of guilty	0,53
Irritability	0,42

Applicating the system of the multiaxial classificator each patient will be classified to it's own "natural" psychopathological constellation. The approximate value to the patterns of the 44 psychopathological types will be determined in the following step. The characterizing of the whole psychopathologic syndroms of the patient contents the broad spectrum of the personality patterns.

sample 2:

day	characterization of psychopathological complex syndroms
1st day	Anxious-hyperaesthetic-hyperkinetic compulsion
2nd day	Anxious-hyperaesthetic-hyperkinetic-obsessive
3rd day	Anxious-hyperaesthetic-hyperkinetic-paranoid
. . .	. . .
30th day	Ataractic-normaesthetic-normokinetic-sensitive
. . .	. . .

DISCUSSION

A multiaxial classificator on psychiatric decision support was constructed on

- Generating of psychopathological types of items correspon- ding with axial syndroms of mental disturbances

- Discovering of hierarchical patterns in the distribution of psychopathological symptoms expressing the internal rules of the pathologic process

- Using the multiaxial classificator each finding can be considered to individually specific steps of the basic psychopathological process characterizing the natural structure of the different levels of the personality (drive, mood, cognition).

This method has the advantage, that with the demonstra- ted procedure of classification and the distribution to a specific type more informations can be received. Hempel 'and Koselowski (in press) illustrated a relatively successful application of these cluster results by course of patients with a differentiated drug treatment. It was possible to demonstrate specific cluster shifts all over the treatment time, at which assertions could be made to each point of the treatment.

This approach may provide an objective if some what inorthodox method for uncovering the complexities of the disease and the evaluation of its predictive validity by computerized systems.

SUMMARY

The development of operational criteria in psycho-pathological states enabled the identification of parameters on differentiating mental disorders and the reflection of basic psychopathological processes. Statistical analyses performed by multivariate techniques (Cluster Analysis, Factor Analysis, Configuration Frequency Analysis) have been defined syndromatic characteristics and classes of mental disturbances. Using new assessment instruments we found out more approaches for classification of individuals into "natural" and not preestablished groups of psychic diseases.

REFERENCES

1. J.J. Bartko, W.T. Carpenter and J. Strauss, Statistical Basis for Exploring Schizophrenia, Am. J. Psychiat. 138:941 (1981).

2. P. Berner, "Psychiatrische Systematik" Huber, Bern (1982)

3. R.E. Kendell, R. "Die Diagnose in der Psychiatrie" Enke, Stuttgart (1978).

4. G.-E. Kühne, H.D. Hempel and G. Koselowski, Toward the development of operational criteria of differentiated mental states, Symposium "Psychopathology and Nosology", Vienna, November 11-14, 1984.

5. G.-E. Kühne and J.U. Grünes, "Das Strukturierte Psycho-pathologische Erfassungssystem (SPES)" VEB G. Thieme, Leipzig (1983).

6. E.S. Paykel, Have multivariate statistics contributed to classification? Brit. J. Psychiat. 139:357 (1981).

AUTOMATION OF LABORATORY EXPERIMENTS USING

AN INTEGRATED LOCAL AREA NETWORK NETEX

Leszek Borzemski

Institute of Control and Systems Engineering
Technical University of Wroclaw
50-370 Wroclaw, Poland

INTRODUCTION

Local area networks (LANs) are of increasing interest in today's computer systems application environments. They can be seen as networks for possibly high speed data transfer among a group of stations via a common interconnecting medium within the geographical bounds of a single building, building complex, university campus or plant area. This paper presents the NETEX (NETwork for EXperimentation) LAN which has been developed and has been used in automation of laboratory experiments at the Institute of Control and Systems Engineering (ICSE), Technical University of Wroclaw (Borzemski et al., 1986a).

The functional features of the computer systems in the field of laboratory automation are similar to those elaborated for process control applications in industry, besides that the laboratory systems are characterized rather by shorter distances, limited number of stations, parallel wiring and much more programable intelligency required because they often must suit for different experiment conditions. On the other way modern laboratories on multi-site experiments comprise groups of measurement and actuation devices which have to be connected together by some communication means to perform a total experiment which has been partitioned into several local experiments. The structures of the computer systems to support such experiments may move from centralized to totally dispersed set of cooperating processors. This text presents an approach to the solution of automation of multi-site experiments where so-called CAMAC stations are connected to the NETEX LAN and are able to perform all local experiments whereas total experimentation is done using the services of local networking within NETEX. The system is not devoted to any specific laboratory, e.g. in physics or chemistry; it supports CAMAC instrumentation devices which can be found in several research and student laboratories. Moreover, due to the architectural features of NETEX, other measurement and actuation devices plugged into the network are also acceptable.

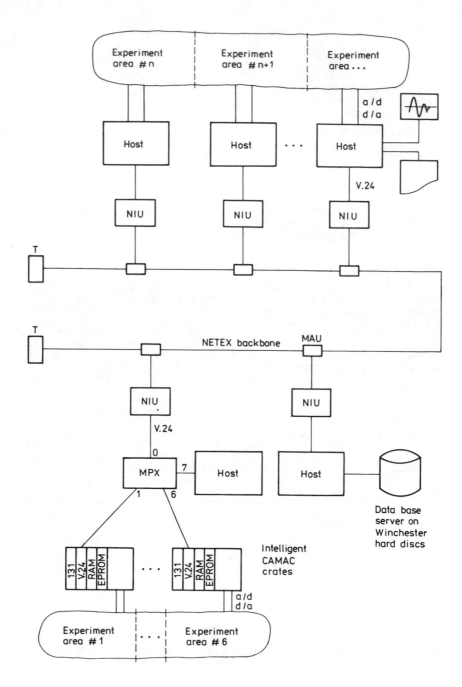

Fig.1 Experiment areas linked by the NETEX local area network

156

NETEX ARCHITECTURE

The structure of the NETEX is shown in Fig.1. The network hardware consists of a number of stations tapped on to a shared coaxial cable. NETEX utilizes baseband transmission technique with the data rate of 2 Mbits/s. At present, the length of the cable is about 400m but the transmission over 1 km cable is expected to have yet enough reliability. Additionally, the extension of the network can be realized using repeaters allowing the system to grow incrementally throughout an establishment.

Up to 100 stations can be attached to the network through the Network Interface Units (NIUs) and the Media Attachment Units (MAUs). MAUs, containing a line driver and line receiver, are tightly coupled without cutting to the cable. Special pucture taps penetrating cable and contacting center conductur are used. Due to this technique of tapping the stations can be attached in any location along the cable.

NIUs perform the functions of station-passive transmission medium interface implementing an access method following the Carrier Sense Multiple Access with Collision Detection (CSMA/CD) technique. This piece of microprocessor-based hardware consists of two parts: network - oriented and station-specific. The former performs functions for medium access control, the latter controls the exchange of data between host and the network-oriented part of the NIU. At present, the NIUs are supplied with serial RS-232C (V.24) interface with 9600 bits/s transmission speed.

The NETEX architecture follows all IEEE 802.3 specifications (IEEE, 1985b) but some medium attachment and baseband medium requirements. In conjunction with the CSMA/CD access method the IEEE Logical Link Control (LLC) protocol (IEEE, 1985a) has been used within the data link layer as defined by the ISO Open Systems Interconnection Reference Model (ISO, 1983) assumed to be our reference in building NETEX architecture. Following this, the NETEX provides data - link - connectionless as well as data-link-connection-oriented services for the Network Layer (Layer 3) which actually is empty. The Transport Layer (Layer 4) provides datagram-type data transmission functions using on request connectionless or connection-oriented LLC service. This protocol has been prepared on the basis of ISO studies on transport protocols and is NETEX specific. The upper two layers-Session and Presentation Layers (Layers 5 and 6) providing basic distributed functions - are also specific to our network and provide services in:

(i)   defining of structures for data sent between the stations,
(ii) sending data between the stations,
(iii) synchronizing actions of the cooperating stations.

The Application Layer (Layer 7) supports both common and specific application services. The common services are independent of the nature of the application and in the NETEX network they are fully provided via two primitives send and receive case on the basis of an integrated local area network approach (Borzemski et al., 1986b). They are used for sending and receiving data of declared type between processes running within the network. These primitives provide fully deterministic, symetric, "rendez-vous" - type communication mechanism. At present, a set of callable routines supports reliable transmission of different type of data between programs written in FORTRAN, BASIC, Pascal and Intel 8080/Z80/CAMAC assemblers, running on 8-bits microcomputers with CP/M 2.2, IBM PC/XT compatibles with MS DOS 3.10 and CAMEX (CAMAC crate executive).

The specific services, built on the basis of previously mentioned services or directly on transport services, provide capabilities to satisfy the needs of particular applications. In the NETEX they are available as server-stations or within general purpose stations. The servers are: CAMAC station and back-end database. The services which are general purpose stations resident are the following: message handling, file transfer, electronic mail, remote file access and job manipulation. Besides of these typical specific application services they are network services designed to support some chosen distributed processing procedures, joining both   common and specific application services.

CAMAC STATION

Process control and laboratory data acquisition systems are generally distributed. The communication is then needed on the board-level bus structures as well as between different plant/laboratory locations. At ICSE there are scientific works on the methodology and techniques of multi-site experiments performed for the needs of solving of complex tasks in system identification and pattern recognition (Koszałka and Borzemski, 1984). Several examples have been found in which distributed computation could be efficiently applied. To support data acquisition and object actuation for these needs we have developed so-called CAMAC station approach which gives experiment/process control capabilities in NETEX network.

The more common bus structures are Multibus, STD Bus, VMEbus and CAMAC (ESONE, 1972). CAMAC bus structure has been used because a broad spectrum of modules is supplied here and we have done much effort in order to develop the operating and application CAMAC software at ICSE.

The networking between plant/laboratory locations can be done by means of LANs with deterministic or random characteristics. The determinism provided by some access mechanism e.g. a token-passing algorithm is important, as it allows successful operation with up to 99 percent load. This feature is required when control devices must have the possibility to transmit an information to some other devices within known and limited time. There is the opinion held by some that LANs utilizing CSMA/CD access method i.e. random algorithm are not safe enough for process control because of collisions during heavy load and non-determinism in accessing the transmission medium. This can be overcome by choosing this kind of networks for applications without rigid real-time requirements, by keeping light load and by designing appropriate system architecture overcoming the objections. In NETEX we applied an approach of dividing the network into a "backbone" network and smaller process control dedicated subnets which are built on the basis of instrumentation bus structures. Thus, CAMAC crates are clustered into process control and data acquisition subsystems -

CAMAC stations - consisting up to six CAMAC crates with autonomous intelligent controllers e.g. type 131 , local microcom-
puter-based host and a multiprocessor-based multiplexer MPX.
Process control capabilities are also added to NETEX by a/d,
d/a interface cards of microcomputers plugged directly into
the network.

The station is linked to the network via NIU through one
of the MPX port which supports NETEX transport protocol services
and conversion between it and internal CAMAC station protocol.
Other ports are used by a station host and CAMAC crates - all
at the 9600 bps rate. MPX transfers data between any pair of
ports. Four transmissions can be served parallely by the MPX.
The highest priority has the transmission which actually is
going through NIU's port. Each crate and station host can be
individually addressed within the network, using concatenated
address formed from the individual address of the CAMAC station
in the network and the number of the port through which the
crate host is linked. Each crate works completely independently
under control of CAMEX executive. All controls, data acquisition,
local processing and storing are supervised by an application
program downloaded before its running from local or remote host.
All hosts can transmit data any time whereas CAMAC creates can
send data only on request. This scheme avoids the network
overloading.

At present, the transfer of sequential files, containing
data or programs in the format specified by host or crate, is
available. Then the files are in standard format for CP/M and
are converted during down/uploading to/from crate RAM. The same
service allows to transfer RAM contents between two crates via
MPX or via two MPXs if transmission between two CAMAC stations
is required. Unfortunately, current version of file transfer
does not support presentation functions, therefore they must
be realized by each pair of communicating programs locally,
depending on required features. The most frequent use of this
service is to transfer programs and data prepared under cross-
assembler of CAMAC assembler within host environment or directly in CAMAC environment. Then the files in hosts external
memories contain the code and data (that are not readable in
this environment) that are dowloaded/uploaded from/to a host
to/from crate RAMs.

CONCLUSION

We believe the development of LANs is an important step
in providing a service for experimentation. In the paper some
features of NETEX LAN have been presented. The first version
of NETEX has been fully operational in late 1985 and now is
being rapidly extended by CAMAC station, 16-bits microcomputers
and back-end database.

REFERENCES

Borzemski,L., Grzech,A., Kasprzak,A., and Koszałka, L., 1986a,
        Development of distributed applications for experiment
        and education support based on the NETEX local area network,
        in: "Computer Network Usage: Recent Experiences," L.Csaba,
        K.Tarnay, and T.Szentiványi, eds., Elsevier Science

Publishers B.V. (North-Holland), Amsterdam.

Borzemski,L., Kobylański, T., and Sas,J., 1986b, Synchronization and interprocess communication in the NETEX local
area network, in: Preprints of the 4th IFAC/IFIP Symposium
on Software for Computer Control, SOCOCO, May 20-23,1986,
Graz, Austria.

ESONE, 1972, CAMAC - A Modular Instrumentation System for Data
Handling, EURATOM Report EUR-4100e.

IEEE Std 802.2, 1985a, Logical Link Control, IEEE, Inc.

IEEE Std 802.3, 1985b, Carrier Sense Multiple Access with
Collision Detection (CSMA/CD) Access Method and Physical
Layer Specifications, IEEE, Inc.

ISO, 1983, Basic Reference Model for Open Systems Interconnection
ISO 7498.

Koszałka,L., and Borzemski, L., 1984, Computer-aided system
identification using a data base, in: Proc. 3rd Int.Conf.
on Systems Engg., September 5-7,1984, Dayton, USA.

FLEXIBLE CONTROL SYSTEM FOR EXPERIMENT AUTOMATION

WITH HIERARCHIAL STRUCTURED SOFTWARE

Sigurd Guenther

Laboratory of Neutron Physics

Joint Institute for Nuclear Research
Moscow, USSR

INTRODUCTION

The experiment and laboratory automation deals not only with data acquisition, but also with man machine communication or communication between machines themselves and control and monitoring of experiment parameters and status. Larger and more complex experiments require larger subsystems dealing with the experiment control, which must be very flexible. On one hand flexibility is required to write various measuring programs. This problem is normally solved by implementing special control languages. On the other hand, the system often has to deal with a very heterogeneous process control interface and experiment environment. Raising or changing requirements, system extensions and improvements can cause significant changes in hardware. To reduce the influence of the software a flexible modular software system is necessary. In this paper a proposal for a software architecture is presented.

HARDWARE STRUCTURE

The flexibility of the hardware can be achieved by applying modular structured intefaces such as CAMAC. The typical structure of an autonomous experiment control system with CAMAC or CAMAC like process interfaces could be divided into 3 major spheres (fig.1):
   - process I/O modules including the local bus,
   - measurement and control equipment (with transducer and actuators),
   - (sub)processes that have to be controlled.

SOFTWARE STRUCTURE

One way of minimizing the amount for software development caused by changes of I/O modules or process control equipment and system extensions is to design a software system with a hierarchical layered structure [1].
   Layered architectures, which have proved themselves good in the field of computer communication, have many advantages [2,3] :
   - independence of different layers,
   - simplier design and maintenance,
   - support migration of control functions from the control language level into the lower program level or into the hardware (intelligent I/O modules, co-processor, special controllers) to gain speed of information

processing.

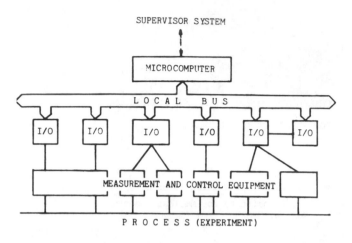

Fig. 1 Hardware structure of a control and monitoring system

Here each "hardware sphere" is represented by a software layer
(fig. 2). The function of a layer is composed by software modules, which
are linked together during system generation. A software module consists
of a parameter list containing the address of the program, module status
and specific control parameters (time constants, parameter limits, etc.)
and the program itself (fig. 3).

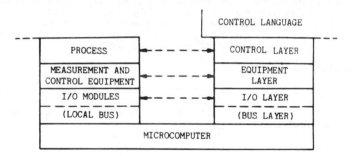

Fig. 2 Program layer structure

LAYER FUNCTIONS

I/O Layer

In this layer the influence of microcomputer peripheral modules is
handled, e.g. code conversion, timing. I/O device arbitration proceeds
here to reduce the lock time. The bus protocol is also carried out here.
In some cases it might be reasonable to put this protocol into a separate
layer.

162

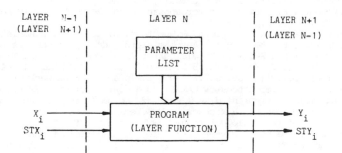

Fig. 3   Software module structure X_i, Y_i - process control
information; STX_i - program control information;
STY_i - module status; i - number of the process

Equipment Layer

Information processing in this layer has to ensure that data are
valid and effects of the measurement and control equipment, transducers
and actuators are "invisible" for the control layer (device timing, code
conversion, normalization of data, linearisation, static and dynamic
limit tests, etc.).

Control Layer

Here simple read or write functions, more time consuming control
functions, as DDC algorithms, digital filtering, or special control functions (2 of 3 selections, signal channel reconfiguration, etc.) are carried out.

Control Language Interface

In this layer the input parameters from control language have to be
scaled and tested.

EXAMPLE

For a neutron scattering experiment at JINR a data aquisition system
was designed, which consists of several loosely coupled subsystems [4,5]. On
On a monitoring and control subsystem (realized in CAMAC standard) a program system of the proposed structure is implemented. Its main functions
are to control magnetic fields, temperature, vacuum, to measure necessary
parameters, to monitor status of the experiment and device. The system
is controlled by an intelligent CAMAC crate controller based on Z80 CPU [6].
All programs are written in assembler. The software is embeded into the
control language MCL (Multi-Control-Language) developed for this experiment [7]. It has an open instruction set and provides great flexibility.

REFERENCES

1. W.P.Gertenbach, A hierarchical organization of real-time process
   control systems. In "Proc.of REAL-TIME DATA 79", West Berlin, Oct.1979,
   North-Holland Publ.Comp., 1980, p.p. 443-446.
2. A.S.Tan enbaum, Computer networks. Prentice-Hall Englewood Cliffs,
   N.Y. (1981).
3. J.Martin, Design and strategy for distributed data processing. Prentice
   Hall, Engelwood Cliffs, N.Y. (1981).

4. S.Guenther, O.I.Elizarov, B.Michaelis, H.Rodiek, Experiment Automatisierung unter Einsatz von Mikrorechnern. Wiss. Berichte der TH Leipzig, 1983, No.1, p.p.135-138.
5. W.Banse, J.Bätge, R.Bilkenroth, S.Guenther, B.Michaelis, H.Rodiek,
   W.Schwenkner, Ein Beitrag zur Automatisierung des Spektrometers polari-
   sierter Neutronen SPN-1 am Impulsreaktor IBR-2 des VIK Dubna. Wiss.
   Zeitschrift der TH Magdeburg 28 H. 1, p.p. 42-46, Magdeburg (1984).
6. S.Guenther, O.I.Elizarov, G.P.Zhukov, B.Michaelis, W.Schwenkner,
   K.-H- Schultz, Microprocessor CAMAC crate controller with 32 kbyte
   internal memory. Preprint JINR 11-84-482, Dubna (1984) (in Russian).
7. S.Guenther, M.Loebner, B.Michaelis, W.Schenkner, K.-H. Schultz,
   Application of the dialogue language MCL for automation of an autonomous multi channel analyser. In: "Proc.of XII International Symposium
      on Nuclear Electronics", Dubna, 2-6 July 1985, JINR Dubna (1986)
   (in Russian).

COMPUTERIZED TEST-SCHEMA FOR HIGH-ACCURACY MEASUREMENTS

Rinaldo C. Michelini and Giovanni B. Rossi

Istituto di Meccanica applicata alle macchine

Università di Genova - Italy

SUMMARY

An interactive measurement procedure used to perform the model identi
fication of a multimass vibrating test rig is discussed. The objective of
this (secondary) observation chain is the collection of additional informa
tion on the test rig dynamical behaviour, in order to improve the accuracy
of the measurements carried by the (principal) observation device. The re
sult is obtained   recalling an original CAT code, purposely developed for
the class of applications as the one described in the paper.

INTRODUCTION

Dynamical measurements represent a rapidily increasing field both
for industrial and for laboratory applications. The present paper specifi-
cally deals with the observation schema developed for the measurement of
the residual three-dimentional motion that appears as environment disturbance on an inertial testing rig realised by the Italian Metrological Institute Gustavo Colonnetti (IMGC) for the very accurate evaluation of the
lattice constant of silicon[1].
Referring to the study that follows, the inertial rig is assimilated
to a couple of masses, suspended over elastic bearing elements, endowed
with proper viscous   damping, structured with the purpose of filtering
away the vibration transmitted from the ground (see Fig. 1a). The residual
motion represents   a disturbance that adds to the reference frame, introducing relative displacements between the two terminals of an X-ray inter
ferometer.
The accurate evaluation of the residual motion, resolved into the related three-dimensional components, is a powerful information for the final trimming of the principal measuring chain, yielding to the conditionning compensation. The analysis, carried with a detailed insight of the phy
sical phenomena, is, moreover, helpful for possible redesign of the test
rig with due regard to its damping properties.
For both purposes, an auxiliary (independent) observation schema is

165

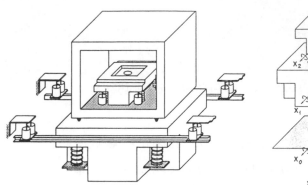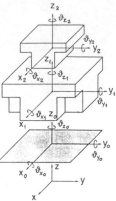

Fig 1

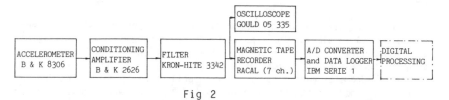

Fig 2

needed, in view of the lack of stationarity of the forcing   disturbances
transmitted from the ground of a densely populated industrial town (Torino),
where the IMGC is situated. A suitable characterisation of the residual mo
tion could be obtained, with the measurement of the statistical indices as
sociated to the driving noise and with the identification of the model parametrisation,once the system transfer functions are evaluated.

THE MEASUREMENT SET-UP

        The measurement and identification problem involved is highly challen
ging, due to the following main difficulties:
- the comparatively high number (twelve in the present schema) of the degrees
  of freedom effectively interesting the system dynamics;
- the low intensity of the observed signals (the accelerometers outputs
  are at the μg level);
- the comparatively low number (six in the actual schema) of   points simultaneously under observation;
- the dependance on environment forcing disturbances, highly cross-correla
  ted and, possibly, not persistently exciting.
        So an interactive experimentation project has to be developed to ob-
tain sufficient estimates of the (additional) degrees of freedom out of the
subset of the points direcly observed. The measurement chain is shown in
Fig. 2.

Reference model

The modelling of the test rig dynamical behaviour is necessary for:
(a) the conditionning of experimental observation with the a priori system
hypotheses; (b) the planning of the interactive experimental sequence
through which the identification problem is satisfied.
The following hypotheses are presumed:
- a simplified geometry is assumed, leading, to the lumped parameters model
  sketched in the Fig. 1 b;
- the motion is limited to "small-oscillations", so to authorise three-di
  mensional linearised displacements;
- the reference solid-frame is assumed rigid, with forcing actions not influenced by the reflected motion of the environment-frame;
- the model parametrisation allows for constant and linearised characterisa
  tion of the forces transmission.

The assumptions, lead to reflected energy uncoupling and, actually,
only the last one is critical and has to be validated only in statistical
sense, with acceptable uncertainty bounds.

The elements of the test rig are referred as shown in the Fig. 1 b,
and dynamical input-output relationships result as follows:

$$\tilde{q} = \tilde{Q}(s)\, \tilde{q}_0 \quad ; \quad \tilde{Q}(s) = (Ms^2 + Cs + K)^{-1} (C_0 s + K_0)$$

$\tilde{q}$ =local generalised displacements
$\tilde{q}_0$ =reference frame displacements.

It is worth noting that: (a) the model being of the second order, the
system has n_0 inputs (corresponding to the number of the d.o.f.) and $2n_0$
outputs; (b) the angular displacements are conveniently transformed after
a multiplication with a characteristic length, for dimensional homogeneity.

Experiment planning

Finally, to optimize information obtainable from the test sequence,
under the constrain of no more the six measurement channels, further assum
tions are made:
. the components ϑ_z are neglected;
. the centre of gravity, due to the symmetry, has negligeable ϑ_x and ϑ_y
  components, along the z axis;
. the reference xyz are directed along directions that are principal of
  inertia.

The following independent-information sub-sets are, accordingly, defined:

$$S_I = (z_0; z_1, z_2) \quad ; \quad S_{II} = (x_0, \vartheta_{y_0} ; x_1, \vartheta_{y_1} , x_2, \vartheta_{y_2}) \quad :$$

$$S_{III} = (y_0, \vartheta_{x_0} ; y_1, \vartheta_{x_1} , y_2, \vartheta_{x_2})$$

The resulting experiment planning is shown in the annexed Table 1.

Table 1

No	MEASUREMENT POINTS	ESTIMATED VARIABLES	SUB-SYSTEMS	REMARKS	No	MEASUREMENT POINTS	ESTIMATED VARIABLES	SUB-SYSTEMS	REMARKS
0				Preliminary test for measurement schema calibration	4		x_o, y_o, z_o x, y, z	$S_I, S_{II},$ S_{III}	$\ddot{z}$ biased by $\vartheta_{x_o}, \vartheta_{y_o}$ x, y, z biased by ϑ_x, ϑ_y
1		x_o, y_o, z_o $\vartheta_{x_o}, \vartheta_{y_o}, x$	S_{II}	$\ddot{x}$ biased by ϑ_y	5		x_o, y_o, z_o $z, \vartheta_x, \vartheta_y$	$S_I, S_{II},$ S_{III}	$\ddot{z}_o$ biased by $\vartheta_{x_o}, \vartheta_{y_o}$
2		x_o, y_o, z_o $\vartheta_{x_o}, \vartheta_{y_o}, y$	S_{III}	$\ddot{y}$ biased by ϑ_x	6		$z_o \vartheta_{x_o}, \vartheta_{y_o}$ x, y, z	$S_I, S_{II},$ S_{III}	x, y, z biased by ϑ_x, ϑ_y
3		x_o, y_o, z_o $\vartheta_{x_o}, \vartheta_{y_o}, z$	S_I	$\ddot{z}$ biased by ϑ_x	7		$z_o \vartheta_{x_o}, \vartheta_{y_o}$ $z, \vartheta_x, \vartheta_y$	$S_I, S_{II},$ S_{III}	

TEST OPERATIONS AND RESULTS

The identification of a dynamical model, as the one considered, within actual running conditions, that is, with correlated driving inputs, de serves the extended application of the individual options offered by the computational code IMIMS , further requiring the addition of modules that directly manage the interactive identification schema. The mean features of the code and the related procedural aspects are not object of discussion of the present paper; hereafter few typical results are presented as characterising examples of the test operations. The arrangement improving of the closure of the computerised information set-up depends on the model transfer properties, and good insight on these properties is provided by the evaluation of the coherence figures and of the frequency responses. Few comments are reported in the following.

Coherence patterns

Typical coherence plots are shown in the Fig. 3 a. The following aspects are pointed out:
- coherence between two signals from the same mass location (trial 0): an evaluation of the trustfulness of the testing conditions is obtained with regard to the SNR.

168

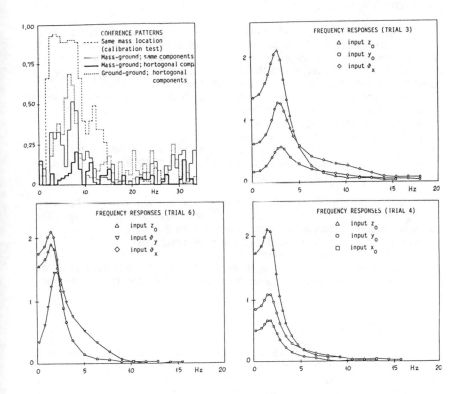

Fig. 3

It is shown that, in the range 1 to 15 Hz, the coherence level is acceptable, while for higher frequencies it decreases rapidly.
- coherence between a ground signal (x_o) and the one corresponding to the
same direction of the first mass (x_1).
The coherence ratio lowers, essentially due to estimate biasness and system multidimensionality.
- coherence between a ground signal (y_o) and one corresponding to a normal
direction of the first mass (x_1). The coherence ratio lowers even more,
since the two signals are (originally) less correlated. However the feature is not critical due to the correlation between inputs.
- coherence between orthogonal ground signals (x_o and z_o). The coherence
between two input signals is comparable with the one measured between an
input and an output signal . To that purpose, the test conditions are
poor, as the driving inputs are by no means independent.
    The coherence functions are estimated by the IMIMS code, operating
with overlapping periodograms (up to the 50%), preprocessed through the
usual Hanning-windows.

169

The frequency responses

Few example frequency responses are shown in the Fig. 3 (b-d); the functions are computed with a parametrical identification procedure, that emploies the minimum square error criterion. Let consider the vibrational behaviour in the vertical direction , that is function of z, ϑ_{x1} and ϑ_{y1} (considering the experiments 3, 4 and 6 on the reference table).

The preprocessing performs the suppression of the continuous component and the attenuation of the lower frequencies with a first-order high-pass filter, for the reshaping compensation.
The following aspects are pointed out:
- Fig. 3 b: The dependance on z_o is the strongest and the structural frequency is readily recognised. Another pole near 10 Hz appears from parametrical analysis, which is probably due to the influence of the second mass.
- Fig. 3 c, d: A comparison between frequency estimates obtained through independent tests, shows acceptable variance only for the z_o input. The lack of repeatability with other inputs may be due to:
  . very low singlas level (especially for ϑ components)
  . high coupling between driving noises.

These remarks suggest, as further development, the analysis of a possibly more robust model with independent motion along the three orthogonal x,y,z axes.

CONCLUDING COMMENTS

The characterisation of the attributes relevant for defining the dynamical behaviour of the testing-rig, within in field working conditions, requires the processing of the experimental data preserving the related time-structure with an interactive identification procedure, since only six measuring channels are simultaneously operating.

The experimentation set-up is, therefore, designed to provide the appropriate information redundancy, conditioning the time-signals, over "sufficiently" extended observation windows, with the functional relationships hypothesised as characteristic constraints of the rig dynamics. Through the interactive procedure, sequences od independent information sub-sets are singled out with the simplifying assumptions, recalled for the experimental planning, and the required rig-attributes are, actually, evaluated and updated.

REFERENCES

1. G. Zosi, Avogadro's constant and X-ray interferometry, ACTA Metrologica Sinica, n. 5, pp. 228-243 (1984)
2. H.G. Natke (ed.),"Identification of vibrating structures", Springer Verlag, Wien (1983).
3. C.W. Silva,"Dynamic Testing and Seismic Qualification Practice", Lexington, Toronto (1983).
4. G.A. Bekey and G.N. Saridis,"Identification and System Parameter Estimation" Proc. of Sixth IFAC Symposium: Pergamon Press (1982).
5. P. Eykhoff, Identification Theory: Practical implications and limitations, Measurement, vol. 2, NO 2 (1984).

HIGH QUALITY ELIMINATION OF LIGHT FLUCTUATIONS IN SENSORS

Rüdiger Haberland

Lehrstuhl Feinwerktechnik
Universität Kaiserslautern
D-6750 Kaiserslautern, Federal Republic of Germany

OVERVIEW

Photo Sensors which are measuring the amplitude of
incoming light are most often degraded because the intensity of the light source is not stable enough. A novel compensation circuit utilizing a pulse - width - modulator as
a divider circuit is compensating for the fluctuations by
dividing sensor light level by a reference light level. The
achieved accuracy in this quasi-analog divider is much
higher than in any other analog divider thus resulting in
an ecceptionally stable sensor.

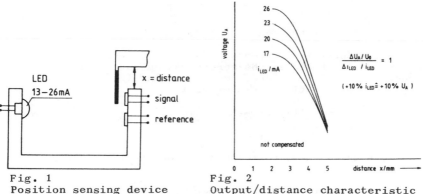

Fig. 1
Position sensing device

Fig. 2
Output/distance characteristic

1. Experimental Setup

For this test a very simple position sensitive device
was chosen. (Fig. 1) It consists of a light emitting diode
and two photo diodes mounted in a metallic u-shaped part.

The marked nonlinearity is not of interest here because the sensor is intended as a position null detector.

2. The pulse width - modulator (PWM)

The first current is charging a capacitor from zero to the reference voltage, the charging time is t_1.

The second current is charging the same capacitor for the time t_2. At the beginning of each charging cycle the capacitor is short circuited.

An output is generated - positive voltage U_A - during t_1. The output is switched to - U_A during t_2.

The mean value $\dfrac{U_A(t)}{U_A} = \dfrac{t_1 - t_2}{t_1 + t_2}$ is thus giving the

dividing principle.

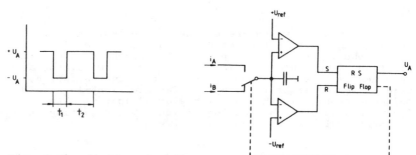

Fig. 3 The divider circuit

3. Generation of the function i_1/i_2.

If we have two currents i_1 and i_2, i_1 representing the reference light and i_2 the sensor raw output and if we generate: $\Delta i = i_1 - i_2$ and $\Sigma i = i_1 + i_2$ and if we feed this into the above outlined PWM we yield

$$\frac{U_A(t)}{U_A} = \frac{\Delta i - \Sigma i}{\Delta i + \Sigma i} = \frac{i_2}{i_1}$$

the function we need to compensate for the light intensity fluctuations.

There are two different possibilities to generate the sum and difference values.

(a) The currents from the two photodiodes are first converted to voltages and then the sum and the difference is computed in the well known analog summing circuit. Thus we need four operational amplifiers and some matched resistors.

(b) The photodiode configuration is switched from parallel in the charging cycle t_1 to antiparallel in the charging
cycle t_2.

4. Typical error mechanisms of the PWM - divider

The basic PWM as depicted in chap. 2 is not susceptible
to instabilities of the charging capacitor. This is true not
only concerning zero stability but also scale factor stability.
The most prominent error source is arising from the two output voltages U_A and $- U_A$. If their magnitude is not the same
a bias is generated. If their magnitude is the same but drifting a scale factor drift results. If their drift is asymmetric, bias drift will result. As reference sources with
temperature coefficients down to 1 ppm are easily obtainable
there is no problem bringing down this part of the drift
coefficient to any desired level. Another drift source is
the temperature coefficient (TC) of the switching time of
the FET - switches as the switching time is between 4 and
400 (approximately) nanoseconds and the switching time TC
is approximately 0.5%/K. The thermal difference between the
switches defines the actual level of mismatch between TC
giving a factor between 10 and 1000 better performance than
the original TC.

If a performance of 10^{-5} is wanted, switching-time-dif-
ference-drift is fixing the operating frequency around
1..10 kHz.

The stability of stray capacities is another limiting
factor but this influence can be made very small, too.

The sum and difference generating circuits can be
designed to a better accuracy than the switching network.
The design of the switching network itself has to be done
carefully to avoid any undefined switch position at any
time.

5. Results

Using the second possibility of parallel - antiparallel
switching the photodiodes we obtained in the setup of fig. 1
the following results.

The sensitivity to fluctuations in the light-level -
light-level measured as current in the LED - was decreased
by a factor of 7500.

So for 1% change in light level there is only $1.5 \cdot 10^{-6}$
output change. This is supposed to be accurate enough for
most applications but further improvements are possible.

With this result the very simple position sensor of
fig. 1 can be used with an accuracy of a few nanometer,
provided the other error mechanisms are by a careful construction brought down to a comparable level.

6. Typical applications

In many instruments photo sensors are used, for example for position sensing in balances, operating typical at a light level of 50% the maximum light level as a null-indicator, so that any light level fluctuation will be misinterpreted as a signal.

A comparable mechanism is valid in distance sensing fiber optic sensors.

7. Pulse operation

If it is necessary to eliminate ambient light fluctuations it is possible to modulate the light as in other sensors.

Two different modes are possible: a. Low frequency modulation of maximum half the operating frequency of the PWM and b. synchronous operation of light modulator and PWM similar to the commutating auto-zero technique used in operational amplifiers.

8. Other useful applications

In the light sensing area the position sensitive devices and the difference photodiodes and the quadrant-diodes demand for the same dividing principle. Today this is implemented with high cost and bad accuracy with conventional analog dividers. This new principle will give much better results at lower cost.

True linear readout of differential capacitors by a similar technique has been reported earlier. Any other application that is utilizing a divider can be implemented by this scheme, too. The possible operating frequency may go up to above 10 MHz.

9. Conclusion

The principle of utilizing a pulse-width-modulator as an analog divider gives much better performance in light sensing devices and many other sensors. High quality of the divider ($1o^{-4}$ - 10^{-6}) and low cost are opening up a wide field of sensor improvements.

10. REFERENCES

1. P 35.19.162.0
2. P 35.31.378.1
3. P 35.24.530.1
4. R. Haberland, Kapazitive Beschleunigungs-Sensoren in Sandwich-Technik, Sensor 85, Karlsruhe

THE THERMOPHYSICAL LABORATORY DATABASE

Emília Illeková, Ľudovít Kubičár, Karol Vass[*]

Institute of Physics EPRC SAS
842 28 Bratislava, Czechoslovakia
*Information Centre SAS, Obrancov mieru 49
814 38 Bratislava, Czechoslovakia

INTRODUCTION

The technique of measurement of thermophysical parameters has undergone recently a rapid development. This was
due to development of materials research and new technological procedures. At present there exists a number of methods
for measuring thermophysical parameters. One of the most
efficient methods is the pulse method. The pulse method with
a planar heat source for measuring specific heat, thermal
diffusivity and thermal conductivity of bulk materials and
thin metallic foils was described in detail by Kubičár and
Illeková /1984/.

Productivity and effectivity of the measurement has been
improved by automation of the experiment. The technique of
information processing has been extensively applied in the
storage and retrieval of information obtained from the measurement. Routine has been made of databases incorporating
standards, specifications of measuring process and sample,
material property data and the like.

The present paper deals with the formation of laboratory
database of thermophysical parameters obtained by the pulse
method.

THE ROLE OF A LABORATORY DATABASE

The automation of experiment, the information processing
and data processing in the material science gave rise to the
gigantic international databanks centralizing physical and
other quatities. In experimental laboratories these take form
of laboratory databases which produce a new technique of
information processing to get the physical quantity from the
measured signals.

The principle to obtain the physical information is
always the same: Some physical model is assumed to hold true.

The process investigated is idealised. In real measurement the fulfilment of the boundry and initial conditions must be evaluated and principal correction factors are determined to calculate the real physical quantities.

Every laboratory database consists of a set of measured material property data /to satisfy the material research requirements/, a set of experimental specifications /to satisfy the metodological requirements/ and the technique, almost the procedures, working on these two sets of data /to satisfy the users/. The quality of the set of specifications in the laboratory database gives the weight to the physical quantity produced by the laboratory.

THE PROCESSING OF THERMOPHYSICAL DATA

The measuring of thermophysical parameters is based on time and space analysis of temperature fields within the sample $T/x, t/_Q$ generated by a heat pulse. The application of this principle allowed to construct an automatized apparatus for measuring thermophysical parameters in the temperature region /80 - 1000/ K.

Standardization and calibration procedures were performed in the automatized measuring system on the bulk samples of ruby and polymethylmetacrylate and on Ni foils. A set of tests was carried out in order to determine correction functions for real experimental conditions.

A large number of information is being automatically obtained, processed and recorded in the course of measurement. These are the specifications of measuring process and the experimental data. Information is being compressed and stored in a SMEP computer system laboratory database.

Specifications

The specifications of measuring process are the following:

Heat pulse characterization. Heat pulse originats the investigated dynamic temperature field. It depends on the quality of heat source, on geometry of the source and the sample, on thermal contact between these and on heat losses. Heat losses are temperature dependent.

Thermodynamic state of measuring system before and after an introduction of a heat pulse into it. The measuring system as well as the stability of the system influence the background of the process investigated.

Geometry of the sample. A fulfilment of boundary and initial conditions of the temperature field studied determines the principal correction factors of thermophysical parameters. Their values depend on mutual geometric and physical properties of sample, heat source and sample holder.

Material property data

Measurement is organized in measuring cycles. In an elementary measuring cycle the measuring process is realized, a lot of information /about 100 data/ is acquired, a local data processing and reduction of data is performed and then the representative set named one point or material property data vector /about 5 data/ is stored /Kubičár and Illeková, 1984/. One measurement is usually represented by a package of about 100 points.

Absolute accuracy of the method is limited by the completeness and the accuracy of the presented specifications, Relative accuracy of measurement is better then 1% for all three thermophysical quantities.

THE DATABASE

A design database is an organised and integrated collection of design data. It provides a natural representation of these data and its relationships and can be shared by a variety of users. It facilitates the development of application programs which enable access to data, protects logical data structure from change, reduces data proliferation, controls accuracy and consistency and avoids the availability of multiple versions of data at different stages of updating.

To achieve this objectives the thermophysical laboratory database has the following architecture: Collection of all material thermophysical property data is stored in a set of files DATA; specifications of every type of measuring process are stored in specialised files named HEADs. The relationships between all levels of stored data and the maintaining and application routines are arranged in one library system.

HEADs

HEADs of bulk materials and amorphous metallic foils already exist. Other HEADs, e.g. of standards, the room temperature thermophysical parameters, temperature dependences of these, relaxation processes viewed through thermophysical parameters, are being designed.

Files HEAD consist of sequentially organised fixed-length records divided into fields. Each field within the record is defined by its ident and data type. Position of the field within record is transparent to the user. These fields consist of: bibliographic data /chemical composition, technological data and thermal history of sample, date of measuring, length of corresponding record of the material property data stored in the file DATA, the value of key of that record, definition of fields in that, input method constants, etc./, parameters of measuring system /dimensions of sample, parameters of holder, technical parameters of heat source and of thermocouples, etc./, parameters of measuring regime /parameters of heat pulse, heating rates, atmosphere, etc./ and remarks about anomalies observed in the course of measurement.

DATA

The index-organised file DATA consists of independent records of variable length containing repeated n-folds of independent material property data /e.g. temperature, time, specific heat, thermal diffusivity, thermal conductivity, weights of these quantities/. The file DATA is protected against rewriting and deleting.

Manipulation tools

The physical and logical independence of stored data, fast access, detection of data inaccuracy and automatic error recovery, good capability of reorganization of changing needs and others are covered through high level programmer interface represented by procedures called NEP, PBE and ZSORT working on files HEAD. All three procedures are designed as tasks working on common data files described by Record-Description-Tables /RDT/. NEP is Input-and-Edit-Program driven by RDT. It reads data from terminal, online checks data type and range permitting their immediate correction and writes the data into data file. Other functions are: retrieve, displaying and updating of stored data. The communication with NEP program is in a simple user-oriented language. Data are displayed on a screen formated according to the user needs. The same field idents are used to retrieve, select and sort the data by simple-driven ZSORT. Complex arithmetical and logical expressions are permitted to specify selection criterion. Data selected from data file are formated and printed by print-By--Example program /PBE/, which uses an example of an output in its real form. Simple description is added at the end of the example, assigning strings from example to the field idents.

Procedures for statistical and scientific calculations and graphics are applied on DATA files.

REFERENCES

Ľ. Kubičár, Pulse method for measuring thermophysical quantities of solids, in "Proc. of IME O Xth WORLD CONGRESS", Budapest /1985/, in press.
Ľ. Kubičár and E. Illeková, Automatic apparatus for measuring thermophysical quantities controlled by calculator EMG 666, in "Proc. of XIth Int. Symp. on Nuclear Electronics, Bratislava, September 6-12, 1983", Report D13-84-53, JINR, Dubna /1984/, p. 331.

OPTIMIZING TEXTILE DRYING AND HEAT SETTING OPERATIONS BY USING

MEASURED TEMPERATURE PROFILES TO FIT A MODEL

Gert Kreiselmeier

Institute of Textile Technology
Karl-Marx-Stadt, German Democratic Republic

INTRODUCTION

Intelligent measurement may be defined as a measurement process in which machine information processing is used to enhance and interpret measurement data. As Finkelstein /1/ pointed out, we may distinguish a number of levels of intelligent measurement, one of which ist for example inferential measurement. In this case the actual measurand ist the output of a model and it is not directly observed but it is iden-tified from measurements of directly observed input variables of the model.

In this paper there is presented an example of inferential measure-ment that relates to the thermal processing of textile fabrics in large ovens called tenter frames.

A difference equation model was developed to describe the drying behaviour of fabric with the help of technological data that can be rou-tinely gathered. Having estimated the model parameters from observable input/output measures which provide a representation of the particular tenter frame, the model can be used to predict the influence of various plant parameters on the temperature and moisture profiles.

From this profile and from some calculated efficiency figures one can get an indication of how near the tenter is operating at its optimum. Thus the model is a practical tool for plant technical personnel to re-duce the number of costly plant tests.

PROCESS DESCRIPTION

In thermal processing of fabric there are two operations which are performed on the tenter frame: drying and setting. In pure drying pro-cesses operation aims at a residual moisture. For drying and setting in one step the whole process is extended to reach a desired dwell time, that is, the fabric has to stay throughout a certain period above a certain temperature. The main factors determining the efficiency of a

tenter are the
- temperature of the circulating air
- fabric velocity
- intensity of jet streams (number of revolutions of
  the air ventilators)
- steam content of the circulating air (exhaust air,
  flap position)

To reach the optimum efficiency of the tenter narrow limits must be observed by means of manual adjustment of the plant parameters with regard to the frequent changes of articles which are to be dried or set.

Since one never knows exactly when drying is finished and setting starts, the fabric velocity is in practice adjusted lower than theoretically possible as a precautionary measure.

Wide and narrow fabrics often being run with an identical exhaust air flap position and the operator is frequently not realizing that a lot of energy is lost by exhausting it above the machine. This is air which has been heated at considerable charge.

As a result of the demand from industry for increased production and the need for lower energy consumption, it has become essential to impose scientific and technological control over textile drying and heat setting operations on tenter frames. In particular, considerable energy saving would be accomplished, if the drying or setting conditions, which are in practice often varied subjectively, were adjusted to satisfy the specific requirements of the particular plant and material.

DESCRIPTION OF THE MODEL

The model equations were obtained for steady state conditions by writing
- the heat balance about a differential element of fabric
  moving through zones of constant drying conditions and
- the mass balance for water and air within a zone of the
  tenter do determine the steam content, which is assumed
  to be constant in the zone.

The heat is transferred from hot air in the tenter to the surface of the fabric proportionally to a heat transfer coefficient which is unknown in the first step and which is a model constant. Vaporization of the moisture takes place according to some assumtions for moisture change with time:
- it is zero for fabric temperatures lower than wet bulb
  temperature
- it is maximal for fabric temperatures equal to wet bulb
  temperature but for moisture contents higher than critical
  moisture (first drying stage)
- it is decreasing with an empirically found slope for moisture
  contents lower than critical moisture (second drying stage)

Thus critical moisture is the second model constant. The heat produced to compensate for insulation losses and to cover the content of the exhaust air can be estimated or computed from the mass balance. The set of difference equations is solved by using a digital computer. In the first loop of computing wet bulb temperature is assumed to be 60 degC, this value is improved iteratively due to the given air temperature and the humidity calculated in previous loops.

In developing a semi-empirically grey model it is not necessary
that the structure (set of equations) of the model corresponds to the
exact theory of the considered process, but only the model must predict
the behaviour of the process variables of interest.

FITTING OF MODEL PREDICTIONS WITH MEASURED TEMPERATURE PROFILES

In order to estimate the two model constants, the model output was
compared with fabric temperature profiles measured over the length of
large commercial tenter frame dryers.

The first step in this procedure was to develop a method for surface
temperature measurement along the length of the dryer. There is a well-
known method to achieve his, viz radiation pyrometry.
But the disadvantage of it is that it is too expensive for a routine
method, which at least as many pyrometers as tenter zones are required for.

Thus 8 thermocouples were positioned along the path of the fabric
through the tenter in the ambient air layer close to the moving fabric.
To diminish the influence of the hot air stream, the rear side of the
thermocouple was thoroughly screened with a metal shield. Thermal conduction between the shield and the thermocouple was depressed by inserting
it into a temperature resistent fabric which is the front side of the sensor. The systematic error of the sensor is proportional to the temperature
difference between the surface of the moving fabric and the hot air jet.
It was determined by comparing sensor readings to that of thermocouples,
which were inserted into the moving fabric and was used to correct sensor
fabric temperatures by this difference.

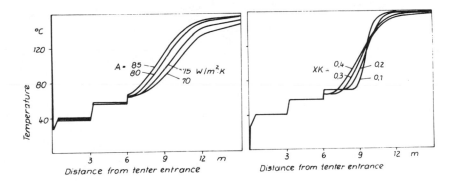

Fig. 1   Influence of the model constant 'heat transfer coefficient'
          (left) and 'critical moisture content' (right) on the tempe-
          rature profile of the model

In the next step model predictions were forced to agree with the experimental data by adjusting the values of the two model constants. Fig. 1 show the influence of the two model constants on the fabric temperature profile.

A graphical, CRT-aided, least-square optimization search routine was used to get the fit. To demonstrate that the model will accurately predict under other process conditions, the same values of the two model constants were used to generate profiles and to compare them with measured profiles at the same operating conditions.

In this way profile measurements were made for 60 tenter/fabric combinations and they were fitted to the model. A certain pair of model constants is supposed to be characteristically for the corresponding tenter/fabric combination.

USE OF THE MODEL

Once this has been established the model can be used to run experiments under unknown process conditions. So it is possible to predict accurately and monitor the drying and setting process over the full tenter lenght. Only the most promising conditions should be tested in the plant. The model is implemented on a personal computer and is in practical use in some textile mills.

REFERENCE

1. Finkelstein, L.; Carson, E.R.
   Intelligent Measurement in Clinical Medicine, in:
   "5th International IMEKO Symposium" Jena June 1986
   Vol. 1, p.56-61

2
INTELLIGENT SENSORS, INTERFACES, MEASUREMENT MODULES AND INSTRUMENTS

SENSOR INTERFACES FOR INTELLIGENT MEASUREMENT

Tilo Pfeifer

Department of Measurement for Automated Production
RWTH Aachen
Steinbachstrasse   Block 53/54
5loo Aachen-Melaten Nord, Federal Republic of Germany

1. Introduction

Linking of sensors and sensor-dedicated microprocessors can
take place either by direct integration of, for example, a
sensor using silicon technology in a microprocessor chip, or
by discrete linking of the two elements. The present contribution will indicate fundamental options and describe an
example of discrete sensor-microprocessor linking which has
already been put into practice.

2. Functional Elements of a Sensor or Sensor System

No clear distinction can be drawn between the terms "sensor"
and "sensor system". In the course of the following observations, "sensor" will in general be taken to mean the basic
sensor with its analogue signal processing. "Sensor system",
on the other hand, will be taken to include all those function modules presented in Fig. 1.

In principle, a sensor system as defined above may also be
described as a measurement chain comprising a sequence of
individual function blocks.

3. From Centralised to Decentralised Structure

As the performance capability of microelectronic devices increases and hardware prices fall, decentrally-structured
measuring systems with distributed intelligence can be used
over an increasingly wide range of applications. The advantages of this type of structure as compared to centralised
processing (Fig. 2) may be summed up as follows:

> - Avoidance of the analogue-digital interface bott-
> leneck, resulting in better time response of the
> system as a whole,

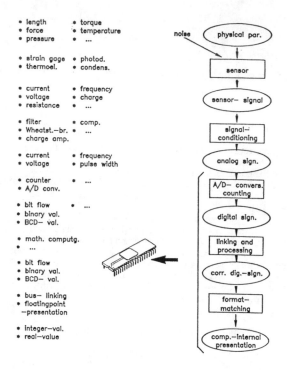

length   torque
force   temperature
pressure   ...

strain gage   photod.
thermoel.   condens.

current   frequency
voltage   charge
resistance   ...

filter   comp.
Wheatst.-br.   ...
charge amp.

current   frequency
voltage   pulse width

counter   ...
A/D conv.

bit flow   ...
binary val.
BCD- val.

math. computg.
...

bit flow
binary val.
BCD- val.

bus- linking
floatingpoint
-presentation

integer-val.
real-value

noise → physical par.
→ sensor
→ sensor- signal
→ signal- conditioning
→ analog sign.
→ A/D- convers. counting
→ digital sign.
→ linking and processing
→ corr. dig.-sign.
→ format- matching
→ comp.-internal presentation

Fig. 1   Function Modules and Interfaces of a Sensor System

- Increased functional capability through decentralised pre-processing or pre-evaluation,
- Hardware simplification through standardised interfaces,
- Reduced interference susceptibility through digital signal transmission.

A microprocessor used in a decentralised structure of this
type must meet the following requirements:

The important functional elements should if possible be
integrated on a single chip. The instruction repertoire of
the microprocessor must be adequate to the measurement processing tasks involved.

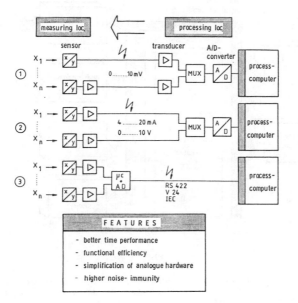

Fig. 2   Creation of a Decentralised Structure

4. Example: Modular Sensor System

An example of the main function blocks of a sensor system
configured according to the principles outlined above is
shown in Fig. 3.

The modular sensor system consists of 3 individual groups of
modules. The core module is represented by the microprocessor, whose significant functions are indicated in the illustration. The matching of the sensor signals takes place on
sensor boards. These contain the analogue signal processing.

The third group of modules consists of the user interfaces.
While a total of six sensor boards can be operated simultaneously by the system, a single user interface is selected
from the available range, according to the problem in hand.

The sensor boards, like the interface modules, are connected
to one another and to the microcomputer module via a unified
mixed analogue and digital internal computer bus. A unified
logical interface underlies the various physical and electrical interfaces of the sensor units. For standardisation
purposes, data exchange is carried out in 7-bit ASCII-code
no matter what physical form the system takes. Data transfer
takes place via a call-answer cycle.

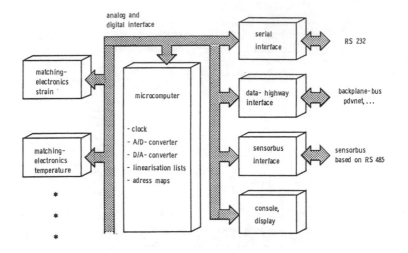

Fig. 3 Elements of the Modular Sensor System

The software modules are split into a driver-software level and a functional level. The driver software comprises all routines serving the various interfaces and components. The higher software level implements the processing functions of the sensor system.

5. Distributed Sensor Systems

Distributed sensor systems in themselves are not new. They have been used for many years in materials processing and chemical engineering. In these applications, however, the structure is frequently centralised and seldom includes several levels. The situation in manufacturing engineering is quite different. Here, a number of levels are generally interconnected. The much-discussed "factory of the future" involves a consistent computer philosophy stretching from the planning level right down to the sensor level. As the sensor itself is in fact encountered in various areas and at various levels, e.g. at incoming inspection, in machine monitoring or for operational data logging, it must be to a high degree system-compatible and networkable. It has to be capable of communicating with field bus systems or with local networks and distributed process control systems (cf. Fig. 4).

Given the expected future structure of industrial manufacturing, developments need to be aimed at making sensor systems capable of simple integration in the manufacturing process and its associated machines.

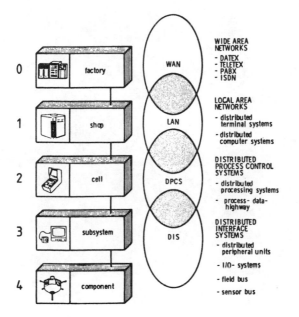

Fig. 4 Distributed Systems for Automated Manufacturing

REFERENCES

1.   M. Komischke and W. Rühle:
      "Intelligente Sensorsysteme"*
      Industrieanzeiger No. 75, 1984, pp. 24-26

2.   T. Pfeifer, U. Schwerhoff and M. Komischke:
      "Sensoren gekoppelt mit Singlechip-Mikrocomputern -
      Möglichkeiten der Vorverarbeitung und Sensorschnitt-
      stelle"**
      VDI-Berichte No. 509, 1984, pp. 399-414

*   Intelligent Sensor Systems
** Sensors Linked to Single-Chip Microcomputers: Pre-Proces-
    sing and Sensor-Interface Options

STANDARDIZED INTERFACES FOR

INTELLIGENT MEASUREMENT

Harald Schumny

Physikalisch-Technische Bundesanstalt
(Federal Institute of Physics and Technology)
Braunschweig, Federal Republic of Germany

1 CONCEPTS OF INTELLIGENT MEASUREMENT

Intelligent measurement is mainly characterized by computer control. That is, instruments or systems either have a built-in processor ("stand-alone") or must be connected to a computer ("computer-based"). We classify intelligent instruments by the following:

* stand-alone   (compact, dedicated)
* computer-based   (modular, flexible)

Stand-alone instruments have all necessary components included in one casing and therefore function autonomously. There is often an interface interconnecting to a host computer by which the instruments run under program control. This is a priori the case (and is necessary) with computer-based systems where the controlling part can be a mainframe or a personal computer (PC). In the latter case the denotation Personal Instrumentation (PI) will be used. The following list shows basic concepts of instrumentation.

- Components have direct access to the processor bus of a system (fastest version). In some implementations stackable or changeable blocks can be used.
- Boards have access to the backplane or system bus of a PC. Free slots can take up acquisition or measuring boards.
- Bus extension means that an external box is connected to the PC by only one slot, but the bus length is limited to about 2 m maximum.
- The IEC Bus (also called IEEE-488) is a peripheral system allowing up to 15 devices to work together with a bus length of no more than 20 m, which seems particularly suited for laboratory automatization.
- Serial connection allows wider distances - depending on technology used, up to 1 km, but only 20 m and slow with RS-232. Communication inside a building or control of a factory floor is possible.
- Front-end as subsystem is the latest concept. It allows the design of powerful, flexible and economical measuring and control systems.

Before discussing interface and bus details, we will mention the ISO Reference Model for Open Systems Interconnection (OSI RM according to ISO 7498). This is now the accepted basis for interface specification. Western countries for example, have agreed upon a harmonization strategy by which uniform European interface standards for Open Systems Interconnection are to be created. Intelligent measurement is a special field where such harmonized standards will be applied.

2 SOME PRINCIPLES AND REQUIREMENTS

Main aspects are: synchronization of all interactions; real-time behaviour by which the response time is restricted to a span no longer than one process or measurement cycle; data rate, noise, distortion, and electromagnetic interference. This means that the electrical characteristics of interface circuitry and transmission lines are of relevance, as are the access method and controlling software protocols.

One of the main principles is to allow an event to interrupt a running program, centrally or peripherally initialized. The first case is deterministic because it underlies software control. Only peripherally initialized "service requests" can be immediately responded to. Identifying an interrupt source is usually done by polling, that is, software controlled reading, but this is only moderate concerning response time. For the fastest transfer of measurement data, a Direct Memory Access (DMA) channel should be implemented. With this, a data transfer rate of several hundred kbyte/s becomes possible.

If a measuring system consisting of several devices is to be configured, the topology used can be important. Most obvious is the classification in point-to-point and multipoint interconnection, the former being the one mostly used. Multipoint systems can have (physically and logically) the topology of a star, a ring or a true bus. Very specific interfaces and protocols are needed to manage the addressing and access procedures in a multipoint measuring system, and to prevent breakdown or even damage. The system control can be centralized or distributed (at least temporarily).

Of particular importance is the way the control information is transmitted. A distinction is made between:

* hardware control - in this case there are additional control lines (8 in the parallel IEC bus, and at least 4 in some serial systems)

* software control - all signalling, handshaking etc. is managed by a software protocol which transfers ASCII or other control characters via the data line.

Software control is the more up-to-date method although it seems to be slower than hardware control. This restriction can be compensated by the often very high transfer rates (e.g. 10 Mbit/s), but interrupt handling and handshaking could become impaird. A combined method where a separate interrupt line joins the one information line may be a solution for short response times.

The electrical interconnection method is another relevant criterion. The two basic techniques are:

* unbalanced - sometimes called single ended, which is the more antiquated version, e.g. used with serial RS-232 interfaces

* balanced - often designated as the differential method, e.g. used with modern RS-422 interfaces.

The latter technique allows galvanic isolation and guarantees relatively high EMI suppression because of its inherent common mode rejection, especially when used together with twisted-pairs. Of added value is the high data rate (up to about 1 Mbit/s) using long transmission media (up to 1km).

The Programming of interfaces is the least standardized and supported part of intelligent measurement. The programming languages mainly used have only control structures for standard I/O, that is e.g. for a printer or simple serial file transfer, but not for process peripherals, interrupt handling or DMA. There are some special process controllers available which provide excellent support for process I/O. One disadvantage with these is that they are not compatible with the up-to-date industrial standard workstations and cannot run the numerous MS-DOS software (e.g. data acquisition and evaluation packages like Asyst or spreadsheet programs like Lotus).

3 "DE FACTO" AND OFFICIAL STANDARDS

The relevant interfaces for personal instrumentation and intelligent measurement are partly defined in written standards, others are de facto or industrial standards. The following is a selection of these:

* System buses   - Multibus, VMEbus, PC Bus
* Peripheral, parallel - TTL, BCD, IEC Bus
* Peripheral, serial   - 20 mA, RS-232, RS-422, Serial Bus.

Table 1 lists the most important system buses and makes plain that IEEE is leading in standardization. A main trend is that new developments follow the Eurocard standard with boards mechanically designed according to DIN 41 494 and indirect connectors (DIN 41 612). The Multibus, VMEbus and PC Bus predominate, all of them processor-dependent, but plug-in boards are available all over the world.

Parallel interfaces using TTL chips are often called "TTL interfaces" and can be considered as accepted industrial standards for fast point-to-point interconnection. They are inexpensive and flexible, but usually require assembler programming. Only in the case of system integration is high-level programming possible. The term "BCD interface" is misleading because it is actually a TTL interface with four-bit ports each used for one binary coded decimal digit.

Of eminent importance for measurement and control is the IEC Bus of which all specifications have been published by IEC 625 and IEEE 488-1975. Both standards are almost identical. They differ only in the connection: IEC specifies a 25-pins connector, IEEE a 24-pins version. A lot of measuring devices and workstations are equipped with such a bus interface. For workstations of the PC or AT types add-in boards are available. There are, however, some problems resulting from the bus specification and the unsatisfactory software situation:

- only 15 devices can work on the bus; the distance between devices
  is only 2 to 3 m; the total bus length is no more than 20 m;
- transmission rate of sometimes less than 100 readings/s, with 16-
  bit computers up to about 10 000 readings/s;
- software has not yet been sufficiently specified; projects and docu-
  ments: IEEE 728-1982 (Code, Format), IEEE P981 (Device Messages).

Serial interfaces have their origin in specifications of the post administrations. Adapting these to data processing has caused a number of problems we now have to live with, but there are also some adaptations and specialized developments worked out mainly by the EIA, IEC, ISO and DIN. Table 2 lists all relevant serial standards including multipoint versions under development.

The most important serial interface is that defined in EIA RS-232-C, in Europe often called the "V.24 interface". But the CCITT paper V.24 is only a list of about 50 signal names for telecommunication. Electrical characteristics are fixed in V.28, connectors are defined in various modem standards. RS-232 is a V.24 subgroup including electrical and connector specifications (25 pins). The main disadvantages:

- designers are free to choose different sub-groups of control lines;
- CCITT and EIA specified a computer-to-modem operation; if there is no modem, then data lines 2 and 3 must be connected cross-over (null-modem); the data rate and distance are largely limited;
- the circuits require +/ 12 V, which is incompatible with the standard 5 V technology;
- the single ended driver and receiver configuration implies cross-talk and noise susceptibility;
- galvanic isolation is ineffectual because of the ground symmetry.

To overcome these drawbacks, CCITT and EIA published the papers V.11 and RS-422, respectively, describing balanced (differential) interface devices. We mentioned the main characteristics in par. 2. Some of the additional advantages are:

- power supply only 5 V; twisted pairs up to about 1 km;
- transmission rate up to 10 Mbit/s.

One consequent step was to make the V.11 chips capable of multipoint operation. The electrical characteristics are now fixed in RS-485 and ISO 8482. The idea is to make available a low cost system e.g. for 32 devices coupled together by twisted pairs of 500 m bus length and with a data rate of about 500 kbit/s. The protocol for such a bus is under discussion. More on networks follows in the next paragraph.

4   NETWORKING

Intelligent measurement equipment is not always of the stand-alone or laboratory environment type, but is used increasingly in complex and extended systems. Networking therefore becomes more and more relevant. The main standardization bodies for LANs are the ECMA and IEEE, the latter runing the following projects:

P802.1 General Structure
P802.2 Logical Link Control
P802.3 CSMA/CD (Ethernet)
P802.4 Token Bus
P802.5 Token Ring
P802.6 Metropolitan Area Network (MAN)
P802.7 Broadband Tech. Advisory Group

Table 1 Important system buses

Bus	Source (Standard)	Processor	Data bits
ECB	Kontron and others	Z80	8
STD	Prolog, Mostek (IEEE P961)	independent	8
STE	(IEEE P1000)	independent	8
G-64	Gespac (single Eurocard)	independent	8/16
ZBI	Zilog	Z80/Z8000	8/16
Q	DEC and others	LSI-11	16
E-/T-	Texas Instruments and others	9900	16
S-100	many (IEEE P696)	independent	8/16
Euro	Ferranti (ISO/DP 6951; BSI)	independent	18
IBM PC	IBM and very many others	8088/8086	8/16
Multi	Intel, Siemens a.o. (IEEE P796)	80 family	8/16/32
AMS-M	Siemens (IEEE P and IEC)	80 family	8/16/32
Versa	Motorola (IEEE P970)	6800	8/16
VME	Motorola, Mostek, Philips, Valvo/Signetics, Thomson and others (IEEE P1014)	68 000	8/16/32
Fast	NBS	independent	32
Future	(IEEE P896)	independent	16/32
Nu	Texas Instruments	99 000	16/32

Table 2 Serial standards; PP: Point-to-point; MP: Multipoint

Name	Standards	Typical line length	Transmission rate	
20 mA	DIN 66 258/1 66 348/1	300 m	2,4 kbit/s	PP
V.24	DIN 66 020/1 66 259/1 RS-232-C; V.28	20 m	19,2 kbit/s	PP
V.11	DIN 66 258/2 66 259/3 RS-422	10 m . 1 km	10 Mbit/s . 100 kbit/s	PP
RS-485	DIN 66 258/3 *) 66 259/4 *) ISO 8482	up to 1 km	up to 1 Mbit/s	MP
ISDN	for ISDN using 1 V			MP

*) under preparation

One strategy is to integrate LANs in a hierarchy extending from system buses over peripheral connections to Wide Area Networks (WANs) which are mostly public nets. The complete hierarchy with some examples is shown in the following.

- Internal system bus   Multibus, VMEbus, PC Bus
- Peripheral, point-to-point   TTL, RS-232, RS-422
- Peripheral, multipoint   IEC Bus, Serial Bus
- Fieldbus   Proway, PDV-Bus, Manchester Bus
- LAN   Ethernet, PC Net, MAP
- WAN   Private and public networks

Fieldbus is a term often used for LANs specialized for process automatization in smaller environments. Some implementations offer a gateway for the IEC Bus and another for LANs. Several IEC standardization projects deal with Process Data Ways (Proways); the West German PDV-Bus proposal has been published as DIN 19 241; chips and boards are available from several sources. The Manchester Bus (or Avionics Bus) is well known as MIL-STD 1553, and is supported by a number of manufacturers. The industrial LAN "MAP" (Manufacturers Automation Protocol) has meanwhile received support from all the relevant firms, since it will define all 7 ISO layers using existing and stable standards. It can be expected that gateways will be avail-able to interconnect both to a field bus and to a WAN.

5   REFERENCES

1. Schumny, H., Schuster,H.-J.: Distributed Intelligence in an Automatic Nuclear Measurement System. CPEM 1980, pp. 210-212
2. Lesea, A., Zaks, R.: Microprocessor Interface Techniques. Paris: Sybex 1981
3. Schumny, H.: Interface Problems in Computerized Measurement. IMEKO Summer School, Dubrovnik 1981, pp. 29-45
4. Schumny, H.: Interface and Bus Standardization. Microproc. and Microprogr. 7 (1981), pp. 266-268
5. Sunshine, C.A.(ed.): Communication Protocol Medelling. Washington: Artech House 1981
6. Stone, H.: Microcomputer Interfacing. Amsterdam: Addison-Wesley 1982.
7. Schumny, H.: Interface Problems in Legal Metrology. Microproc. and Microprogr. 11 (1983), pp. 243-247
8. Schumny, H.: Measurement and Process Control Need Clearly Defined Interfaces. Microproc. and Microprogr. 12 (1983), pp. 115-118
9. Schumny, H.: Technical Applications of PCs. Euromicro Tutorial, Brussels 1985

INTELLIGENT SENSORS FOR ON-LINE TOOL-MONITORING
IN UNMANNED FLEXIBLE MANUFACTURING

Hans-Jürgen Jacobs, Bertram Hentschel, Bernd Stange
Peter Winkelmann
Dresden University of Technology, GDR,
Department for Manufacturing Engineering and Machine Tools

INTRODUCTION

The long-term, future-oriented concept of computer-integrated, flexibly automated manufacture reveals the first convincing results at present. They indicate a fundamental change of production modes.

Here, the last stage of information flow of computer-integrated machining in mechanical engineering will be dealt with, i. e. on-line process monitoring and on-line process control in a technological optimum. There, all those technical and organizational disturbances have to be compensated that cannot be prevented by classical CAD/CAM-Systems.

The stochastic variations from the normal state of the manufacturing process occurring within the time-critical range require technological machine intelligence to be raised in machining cells and flexible manufacturing systems, resp.

Simultaneously, the machine operator with his sensorial and control abilities is to be replaced by technical systems capable to ensure hitherto existing human performances at this place, and even surpassing them, resp. Finally, in order to achieve high flexibility of manufacture monitoring and control strategies of a new type have to be developed. They are to realize the transition from process-variant to process-invariant solutions. The technological intelligence of most monitoring systems available today is characterized by too less flexibility. These systems implement process-variant strategies, i. e. for each new value combination of manufacturing parameters the process normal state must be recognized and stored anew. Thus, achieving defined limits the deviating process state will be re-transferred to normal state again. Or the critical state will be removed, e. g. by tool change, or the dangerous state finished, resp., by "alarm stop". But these solutions do not fit for flexible manufacturing with a batch size of one.

The process invariance to be aimed at for process monitoring involves now to develop such strategies applicable for a maximum value range of a maximum number of process, tool and machine tool parameters. That means the currently used learning cycle for the storage of process normal states of each new operational step can be omitted.

NEW CONCEPT OF A PROCESS-INVARIANT CUTTING-TOOL-SENSOR

The results achieved in the research work at the Dresden University
of Technology, hitherto, shall be demonstrated. Tool monitoring is the
focus of process monitoring. This will be shown by raised technological
machine intelligence of flexible automated turning cells with the main
components of sensor system, control concept and process model for monitoring.

For the monitoring tasks of tool breakage and end of tool life recognition the cutting force acting during machining is used as a measuring
and control value for characterizing the actual process state.

CONCERNING THE SENSOR SYSTEM

The sensor system functioning on the basis of a strain transformer
shall be emphasized among other solutions (fig. 1).
By means of the design of the strain transformer displayed in the figure
five- to tenfold increases of extension due to the cutting force can be
achieved at the semiconductor resistance strain gauge compared to the measured object. This sensor principle is suitable both for further development of automated machine tools and supplementary equipment of existing
machinery.
The decisive condition for a high efficiency of this measuring system
consists in the selection of a suitable location in order to achieve high
measuring effects and restrict hysteresis and temperature influences.

CONCERNING THE CONTROL CONCEPT

In the figure 2 the general concept can be seen. The basis is technical equipment available in the GDR.

The CNC can be connected to a superior DNC computer. It also permits
the connection of further microcomputers as external slave computers.
One slave computer in each case is planned for

. process monitoring - tool breakage recognition

. process monitoring - tool life end recognition

. process control with adaptive control systems.

The monitoring computer for tool breakage recognizes tool fracture,
collisions between tool and workpiece as well as dangerous system vibration
during machining. The module for breakage recognition evaluates the actual
cutting force signals of the sensor and transmits the according status
signals to the PIO interfaces of the programmable controller of the CNC.
In this way the corresponding reaction programmes - implemented in the
CNC - can be activated. Thus, a time-critical monitoring task will be
solved, i. e. if scrap work caused by tool breakage is to be avoided the
feed motion of the tool must be stopped within one workpiece revolution.

Microcomputer "2" carries out the cutting force signal processing for
tool life end recognition. Generally, this monitoring task is not time-
critical and can be realized by a programme loop operating in the second
or minute range. Algorithms for the correlation between actual cutting
force signal and actual tool wear signal are the basis.

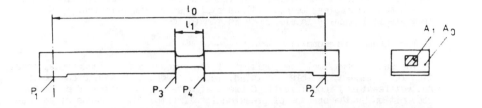

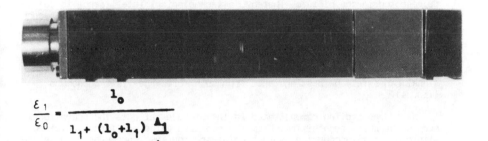

$$\frac{\varepsilon_1}{\varepsilon_0} = \frac{l_0}{l_1 + (l_0 + l_1)\dfrac{A_1}{A_0}}$$

Fig. 1   Strain transformer for monitoring tool behaviour

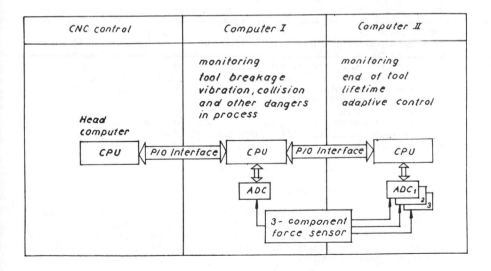

Fig. 2   Machine control with integrated
tool monitoring system

Thus, by means of this control concept the very same cutting force signals are processed two times by means of two specific process models. These two models and strategies, resp., will be explained as the principal item of technological machine intelligence in the following.

1) Tool breakage recognition:

The secure and timely recognition of tool breakage is a fundamental condition for unmanned flexible machining. Today, forecast of tool breakage is not yet feasible, industrially. Current efforts are concentrated on beakage recognition by the process of breaking itself (fig. 3). By means of an analysis of numerous breakage processes three main forms of signal courses could be stated as typical for tool breakage. The cutting force - time behaviour is represented in the picture. Thereby, a technological intelligence is required such that signal courses similar to tool breakage will not be recognized as breakage resulting in false alarm. Situations are refered to such as cutting across shrinkage cavities or hard spots in the workpiece, resp., or force interruptions caused by grooves and bores in the rotating part.

By keeping time conditions during signal processing the actual breakage signal can safely be distinguished from such "breakage-like" phenomena (fig. 4).

The representation demonstrated in the picture reflects the operational mode of breakage recognition. From the measured values of forces obtained within one millisecond the average value and the lower and upper limits are calculated enveloping the cutting force development in normal state. These limits hold for the next millisecond, where new measured values are recorded and evaluated. Consequently, the limits will follow the actual signal course with a slight delay. Thus, force variations due to tool cutout and start of the cut as well as cutting depth variation can be recognized as false alarm. Hence, first conditions for the process invariance mentioned are secured. If the actual force measuring signal exceeds the associated limits they will be kept constant for a certain time. After a time interval to be defined it will be questioned if the signal has returned to the restricted range, meanwhile. In this case, cutting process will be continued.
By this speeddependent time interval process invariance in terms of force variations due to shrinkage cavities and hard spots in workpiece material, by grooves and bores interrupting cutting will be secured. If the signal has not returned to the restricted range after the time interval mentioned toolbreakage alarm will be actuated. The length of the time interval is determined by "expertises".

Altogether however, this strategy will warrant process and tool invariance required for flexible manufacture. "Learning cycles" in the classical sense of current solutions are not necessary anymore. The tool monitoring system only needs to be adjusted to the characteristics of the machine tool. This is performed via software adaptation.

2) Recognition of tool life end:

Tool life end recognition by means of process sensors of cutting force signals is a problem more intricated than breakage recognition, although it is no time-critical problem. Current solutions use force limits depending on the actual values of process parameters. That means a process-variant operational mode is applied.

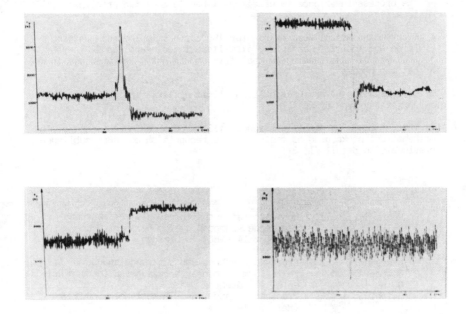

Fig. 3   4 typical behaviours of cutting force components at tool
         breakage and vibrations

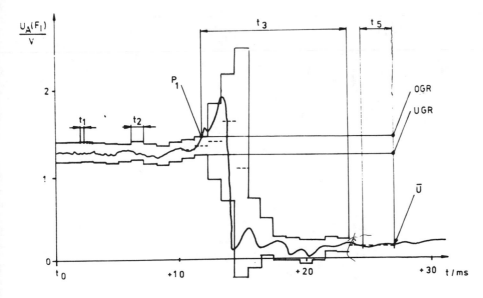

Fig. 4   Operational mode of breakage recognition

The process invariance to be aimed at has to meet two principal conditions:

1. Exact mathematical description for the correlation between cutting force and tool life behaviour at different tool wear forms. Thereby, the type of wear leading to tool life end is not known previous to the cutting process.

2. Determination of process-invariant limit criteria indicating tool life end at any type of wear.

Thus, "double" process invariance will be demanded here in a way. The mathematical background of tool life end recognition has been published previously in detail /1, 2/.

REFERENCES

1. Hentschel,   Spanungskraftverhältnisse als verfahrensbe-
    Maler   gleitende Kenngrößen - am Beispiel Drehen
                     Preprint 14-6-80, TU Dresden

2. Hentschel,   Prozeßüberwachung und Werkzeugdiagnose im Ar-
    Stange,   beitrsraum spanender Werkzeugmaschinen am Bei-
    Guth,   spiel des Drehens
    Geipel   Proceedings Aupro 84, TH Karl-Marx-Stadt
                     (Heft 5, Teil 1)

DESIGN OF AUTOMATIC ON-LINE CONTROL AND TROUBLESHOOTING
SYSTEMS FOR COMPLEX CONTROL PLANTS

A.Yu. Vasilyev and Ye.K. Kornoushenko

Institute of Control Sciences

65, Profsoyuznaya, Moscow 117342, USSR

Control of functioning and diagnostics of present-day
control systems are of great importance in their design and
maintenance.
The solution of the control and diagnostic problem is
complicated by the fact that only the external inputs and outputs of the system are accessible for observations and control.
Thus diagnosing of complex automatic control systems in the
course of their functioning is impossible without the use of
a computer linked by communication lines with the plant.
The suggested principle for on-line fault detection and
localization may be implemented on a wide class of computer-
aided control systems already in operation. A typical structure
containing the minimal number of functional units is shown in
Fig. 1. The keys are provided for connecting the system to the
external leads of the control plant. After passing normalizers
the scaled signals reach the analog-to-digital (A/D) converter
and further they get into the computer via the I/O device.
The result of the diagnosis appears at the display unit.
We describe the procedure of diagnosing discrete close-
loop control system in the course of their operation. An example of a system with a two-dimensional control plant and a
two-dimensional controller connected is shown in Fig. 2. This
is the case of the so-called separate control. Let the plant
be described by the equations

$$a_{11}(D)y_1 + a_{12}(D)y_2 = b_{11}(D)u_1 + b_{12}(D)u_2 \; ;$$

$$a_{21}(D)y_1 + a_{22}(D)y_2 = b_{21}(D)u_1 + b_{22}(D)u_2 \; ,$$

(1)

where D is a step-long delay operator and $a_{ij}(D)$ and $b_{k\ell}(D)$
are some polynomials. Similarly, let

$$c_1(D)u_1 = d_1(D)v_1 \; , \qquad c_2(D)u_2 = d_2(D)v_2$$

(2)

be the description of the controller. Then the closed-loop system with inputs g_1 and g_2 and outputs y_1 and y_2 is described as

$$c_2(c_1 a_{11} + d_1 \alpha_1 b_{11}) y_1 + c_1(c_2 a_{12} + d_2 \alpha_2 b_{12}) y_2 = \\ = c_2 d_1 b_{11} g_1 + c_1 d_2 b_{12} g_2 \ ; \tag{3}$$

$$c_2(c_1 a_{21} + d_1 \alpha_1 b_{21}) y_1 + c_1(c_2 a_{22} + d_2 \alpha_2 b_{22}) y_2 = \\ = c_2 d_1 b_{21} g_1 + c_1 d_2 b_{22} g_2 \ . \tag{4}$$

For the sake of simplicity we omit symbol D with all the coefficients a, b, c, d and α. Specify some concrete form of the polynomials $a(D)$, $b(D)$, $c(D)$, $d(D)$ and $\alpha(D)$. Let, for simplicity, we have

$$a_{ii}(D) = a_{ii}^o + a_{ii}^1 \cdot D \ , \quad c_i(D) = c_i^o + c_i^1 \cdot D \ ,$$
$$a_{ij}(D) = a_{ij} \cdot D \ , \quad\quad d_i(D) = d_i \cdot D,$$
$$b_{ij}(D) = b_{ij} \cdot D \ , \quad\quad \alpha_i(D) = \alpha_i \cdot D \ , \tag{5}$$
$$i, j = 1, 2$$

In the course of diagnosing equations (3) and (4) are treated independently, therefore we may restrict ourselves to consideration of equation (3) only.

Assume that we have to diagnose the system up to its isolated structural parts: the plant itself, each of the controllers, each of α -feedbacks. Consider, for instance, the first hypothesis H_1 "the control plant is faulty". The analysis of this hypothesis requires consideration of the equation obtained from (3) by specifying the polynomials $a(D)$, $b(D)$, $c(D)$, $d(D)$, $\alpha(D)$.

All the coefficients of $a(D)$ and $b(D)$ which refer to the control plant are regarded to be unknown while all other coefficients, known and equal to their "failure-free" values. Let k denote dimensionless time. Specify some values of k and N ($N > 10$) and take down a set of readings from the system to be diagnosed: inputs g_k^i, g_{k+1}^i, g_{k+2}^i, ..., g_{k+N}^i and outputs y_k^i, y_{k+1}^i, ..., y_{k+N}^i, $i = 1, 2$. Substituting these into equation (3) with due regard for the above, we

204

obtain, for coefficients of polynomials $a(\mathcal{D})$ and $b(\mathcal{D})$, an overdefined system of uniform linear equations of the form

$$M(a,b) \cdot (a,b)^T = 0, \qquad (6)$$

where $(a,b) = (a_{11}^0, a_{11}^1, a_{12}, a_{21}, a_{22}^0, a_{22}^1, b_{11}, b_{12}, b_{21}, b_{22})$

Assume the hypothesis H_1 to be true i.e. the control plant is indeed faulty and characterized by some set of "faulty" values of coefficients $(\tilde{a}, \tilde{b}) = (\tilde{a}_{11}^0, \tilde{a}_{11}^1, \tilde{a}_{12}, \tilde{a}_{21}, \tilde{a}_{22}^0, \tilde{a}_{22}^1, \tilde{b}_{11}, \tilde{b}_{12}, \tilde{b}_{21}, \tilde{b}_{22})$.

Provided there are no observation processing errors and equations (1) and (2) offer accurate mathematical models of the plant and the controller, respectively, system (6) has both zero and nonzero $(\tilde{a}, \tilde{b})$ solutions. Inevitable presence of disturbances results in inconsistency of system (6). However the "degree of inconsistency" depends on the magnitude of such disturbances. Quantitatively this of degree inconsistency of the system of linear equations is defined via singular values of the matrix of this system. Thus for instance, if $\delta_{max}(a,b)$ and $\delta_{min}(a,b)$ are the maximal and minimal singular values of matrix $M(a,b)$, respectively, then the "degree of inconsistency" of system (6) is defined by the relationship $\mathscr{æ}(a,b) = \delta_{min}(a,b) : \delta_{max}(a,b)$. The core of the diagnosing procedure is based on the fact that the "degree of inconsistency" of the system of linear equations corresponding to the first (correct) hypothesis is gnerally much less than the "degree of inconsistency" of the system of linear equations corresponding to an incorrect hypothesis. If the hypothesis is true and "faulty" coefficients of $\tilde{a}$ and $\tilde{b}$ significantly differ from the corresponding faultless values, we obtain

$$\mathscr{æ}(a,b) \ll \mathscr{æ}(c_i, d_i), \quad \mathscr{æ}(a,b) \ll \mathscr{æ}(\alpha_i), \quad i = 1, 2.$$

Where, for instance, $\mathscr{æ}(c_1, d_1)$ is the quantitative estimate of the "degree of inconsistency" of the system with respect to the coefficients of controller c_1

$$M(c_1, d_1) \cdot (c_1, d_1)^T = 0$$

obtained from (3) in a way similar to that for system (6). If the difference between $\mathscr{æ}(a,b)$, $\mathscr{æ}(c_i, d_i)$, and $\mathscr{æ}(\alpha_i), i = 1, 2$ is small enough we may artificially increase the "inconsistency degree" of the corresponding system of equations.

The diagnosing procedure is software realizable and requires no additional hardware expenditures. A generalized block diagram of the diagnosing algorithm is presented in Fig. 3. The algorithm realizes the procedure of N-fold verification of the hypothesis on faulty operation of each component of the system. The information malfunctioning of an i-th component is contained, at each step of diagnosing, in the degree of singularity $\mathcal{X}_i$ of the diagnosing matrix. The degree of singularity is defined with the use of the well known singular matrix decomposition algorithm which features sufficient performance and high stability. With the depth of diagnosing up to 10 units of the system the computer time taken by the procedure for faulty unit search is about a second on a ES-1033 computer and 10 to 15 seconds on a microcomputers ELEKTRONIKA 60.

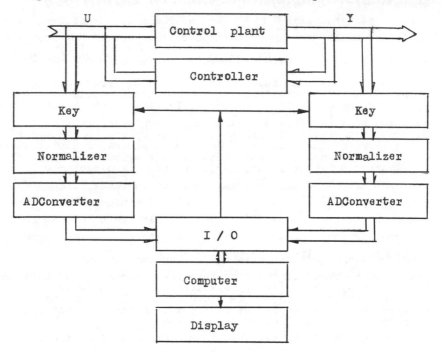

Fig.1 A structural diagram of a measurement system

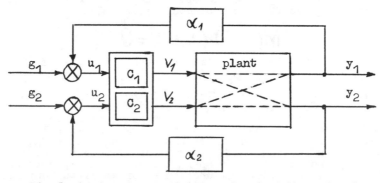

Fig.2 A structural diagram of a two-dimensional
control system

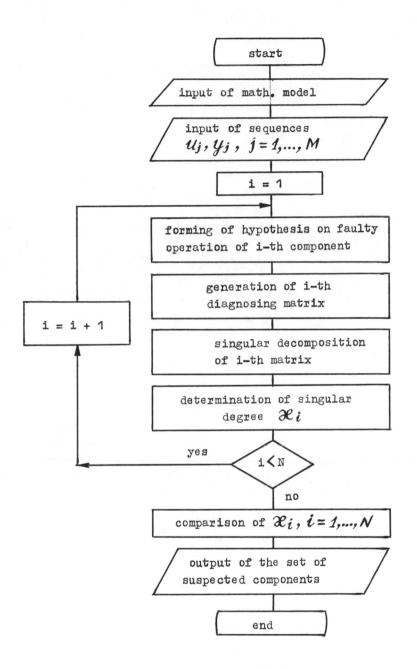

Fig.3 A flowchart of the diagnosing algorithm

A CONTINUOUS CAPILLARY VISCOMETER FOR GASES

Tilo Hipp and Heinz Kronmüller

Institut für Prozeßmeßtechnik und Prozeßleittechnik
Universität (TH) Karlsruhe
D-7500 Karlsruhe

1. INTRODUCTION

Within the frame of a multicomponent gasanalyser a viscometer for gases was designed for process instrumentation. The multicomponent gasanalyser detects molecular quantities inherent to any molecule of the mixture. Such quantities are the mass m, the degree of freedom f and the effective cross section σ^2. Corresponding macro-quantities are the density ρ, the specific heat c_v and all transport phenomena depending on the free path l like viscosity η, thermal conductivity or diffusion. Given the molecular fraction x_i of component i $x_i = n_i / \Sigma n_j$, n_j mol of component j, the macro quantity of mixture is /1,2/:

$$\rho = \Sigma \rho_i x_i \qquad c_v = \Sigma c_{v_i} x_i \qquad \eta = \Sigma \eta_i \frac{x_i}{\sum_j A_{ij} x_j} \qquad \begin{array}{l} A_{ii} = 1 \\ A_{ij} \neq A_{ji} \end{array} \qquad (1)$$

The resulting viscosity η depends nonlinear on the viscosity η_i of the components. The parameters A_{ij} are functions of the σ_t^2 and m_i. On the view of kinetic gas theory the viscosity η of a pure component is given as /1,2/

$$\eta = \frac{1}{3} N m \bar{v} \, l \sim \frac{m \bar{v}}{\sigma^2}; \qquad \text{N molecules per volume, } \bar{v} \text{ mean of velocity.} \quad (2)$$

2. PRINCIPLES OF THE NEW VISCOMETER

Compared to liquids gases have very small viscosities in the range of a few 10^{-5} Pas. For process instrumentation a minimum of moving parts or mechanics is desired. Following this you will be at a capillar viscometer /3/ (fig. 1).

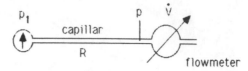

Fig. 1 capillar viscometer

A pressure source p_1 feeds the capillar with the resistance R. The volume flow $\dot{V}$ depends on the pressure difference p_1-p and the resistance R. R is proportional to the wanted viscosity η;

$$p_1 - p = R \cdot \dot{V} \qquad R = \frac{8l}{\pi r^4}\,\eta = a\cdot\eta \qquad\qquad \begin{array}{l} l\ \text{length,}\\[4pt] r\ \text{radius of the capillar} \end{array} \qquad (3)$$

Problems raise with measuring the very low flow $\dot{V}$. The new concept replaces flow metering of $\dot{V}$ by dynamic measuring p in a fixed volume (fig. 2).

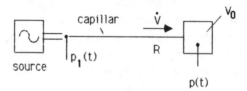

Fig 2 principle of the new viscometer

The most simple mathematical model of the device is given by the state equation of gases and the energy conservation theorem. We linearize and get with

$$p_1(t) = p_0 + \Delta p_1(t), \quad T(t) = T_0 + \theta(t), \quad V = V_0 + \Delta V(t), \quad \frac{d}{dt}\Delta V(t) = -\dot{V}:$$

$$pV = cT \quad\leadsto\quad \frac{\Delta\dot{p}(t)}{p_0} - \frac{\dot{V}}{V_0} = \frac{\dot{\theta}}{T_0} \quad\circ\!\!-\!\!\bullet\quad \frac{s\,\Delta p(s)}{p_0} = \frac{\dot{V}(s)}{V_0} + \frac{s\,\theta(s)}{T_0}$$

$$C_v\,\rho\,V_0\,\frac{\partial\theta(t)}{\partial t} + \alpha A\,\theta(t) = p\cdot\dot{V}(t) \approx p_0\cdot\dot{V}(t) \quad\bullet\!\!-\!\!\circ\quad \theta(s) = \frac{p_0\,\dot{V}(s)}{\alpha A + C_v\rho V_0 s} \qquad (4)$$

$\alpha A\theta$ gives the heatflow from the volume V_0 to the walls of V_0. This flow is proportional to wall area A and to the temperature difference θ between gas and wall. Eliminating $\theta(s)$, the system function of Volume V_0 becomes:

210

$$\frac{\dot{V}(s)}{\Delta p(s)} = \frac{V_0}{p_0} \left\{ \frac{1 + T_{th}\, s}{1 + \kappa\, T_{th}\, s} \right\} \cdot s \qquad (5)$$

with the capillar equation (3) the system function of system fig. 2 is:

$$G(s) = \frac{\Delta p(s)}{\Delta p_1(s)} = \frac{1}{1 + s\, T_p\, \dfrac{1 + s\, T_{th}}{1 + s\, T_{th}\, \kappa}} \qquad
\begin{aligned}
T_p &= \frac{V_0}{p_0}\, R = \frac{V_0}{p_0}\, a\, \eta \\[4pt]
T_{th} &= \frac{c_v\, \rho\, V_0}{\alpha A} \qquad \kappa = \frac{c_p}{c_v}
\end{aligned} \qquad (6)$$

κ is the ratio of specific heat c_p and c_v.

One concept to evaluate η from the system function is to give a harmonic input $\Delta p_1(t) = \Delta p_1 e^{j\omega t}$ to the system and to choose $T_p/T_{th} \gg 1$. Then we get

$$G(\omega) = A(\omega)\, e^{-j\varphi(\omega)} \approx 1 / (1 + j\omega\, T_p) \qquad (7)$$

The phase φ may be measured by lock-in-amplifier techniques. The phase

$$\varphi = \arctan \frac{V_0}{p_0} \cdot a \cdot \eta \cdot \omega \quad \text{supplies the highest sensitivity} \quad \frac{\partial \varphi}{\partial \omega} \quad \text{at}$$

$$\omega \cdot T_p = 1 \quad \text{or} \quad \varphi = \pi / 4 . \qquad (8)$$

The result is not affected if the pressure transducers Δp_1 and Δp change their sensitivity. The frequency ω is controlled to keep condition (8). In this case the viscosity is given, coming up from a standard gas η_0, ω_0 :

$$\eta = \frac{\eta_0\, \omega_0}{\omega} \qquad (9)$$

Following the design concept $T_p / T_{th} \gg 1$ a time delay of about one minute raises. Detailed investigations showed, that the heat transfer in the volume V_0 has to be considered with distributed parameters, as well the volume of the capillar. Hipp /4/ came along with a good compromise beetween time delay, independence to T_{th} and κ by controlling the real part of the system function instead of the phase $\varphi = \pi/4$:

$$\mathrm{Re}\{G(j\omega)\} = G(0) / 2 . \qquad (10)$$

3. THE NEW VISCOMETER

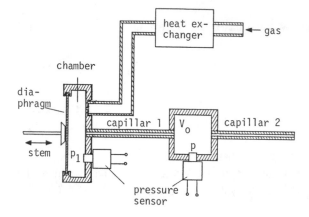

Fig. 3 mechanics of the viscometer

Fig. 3 shows the mechanical part. The gas to be measured flows continuously through a thermostat and then to a chamber closed to the atmosphere by a thin metallic diaphragm. In this chamber the pressure sensor p_1 is mounted. Then the gas flow passes the capillar and then the volume V_0. The pressure p in the volume is measured by the second pressure transducer. Finally the gas flows through a second capillar to the open air. All dimensions are kept as small as possible to prevent a considerable time delay. For example one capillar has a length l = 90 mm, a diameter D = 0.15 mm and the volume V_0 = 700 mm^3. The harmonic pressure signal $\Delta p_1(t)$ with an amplitude of about 10 mbar is generated by a moving coil in a permanent magnet, which acts on the stem of the diaphragm. The pressure sensors are of the miniature piezo-resistive type, range 400 mbar.

A sine-generator supplies sine waves up to 3 Hz in digits of 0.01 Hz. A micro-processor samples the signal Δp_1 and Δp. The frequency-response $\text{Re}\{G(j\omega)\}$ is calculated by lock-in-techniques. Is there a difference $\text{Re}\{G(j\omega)\}-G(0)/2$, the microprocessor controls the frequency of sine-generator until the difference vanishes. Compensation of the phase or the real part of $G(\omega)$ takes less than 5 periods of the sine wave.

Besides the time needed for control, the time delay is affected essentially by the time needed for gas exchange in capillar and volume V_0. This time is given by

$$t_F = \frac{2}{\pi} \frac{1}{\Delta p_1 / \Delta p_0} \cdot T \; ; \quad \omega = \frac{2\pi}{T} \tag{11}$$

For example with $\Delta p_1 / \Delta p_0 = 0.2$, it takes 3 periods to get a new sample into the system.

The expenditure to control phase or difference of $Re\{G(j\omega)\}-G(0)/2$ is not negligible. So Hipp /4/ tested phase measuring with lock-in-technique with a suitable fixed frequency, followed by worsening.

4. RESULTS

Table 1 shows the results with 2 viscometers VM3 and VM4.

Gas	VM3			VM4		
	$\dfrac{f_{0.5}}{Hz}$	η_m	$F_r/\%$	$\dfrac{f_{0.5}}{Hz}$	η_m	$F_r/\%$
He	1.627	2.034		0.638	2.034	
H_2	3.556	0.931	1.07	1.400	0.927	0.65
CH_4	2.794	1.184	2.70	1.105	1.174	1.87
CO_2	2.017	1.641	5.12	0,804	1.614	3.53
N_2	1.784	1.855	0.81	0,700	1.854	0.76
O_2	1.536	2.155	1.02	0.603	2.151	0.88
Dimensions:	VM3: $D = 0.15$ mm, $L = 90$ mm, $V = 395$ mm^3					
	VM4: $D = 0.15$ mm, $L = 90$ mm, $V = 710$ mm^3					

Table 1 results with type VM3 and VM4

Table 1 gives data compared with theoretical relation (9). The errors may be reduced by empirical fixpoint adjustment by the factor 2. An outstanding error was measured with CO_2.

Fig. 4 shows quite a different κ for CO_2 and CH_4. The error diminishes with a large volume V_0 along with the ratio T_{th}/T_p, for example VM4.

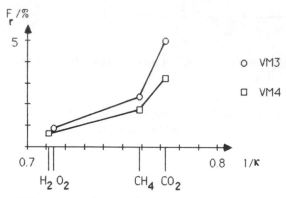

Fig. 4 error of the measured viscosity

5. CONCLUSIONS

The conformity between mathematical model and experimental results is excellent. All system parameters of any influence on the result are given. In order to get a small time delay one has to find a compromise between the errors mainly caused by variations of κ. An example for such a compromise is the type VM4 with gas exchange time of about 10 s and output frequencies in the range of $1/f = 0.5 ... 2$ s.

6. REFERENCES

1   Tsederberg, N.V.: Thermal Conductivity of Gases and Liquids. MIT-Press 1965

2   Jeans, J.H.: Dynamische Theorie der Gase. Friedr. Vieweg & Sohn, Braunschweig 1926

3   Hengstenberg, J.; Sturm, B.; Winkler, O.: Messen, Steuern und Regeln in der chemischen Technik. Bd. II, Springer, Berlin-Heidelberg-NewYork, 1980

4   Hipp, T.: Ein kontinuierliches dynamisches Kapillarviskosimeter für Gase. Dissertation am Institut für Prozeßmeßtechnik und Prozeßleittechnik, Universität Karlsruhe 1985.
     Published in : Fortschritt-Berichte VDI, Reihe 8: Meß-Steuerungs- und Regeltechnik Nr. 89.

INTELLIGENT SIGNAL TRANSMITTERS

Manfred Seifart

Technical University Otto von Guericke
3010 Magdeburg
German Democratic Republic

INTRODUCTION

In industrial instrumentation and process control the availability of low-cost microprocessors and single-chip-microcomputers (SCMC) leads to intelligent multiple-transducer data acquisition structures and systems. Such systems consist of remotely located intelligent modules, which are controlled by a central unit, for instance by a personal computer.

This paper describes a SCMC-based signal transmitter designed for remotely located intelligent measuring data acquisition using the party-line technique. This concept has several advantages over existing systems with analog signal transmission. It leads to a remarkable reduction of cable costs, because all signal transmitters of the system are connected to a single two-wire-bus (field bus), which transmits the signal bidirectional up the 19200 baud.
Two slightly different signal transmitters have been designed. The first type is equipped with two analog sensor inputs with a separate amplifier in each channel. The second type is designed with 8 inputs (7 individually programmable analog or binary) and 4 additional binary outputs for process control. Special attention was given to high flexibility. The following text refers to this type.

BASIC CONCEPT

Starting from practical demands the intelligent signal transmitter has been equipped with 7 analog/binary direct sensor inputs, one counter input and output, 4 binary outputs, one analog signal output which can be scaled to any range from 0-20 mA or 0-10 V and with an isolated serial output/input (UART). Signals of ≤ 256 transmitters may be transmitted bidirectionally by means of this serial output/input over a 2-wire-bus (field bus) up to 19200 baud. This bus also serves as interface to further signal processing devices.

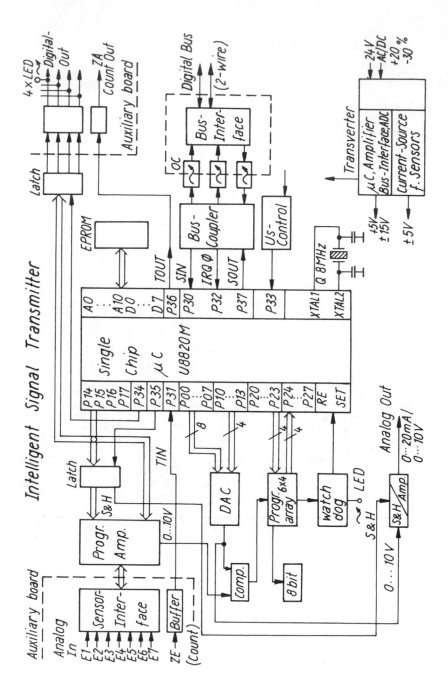

Fig. 1   Block diagram of the signal transmitter

216

Parameters of the intelligent transmitter such as range
(offset, full scale), alarm levels, configuration of the
transmitter, are adjustable remotely or by means of a programmable array and EPROM. The unit processes analog signals
with a resolution of 12 bit. Communication on the serial bus
enables remote-controlled diagnosis (access to all registers
of the SCMC). The SCMC allows preprocessing of sensor signals,
which is easily software-adaptable to the demands of special
applications. Examples are routines for linearization of
sensor characteristics, correction of offset and drift errors,
averaging, elimination of noise (filtering) and control of
limits. The SCMC in combination with the EPROM leads to some
additional flexible properties: ability for the solution of
new tasks by interchanging the EPROM, selection of pre-fabri-
cated basic signal processing routines by proper wiring the
programmable array or remotecontrolled by means of the function
field in the communication protocol, self-diagnosis (program
run, supply), signalling of errors.

CIRCUIT DESIGN

Fig. 1 shows the block diagram of the signal transmitter.
Main elements are the SCMC, the programmable sensor signal
amplifier with multiplexer, the ADC and DAC, the bus coupler
with isolated output stage, the programmable array (matrix)
with 6·4 programmable wired connections and a transverter for
isolated power supply.

The ADC is based on the successive-approximation principle. The DAC C 565 D is time-shared both for the AD- and
for the DA-conversion. During the AD-conversion a sample-
and-hold circuit holds the analog output signal.
The adress of the transmitter is selectable by a DIL-switch
in connection with the programmable array and a decoder. The
programmable array allows the user to select and combine
special statements, program starts and software conditioning
without variations on hard- and software.

The number of channels, the organization of the analog/
binary inputs and the selection of sensor types are adjustable
to the special demands of application by means of an inter-
changeable auxiliary board (80·120 mm^2). The hardware of the
main board (160·120 mm^2) remains unchanged for the various
applications of the transmitter.

SOFTWARE CONCEPT

The program memory (2-4 Kbyte) of the transmitter
contains a library of various input/output types and ranges
and usual process functions. This enables the user to field
program features required for his specials application. The
program consists of the following three modules: 1. Initiali-
zation, 2. Bus-communication and 3. Process software. The
process software realizes acquisition of analog and binary
signals, their arithmetic and logic processing as well as
output of data and control signals.

Transfer of data on the serial bus is asynchronous in
Manchester II-Code or NRZ-Code with a communication protocol
similar to the PDV-bus. The adress byte is followed by one
function byte and data bytes with cyclic redundancy check or
parity check. Only a complete protocol received without errors
is processed in the SCMC.

APPLICATION, ADVANTAGES

The universal concept of the intelligent transmitter
leads to a broad field of applications. The flexible system
structure involves several advantages concerning project work
and operation of automation systems. Some advantages are:
- saving cable costs by party-line-principle
- high reliability of communication and signal acquisition by
  digitizing near the sensor
- fixed hardware structure for different tasks, quick altera-
  tion of the application program (EPROM, programmable array,
  function field)
- combining and preprocessing of sensor in the transmitter.
The supply voltage is 24 V, power requirement amounts 3,5 W.
By applying CMOS-circuits and CMOS-SCMC the power requirement
can be further decreased by a factor of 5-10.

REFERENCES

Allen, Ch. 1983, How distributed should your control-system
        be?, Measurement and control, 16:174.
Bradshaw, A.T., 1984, Smart pressure transmitters,
        Measurement and control, 17:353.
Seifart, M., 1983, Mehrkanal-Meßwerterfassungssystem mit
        seriellem Bus, Wiss. Zeitschrift TH Magdeburg, 27:81.

DESIGN OF INTELLIGENT TEST SYSTEMS

Gu Wei-jun, Shan Guang-zhi

Tianjin Radio Factory No.1
Tianjin
The People's Republic of China

INTRODUCTION

Several instruments can be connected to each other through GPIB to build an automatic test system /ATS/. GPIB is very useful and is in common use because instruments produced by different manufacturers can be used in an ATS, as long as they are all equipped with GPIB. The ATS based on GPIB is very flexible, too. Its behaviour may change greatly, if the control software is changed.

The instrument manufacturers benefit greatly from GPIB, because theirs is to produce instruments for general purposes without caring about the varieties of user's systems. On the other hand, in order to use ATS based on GPIB the users have to pay a lot of money to buy some basic instruments and computers. What's more, the users have to design the control software themselves. Their technology level must be high enough to do such work. In fact, many user's test problems are unique and constant. So it is not worth the money for them to buy such a complicated system, especially in developing countries.

AN "OEP" INTELLIGENT INSTRUMENT

We have developed a new type "Open-end program" /OEP/ intelligent instrument. In an OEP instrument the software consists of two parts. One is the monitor of the instrument itself. The other part is the software of the ATS. The two parts are all fixed in the instrument. The microprocessor in an OEP instrument completes not only the function of the instrument itself but the function of the test system as well. Therefore in an ATS based on an OEP instrument the controller is not necessary. By pressing some buttons on the panel, the user can control such an ATS easily even if he has not any knowledge of computer hardware or software.

Of course, the ATS based on OEP instrument is not as flexible as the one based on GPIB. But it is not a problem for many users. Besides, an OEI instrument can be equipped with GPIB,

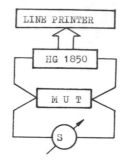

Fig.1.

too. Under the control of a computer it can complete more
complicated tests.

Hg 1850 voltmeter is a typical OEP instrument. It is a
precision digital voltmeter based on microprocessor. Half of
its byte EPROM contains the monitor of the voltmeter and the
other half is reserved for the system software. On its panel
there are two buttons named "user" and "return", respectively.
When the "user" button is pressed, HG 1850 begins its work
on ATS. Whenever the "return" button is pressed, HG 1850
will return back to its original voltmeter state immediately.
There are several softwares for different systems in HG 1850.
This paper is going to deal with two of them: Automatic
Calibration Program for meters and for thermocouple.

AN ATS FOR CALIBRATING METER

The method for calibrating voltmeters is shown in Fig.1.
MUT is the meter under test. S is a continuously adjustable
precision voltage source. S is used to drive the meter under
test. HG 1850 voltmeter measures the voltage applied on the
meter. At the same time HG 1850 controls a line printer to
print out the calibration report. After HG 1850 has entered
into automatic calibration program for meter, it will inter-
rogate about the peculiarities of MUT. With the aid of keyboard on panel the operator can put in date, ambient temperature, test frequency, MUT's serial number, grade, range
number, the full scale voltages of every range, the number
of tested points in basic range and scales of every tested
point. The man-machine interaction enables HG 1850 to suit
any meter.

After the man-machine interaction, HG 1850 begins calibrating the basic range. The scales of tested point are displayed on the panel to call the operator's attenation. The
operator should adjust the output voltage of S and make the
pointer of the MUT go up. When the pointer arrives at the
shown scales accurately, the operator should press another
button named "manual" and the measured voltage, which we
call "up-value", will be stored into RAM. The meter is calibrated at every tested point one by one, until all the points
are measured. Then the operator will turn down the output of
S and measure the voltage at every tested point in an opposite

220

direction. All the "down-value" will be stored into RAM, too.
Because of the existence of error, at any tested point the
nominal value, the up-value and down-value are different from
each other.

The calibrating procedure of higher ranges is similar to
that of the basic range. But there are only four tested points
for every higher range. Apart from the least and the full scales,
HG 1850 should find another two points according to the
following principle: the error is the largest at these two
points in the basic range. After having calibrated all the
ranges HG 1850 begins calculating the correction value for
every tested point of every range and the maximum error of
the meter under test. Then HG 1850 should determine whether
the meter under test is up to standard or not. The line printer
will print out the final calibration report and as many copies
as required. Then HG 1850 stops and waits for the operator's
instruction. The operator must decide either to calibrate a
next meter or to retire from the user's program and to return
to the original voltmeter monitor.

AN ATS FOR CALIBRATING THERMOCOUPLE

The calibration of thermocouple is another example. Fig.2.
shows the principal connection of this method. HG 1850 controls
a scanner and a line printer. A standard thermocouple and couples under test are placed into an oven and their cold junctions
is measured with a platinum resistor. With the aid of the
scanner HG 1850 can measure the electromotive force of every
thermocouple and the resistance of platinum resistor successively.

After HG 1850 enters into the "Automatic Calibration Program
for Thermocouple" the operator should put in the type of
standard couple, the type of the couple under test, the type
of platinum resistor, the correction values of the standard
couple and the platinum resistor. After man-machine interaction HG 1850 measures the resistance of platinum resistor.
By means of curve fitting HG 1850 can derive a temperature
from this resistance. /By the way, the coefficients of the
curve fitting formulas for platinum resistor and some common
couples are stored in the ROM./ After correction, the real
temperature of the cold junction can be obtained. Then HG 1850
selects the fitting formula corresponding to the standard
couple to derive the "cold junction compensation emf" from
the cold junction temperature. The sum of this "cold junction
compensation emf" and the emf measured on the work-junction
of standard couple equals the emf which is the output emf if
the cold junction is put into ice-water. By means of the curve
fitting and correction the real temperature of the oven can
be obtained.

In this way, HG 1850 measures and controls the oven's
temperature continuously. When the temperature rises steadily
and finally arrives at a certain test point, the operator
should press the "manual" button and star a measuring cycle.

In the cycle the temperature of cold junction is measured
once more. According to the type of couple under test HG 1850
derives the cold junction compensation emf from this temperature. Then the emf of every couple under test is measured.
From the sum of these and the cold junction compensation emf

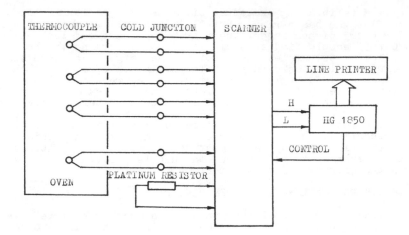

Fig.2.

HG 1850 derive the temperature for every couple. The difference between the temperature and the oven's real temperature indicates the errors of every couple under test at this temperature point. HG 1850 will compare these errors with the maximum tolerance to see if they are up to standard or not. After the test points are all measured the line printer will print out the final calibration report.

It takes only a few seconds to accomplish a measuring cycle, so the temperature of the oven can be maintained unchanged in this period. It is essential to the calibration of the thermocouple. The main task of this program is to overcome the non-linearity of thermocouple and thermoresistor automatically.

SIMPLE SOFTWARE-BASED IEC625 INTERFACE

FOR COMPUTER AIDED MEASUREMENT

Wiesław Winiecki, Piotr Sokołowski

Institute of Radioelectronics
Warsaw University of Technology
PL-00665 Warsaw, Poland

INTRODUCTION

One of the major factors limiting the automation of laboratory experiments, esp. by non-professionals, is the high cost of a computer controlling an automation system, and often rather complicated procedure of operating the system. A low -cost personal computer, applied as a system controller, would relinquish both above limitations. It should satisfy the following requirements:
1) It should enable one to connect it with measuring and peripheral devices via standard interface IEC625-bus.
2) The controller should be operated from a high-level language, preferably BASIC, to broaden the range of its potential users not necessairly familiar with programming. For the same reason interface commands should formally resemble BASIC instructions and should be easy to use in programming.

GENERAL APPROACH

The above requirements may be met most easily when the hardware part of the interface is minimized and software means are used as wide as possible for implementation of interface functions.

The first task to be solved is functional analysis of the IEC625-interface standard [1] (also known as the IEEE488, HP-IB and GPIB), and formulating a list of directives which are necessary to accomplish full control of the measuring system in accordance with this standard. The user interface must to be as simple as possible. The same applies to the directives. "The time which can be saved by using one character cryptic commands is often a lot smaller then the advantage gained by the use of readable easy-to-remember commands. It should be possible to operate the system by only a few commands" [2]. In the discussion of the syntax and meaning of directives, some already existing controllers, e.g. HP-85 [3] and Philips PM4410 [4], has been taken into consideration. But the solution presented in this paper differes from those mentioned above, esp. as

regards simplicity of the use of directives. This has been
achieved by implementation of more general directives — each
one consisting of several IEC625-commands or messages. All the
mnemonics corresponding to the directives has been chosen as to
resemble BASIC commands and describe the action the directives
carry out. An additional advantage of the proposal, which
broadens the circle of the potential users, is that using the
directives doesn't require any special experience in the
standard IEC625. After the detailed analysis, the following
list of the directives has been generated:

INIT   — initializates the hardware part of the interface and
             sends IFC *) message;
REMOTE   — sends REN message, LLO command to actual listeners
             in the system, and UNL-UNT commands;
CLEAR   — sends DCL or SDC command to actual listeners in the
             system, and UNL-UNT commands;
TRIGGER   — sends GET command to actual listeners in the system
             and UNL-UNT commands;
LOCAL   — sends GTL command to actual listeners in the system
             and UNL-UNT commands;
WRITE   — sends data to actual listeners in the system and UNL
             -UNT commands;
READ   — receives data from actual talker in the system,
             stores them in memory and sends UNL-UNT commands;
LINK   — enables actual talker to send data to actual listen-
             ers in the system;
SERVICE   — carries out serial polling when SRQ message is de-
             tected, sends SPE command, addresses all talkers in
             turn, receives their status bytes, identifies talker
             which has sent SRQ message, sends SPD, UNL and UNT
             commands;
PASS CONTROL — passes control over the system to other control-
             ler with command TCT and awaits return of control.

     Any directive may have a parametric form (e.g. TRIGGER
<address 1>, <address 2>,...) or non-parametric (e.g. TRIGGER).
In this latter case, some variables must be reserved to save
relevant addresses and data (e.g. L$=<address 1>,<address2>,..).
All above directives have to be performed according to IEC625
-interface functions. Those functions are usually implemented
using hardware means. In order to minimize the costs of repro-
ducing the controller, software means has been used here for
implementation of interface functions, and the hardware part
acts only as an adaptor of microprocessor signals to IEC625
-standard bus.

     The described approach can be adapted if only the micro-
processor bus is accessible for the user of the microcomputer.

IMPLEMENTATION

     The concept of the system controller presented in this
paper was intended to be implemented on a microcomputer. The
analysis presented above does not specify any particular microcomputer to be used as a system controller. The only requirements that the microcomputer must fulfil are:
- availability of processor signals on the expansion bus,

----
*) all abreviations used are after [1]

224

- possibility of calling machine-code routines from an existing operating system.
Both those requirements are not very restrictive: actually all
the microcomputers the authors happened to work with fulfilled
them, although at different levels of satisfaction. Up till
now the conception of the IEC625 system controller has been
implemented on Sinclair ZX81, Sinclair Spectrum and Amstrad
6128 microcomputers. In the process of working out the software,
two general problems emerged. The first one was the linkage of
the new directives with already existing commands of BASIC
interpreter. In ZX81 and Spectrum, the interpreter is located
in ROM and represents closed entity. The natural extension of
BASIC would require re-building the interpreter by either
doubling the part of the operating system in RAM or adding
"shadow" EPROM to the hardware part. Both those solutions were
discarded: the first one because of RAM limitation, the second
one - because of the complication of hardware. Another approach
to the problem has been chosen, viz. special program construction has been used to enable the user to introduce controller
directives into a program in the form: GOSUB <directive name>
(e.g. GOSUB WRITE). The software part of the controller trans-
lates it, using an appropriate machine-code routine. This construction makes clear and readable the listing of any user
program. In the case of Amstrad 6128 microcomputer this problem
has been satisfactorily solved by its constructors. The controller directives can be easily implemented using so called
RSX (Resident System Extension) facility. All the new commands
have been simply logged-on and can be used directly like usual
BASIC keywords.

The second problem was the way of passing parameters from
and to controller directives. It has turned out to be more complicated. On our view, its succesful solution determines the
quality of the controller. In ZX81 and Spectrum, the USR <ad-
dress> command does not pass any parameters; so in order to
make possible passing adresses and data, it was necessary to
assign certain variables exclusively to controller directives.
Special machine-code procedure was then used to scan the
variable space in operating system in order to find an
appropriate variable. The data received via DIO lines were put
into assigned variables in the same manner. In Amstrad 6128
the directives logged-on using RSX facility pass parameters
both in integer and string format, so adresses and data could
be passed directly in a very easy way. Unfortunately, there
was no straitforward way to store received data in BASIC
variables. Therefore the string or integer variable array of
data is passed by address of its first element.

The experience gained with ZX81, Spectrum and Amstrad has
proved that the system controller is easily implementable. The
code written in Z80 assembly language occupies ca 1.5 kB of
RAM space and the linkage of directives and passing parameters
seems to be done satisfactorily.

EXAMPLE OF APPLICATION

A short program, written for the controller based on Amstrad 6128, to animate a simple measuring system is presented
below. The system consists of the controller and a digital
voltmeter (DVM) V553 MERATRONIK. The controller programs the

DVM, triggers measurement and stores received data in memory and displays them.

```
10  !INIT                          ;initialization of the hardware
20  B$="0"
30  T%="0"                           ;setting up initial values
40  K$=""                             of controller variables
50  DATA$="        "
60  !REMOTE,48                      ;setting DVM into remote state
70  !CLEAR,48                       ;clearing DVM interface
80  !WRITE,48,"HOX4Y5","ASCII"      ;programming DVM
90  !TRIGGER,48                     ;triggering the measurement
100 !SERVICE,80,B$,@T%              ;serial poll procedure
110 !READ,@T%,DATA$,K$              ;receiving data from DVM
120 PRINT DATA$                     ;displaying the data
130 !LOCAL,48                       ;transfering DVM to local state
140 STOP                            ;the end of the program
```

CONCLUSIONS

The presented idea of the software-based IEC625 interface for personal computers enables one to build a simple and inexpensive system controller fully compatibile with IEC625 standard. It makes possible automation of measurements in those cases, where it is unresonable because of the high cost of a professional computer with the IEC625 interface. The software -based implementation of IEC625-interface functions proved to be possible. The limited speed of data transfer over interface bus, in comparison to hardware-based-controllers is the only negative consequence of the proposal. The data can be transferred at the rate of ca 5 kB/sec. But in simple systems which the controller is designed for, it is usually acceptable.

The usefulness of the low cost controllers with software -based IEC625-interface has been proved in the laboratory of the Institute of Radioelectronics, WUT.

REFERENCES

1. IEC publication 625-1: Standard Interface Systems for Programmable Measuring Equipment, IEC, (1979).
2. L. Van Biesen, and P. Bakx, Expert System for Intelligent Measurements, Proc. 5th Int. IMEKO-TC7 Symp. Intelligent Measurement, No 1, Vol.2, 297:300, (1986).
3. HP-85 I/O Programming Guide, pt IV, Hewlett-Packard USA Feb. (1981).
4. J. A. M. Grimberg, Digital Instrument Course, Pt. 4, IEC Bus Interface, N. V. Philips - Gloeilampenfabrieken Eindhoven, The Netherlands. Publ. N° 9498.829.99311.

APPLICATION OF A MICROPROCESSOR

IN A MULTIFUNCTIONAL INSTRUMENT

Fei Zheng-sheng   Keng Ye   Li Dong-bin

Harbin Institute of Electrical Technology

Daqing Road, Harbin, China

INTRUDUCTION

Due to the application of semiconductor rectifiers and switching devices, there are more and more nonlinear loads in the industrial power distribution systems. Therefore, the true rms measuring technique is now seriously concerned.

This paper suggests an intelligent multifunctional instrument which is designed to measure, store and print the values of frequency, power factor, true rms values of voltage and current, active power, reactive power and energy in alternating current circuit.

This meter is based on the sampling theorem and calculation method. The intellectualization of the instrument and GP-IB interface are provided to suit the need of automatic measuring system.

PRINCIPLE OF INSTRUMENT

This instrument consists of an analog circuit and a microcomputer board consisting of Z-80A CPU, PIO, CTC, 10K EPROM and 4K RAM (Fig.1). In the analog circuit, there is a fast A/D converter which uses the successive approximation conversion technique to achieve a 12-Bit conversion in 8 microseconds. The timing sequence of the instrument is shown as Fig.2. Tu is the sampling period for u(t) and Ti for i(t). While measuring the power, we use only one A/D converter in order to reduce the cost. The instantaneous values of the voltage and the corresponding current are sampled during the different cycles. During the same beginning sampling point in two cycles to measure voltage and current respectively as shown in Fig.2, the result is the same as simultaneously sampling both the voltage and cuttent by two A/D converters.

The measurements of frequency and phase in AC circuit may be realized through controlling the timing circuit.

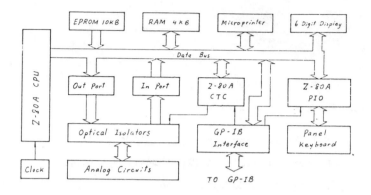

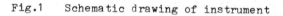

Fig.1   Schematic drawing of instrument

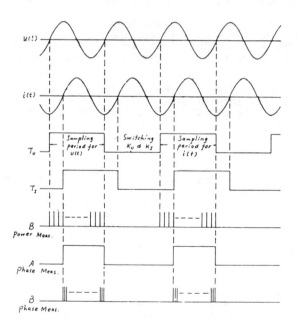

Fig.2   Timing sequence

SAMPLING METHOD

If we suppose the voltage and current are sinusoidal,
they are:

$$v(t) = V_m \sin\omega t; \\ i(t) = I_m \sin(\omega t + \varphi)\Big\}$$
(1)

where ω is the frequency and φ is the phase angle. V_m and I_m
are the amplitudes of the voltage and current. The power may
be written as following:

$$p(t) = V_m I_m \sin\omega t \sin(\omega t + \varphi)$$
$$= \frac{V_m I_m}{2}\left[\cos\varphi - \cos(2\omega t + \varphi)\right]$$
(2)

The sampled value of $p(t)$ at time tj is

$$p(tj) = v(tj) \cdot i(tj);$$
(3)

and the mean value of $p(t)$ in one period may be written as
follows:

$$P = \frac{1}{\Delta T \cdot n} \sum_{j=1}^{n} v(tj) \cdot i(tj) \,\Delta T$$
$$= \frac{V_m I_m}{2}\cos\varphi - \frac{V_m I_m}{2n} \sum_{j=1}^{n} \cos\left(2j\frac{2\pi}{n \cdot \Delta T}\,\Delta T + \varphi\right)$$
(4)

where ΔT is the sample interval and n is the total sample
number.

C.H. Dix had proved [1] that, if $2/n$ is not an integer,
it is $n \gtrless 3$, the sampled error will be zero in theory. In
practice, the signal may be not sinusoidal and the sampling
is always completed in more or less than one or several periods, and the error may be written as following:

$$\Delta P = -\frac{V_m I_m}{2n} \sum_{j=1}^{n} \cos\left[2j\frac{2\pi+\theta}{n} + \varphi\right]$$
(5)

If $\varphi = 0$, θ and $2\pi/n$ are very small, ΔP will be maximum at
$\theta = 2\pi/n$:

$$\Delta P_{max} = -\frac{P}{n+1}$$
(6)

The maximum error in this case is inversely proportional to
the number of samples.

There are several methods(2) ,(3) , (4) which can be
used to reduce the truncation error. Here, we use trapezoidal
integration to calculate the power and rms values. The average
power of $p(t)$ may be numericaly integrated from Fig.3 as
follows:

$$P = \frac{1}{T}\int_{0}^{T} p(t)dt$$
$$= \frac{1}{T}\sum_{j=0}^{n-1} (P_j + P_{j+1})\,\frac{Tj}{2}$$
(7)

Where Tj is the jth sample interval, it may be written as
follows:

$$Tj = t_{j+1} - t_j$$

During regularly spaced sampling before t_{n-1}, there are

$$\Delta T_1 = \Delta T_2 = \cdots = \Delta T_{n-2} \quad ; \quad \Delta T_j = \Delta T \quad (j < n-1)$$

(7) may be reformulated as follows:

$$P = \frac{1}{T} \sum_{j=1}^{n-2} p_j \Delta T + \frac{1}{2T} (p_{n-1} + p_n)(\Delta T + \Delta T_{n-1}) \tag{8}$$

The final sample interval $\Delta T_{n-1} = t_n - t_{n-1}$ may be less than ΔT, it may be $\Delta T_{n-1} = \Delta T - \xi$. then (8) may be written as following:

$$P = \frac{1}{T} \sum_{j=1}^{n} p_j \Delta T - \frac{1}{2T} (p_{n-1} + p_n) \xi \tag{9}$$

If we take n samples during one period, and assume that the voltage and current are sinusoidal wave:

$$\left. \begin{array}{l} v_j = V_m \sin(j x + \alpha) \quad ; \\ i_j = I_m \sin(j x + \alpha + \varphi) \, . \end{array} \right\} \tag{10}$$

where v_j and i_j --- the sample values of V and i at j:

$x = \frac{2\pi}{n}$ --- the sample interval;

α --- the beginning phase angle of v;

φ --- the phase difference between v and i.

P may be rewritten as follows:

$$P = \frac{1}{n \Delta T} \sum_{j=1}^{n} V_m I_m \sin(j x + \alpha) \sin(j x + \alpha + \varphi) \Delta T - \frac{1}{2T} (p_{n-1} + p_n) \xi \tag{11}$$

$$= UI \cos \varphi + \Delta P$$

where

$$\Delta P = -\frac{1}{2T} (p_{n-1} + p_n) \xi = \frac{\xi}{T} UI \cos \varphi \left\{ -1 + \cos \left(2\alpha - \frac{2\pi}{n} \right) \cos \frac{2\pi}{n} \right.$$

$$\left. - tg \varphi \sin \left(2\alpha - \frac{2\pi}{n} \right) \cos \frac{2\pi}{n} \right\} \tag{12}$$

when $\cos \varphi = 1$, the truncation error may be written as follows:

$$\gamma_p = \frac{\xi}{T} \left[-1 + \cos \frac{2\pi}{n} \cos \left(2\alpha - \frac{2\pi}{n} \right) \right] \tag{13}$$

If α and $2\pi/n$ are very small as compared with 2π, γ_p may be written as follows:

$$\gamma_p = \frac{\xi}{T} \left[-\left(\frac{2\pi}{n} \right)^2 - \frac{(2\alpha)^2}{2} + 2\alpha \left(\frac{2\pi}{n} \right) \right] \tag{14}$$

when $\alpha = 0$, we get:

$$\gamma_p = -\frac{\xi}{T} \left(\frac{2\pi}{n} \right)^2 . \tag{15}$$

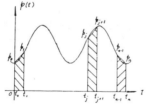
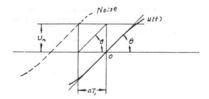

Fig.3   Trapezoidal integration   Fig.4   Zero-crossing drift due to noise

ERROR ANALYSIS

The errors due to timing and discrete signal measurement are the principal component. But the sampling method error may be neglected.

a. The Effect of A/D Conversion Error

If the gain error of A/D converter is α and the offset error is β, then the sampling values of voltage v and current i may be written as following:

$$v_{xj} = v_j + \alpha\, v_{xj} + \beta\, v_m \;, \qquad i_{xj} = i_j + \alpha\, i_{xj} + \beta\, i_m \tag{16}$$

where v_j and i_j are true values of signals, and v_m and i_m are the full scale range. The errors due to attenuator, amplifier and sampling resistance are neglected.

Substituting (16) into the power calculation formula (4) we get:

$$P_x = \frac{1}{n}\sum_{j=1}^{n}(v_j + \alpha\, v_{xj} + \beta\, v_m)(i_j + \alpha\, i_{xj} + \beta\, i_m) \;. \tag{17}$$

Neglecting the errors of bigher order, we may get approximately

$$P_x = \frac{1}{n}\sum_{j=1}^{n}[(v_j\, i_j) + 2\alpha\, v_{xj}\, i_{xj} + \beta\, v_j\, i_m + \beta\, v_m\, i_j] \tag{18}$$

$$= P + 2\alpha\, P_x + \frac{\beta}{n}\sum_{j=1}^{n}v_j\, i_m + \frac{\beta}{n}\sum_{j=1}^{n}v_m\, i_j$$

Then, the error of power may be written as follows:

$$\Delta P_x = P_x - P \leq 2\alpha\, P_x + 2\beta\, P_m \tag{19}$$

where $P_m = V_m\, i_m$, V_m and i_m are the full scale range of voltage and current.

We may rewrite the error formula due to A/D converter as follows:

$$\Delta = 2\alpha\,(Indication) + 2\beta\,(Full\ Scale\ Range) \tag{20}$$

b. The Effect of the Timing

The timing accuracy is determined by the accuracy of crystal-controlled frequency and zero-crossing accuracy of the periodic signal. The effect of the crystal may be neglected due to its high stability.

The zero-crossing error due to the noise effect shown in Fig.4 may be written as follows:

$$tg\,\theta = \frac{U_n}{\Delta T_1} \tag{21}$$

For sinusoidal signal $v(t) = V_m \sin \omega t$, we have:

$$\frac{\Delta U_n}{\Delta T_1} = tg\,\theta = \frac{d\,v(t)}{dt}\bigg|_{t=0} = \omega V_m = \frac{2\pi}{T}V_m \tag{22}$$

Then

$$\Delta T_1 = \Delta T_2 = \frac{T}{2\pi}\cdot\frac{U_n}{V_m} \tag{23}$$

Considering the randomness of the noise, we have:

$$\Delta T_{max} = \sqrt{\Delta T_1{}^2 + \Delta T_2{}^2} = \frac{T}{\sqrt{2}\,\pi}\cdot\frac{U_n}{V_m}\;. \tag{24}$$

It is obvious that the zero-crossing error of signal period is in the opposite proportion of the signal-noise ratio and also the slope of the signal at zero-crossing point. Therefore, we must choose low noise preamplifier and raise the signal-noise ratio, and choose fast response and wide bandwidth comparator.

The phase difference error caused by the indeterminateness of the period is

$$\Delta \varphi = \omega \, \Delta T_{max} = \sqrt{2} \, \frac{U_n}{V_m} \tag{25}$$

Obviously, raising signal-noise ratio is also beneficial to the decrease of the phase difference error.

The power error due to $\Delta \varphi$ may be written as follows:

$$\gamma_p = \pm \sqrt{2} \, \frac{U_n}{V_m} \, tg \, \varphi \tag{26}$$

When the power factor is high enough, we may rewrite the formula (26) as follows:

$$\gamma_p \doteq (\frac{U_n}{V_m})^2 \tag{27}$$

It is shown that the error is in the opposite proportion of the signal-noise ratio. This error will be disappeared in the rms measurement.

ACCURACY

This instrument's accuracy has been calibrated by NIM of China(NIM-The National Institute of Metrology). Our intelligent AC multifunctional instrument HL-8602 has the accuracy as follows:

```
f:      0.2 % Reading 20Hz-10kHz
φ:      0.1 % Reading + 0.05 % Full Scale
Vrms:   0.1 % Reading + 0.05 % Full Scale 20 Hz - 1000 Hz
Irms:   0.2 % Reading + 0.1 % Full Scale  20 Hz - 1000 Hz
P:      0.3 % Reading + 0.15 % Full Scale Cos φ = 1
        0.5 % Reading + 0.3 % Full Scale  Cos φ = 0.5
```

REFERENCE

1. C.H.Dix, Calculated performance of a digital sampling wattmeter, NPL Memo DES 17, 1975
2. F.J.J. Clarke and J.R. Stockton, Principles and theory of wattmeters operating on the basis of regularly spaced sample pairs, J.Phys E:Sci. Instrum, Vol, 15, 1982
3. Tai Hsien-chung, The quasisynchronous sampling and its application in the measurement of nonsinusoidal power,

   Chinese Journal of Scientific Instrument, Vol. 5, No.4, 1984
4. Lu Zu-liang, On the quasi-integer-period-sampling and the approach for improving its accuracy. CPEM'86, Gaithersburg, MD, USA

A PRECISION INTELLIGENT AC-DC SOURCE

Dušan Fefer, Janko Drnovšek, Anton Jeglič

Faculty of Electrical Engineering
Edvard Kardelj University of Ljubljana
Ljubljana, Yugoslavia

INTRODUCTION

An AC voltage source under software control with an internal intelligent DC reference has been developed. The main
purpose of the unit is to be used as a precision secondary or
working standard for power and energy measurements in laboratories and in production lines. Such a multi function source
(AC, DC voltages, current, phase, power) benefits from its
complexity, since the interconnection problems of separate
modules are minimized. On the other hand, integration of precision analog digital and power circuits generates severe new
problems.

BASIC CONCEPT OF AN AC-DC SOURCE

Block diagram of basic elements shows Fig. 1.

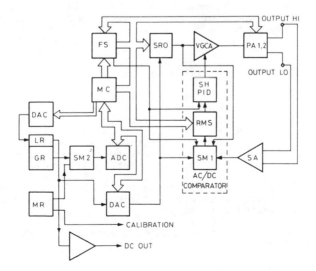

Fig. 1

Internal sine reference oscillator SRO generates an amplitude stable voltage, with very low harmonic distortion and ability of frequency adjustment (internally or sinchronized externally). Output of SRO is connected to voltage gain control amplifier VGCA that is in amplitude control loop with the ability of setting the required output power via power amplifiers PA1,2. For precision amplitude control, the ouput is atenuated by sense amplifier SA and led to a precision AC/DC comparator. Comparator compares true RMS value of output voltage from the sense amplifier SA with a precision DC reference. Error signal controls voltage gain control amplifier VGCA.

An intelligent internal DC voltage source consists of a group of four temperature stabilized Zener diods. A constant mean value of the group reference GR enables to trim the leading voltage reference LR to the appropriate value. Master reference is used for calibration purposes.

DETAILS OF BASIC AC CIRCUITS

SRO is the most critical element of the AC control loop. Since there are severe demands on voltage stability, very low harmonic distortion, frequency stability and a fast response on parameter variations, a digital solution proved to be the most suitable. A digital sine generator is always a compromise between high frequency of output signal and a total harmonic distortion.

Limiting factors concerning speed are DAC conversion times and EPROM access time. Besides, temperature stability and stability with time are to be considered as well.

Analyses show however that a digital sine wave constructed of 36 discrete values per period, each 12 bits long, has a total harmonic distortion below 0,1%, short term stability better than 10^{-4} and standard deviation $1.5 \cdot 10^{-4}$. High stability of AC source is based on direct comparison with the internal intelligent DC reference. AC voltage could be compared with DC by measuring its peak, mean or RMS value. An electronic RMS converter has been used for wide dynamic range, fast response and high speed. VGCA in the AC cmplitude control circuit is affected by the AC-DC comparator output, where a positive or negative signal appears, corresponding to the difference between output and reference voltage. Regulated voltage output from VGCA is amplified by power amplifiers PA1 and PA2.

PA1 and PA2 are for voltage ranges from 0.1 V to 10 V and from 10 V to 100 V respectively, built in VMOS technology. Output transformator for voltages from 100 V to 1000 V is in the feedback loop of PA2 to minimize the output impedance and to fasten the response on load variations. For correcting gain, linearity, drift, frequency dependence etc. an automatic testing sequence is performed, using internally stored correction data. Autocalibration cycle consists of 15 measuring points and takes place after every "switch on" of the whole unit, altering the output parameters, after every 10^3 measuring cycles or beeing requested externally.

Calibration and correction constants are at the same time basis for diagnostics.

DETAILS OF BASIC DC CIRCUITS

An intelligent internal DC voltage source defines the
upper limit of absolute and relative accuracy of the instrument. Three DC reference levels (master reference MR, group
reference GR and leading reference LR) are to improve technological and metrological parameters of a major reference
system within the whole AC-DC source, and equally important,
imrove the reliability of metrological parameters within a
certain calibration interval.

By permanently measuring the differences of a group of
four semi-conductor DC references, information about drift
with time of every particular element is obtained. Random
errors are reduced since in a group of N references the remaining voltage drift is reduced by the square root of their
number. Knowing the prestored mean value of the group and of
single elements, numerical correction is performed via digitally controlled power supply DAC causing a change of current
in a leading voltage reference LR. Numerically corrected leading voltage reference LR is amplified to appropriate voltage
levels thus becoming a voltage reference for a DAC whose output
is already a reference point for AC circuits. Systematic errors
(simultaneous drift of the group, eccessive temperature deviations etc.) as well as unpredicted variations (excursions
exceeding voltage tolerances of references or even a total failure of a single or more elements) are supressed by occasional
autocalibration with an on board master reference. Calibration
of master reference is performed via front panel connection to
a null volt meter and a primary standard. All the differences
are entered as new numerical correction constants.

CONCLUSIONS

A significant improvement of stability, reliability and
maintainability has been achieved by integrating a DC reference
into an AC source under microcomputer control. Further research
is going on for AC and DC divider circuits, to improve the resolution and the accuracy of output parameters, thus obtaining
a high precision multipurpose AC-DC calibrator.

REFERENCE

1. J.Drnovšek, A.Jeglič, D.Fefer, "Systems approach to longterm
   stability of DC references", IMEKO world congress, Praga
   1985, pp. 75-77.

ITERATIVE METHOD FOR CONVERSION OF THE RMS VALUE

IN PRECISION AC MEASUREMENTS

Jerzy S. Olędzki, Jan M. Szymanowski

Department of Electrical Engineering
Warsaw University of Technology
Koszykowa 75, 00-662, Warsaw, POLAND

INTRODUCTION

The ability to determine the rms value of an ac signal
with a high degree of accuracy is of critical importance in
many tasks. Recently designed voltmeter of Fluke (Goyal and
Brodie, 1984) provides accuracy to about 160 ppm. Signal
recirculation method applied there may be considered to be
a special case of the iterative method (Alijev and Schekichanov,
1983). The purpose of this work is to present another version
of the method, which makes it possible to build a relatively
inexpensive converter system that would determine the rms
value of an unknown ac signal with great accuracy and relatively   quickly. The structure of the converter system leaves an
allowance for various auto-adaptive possibilities.

DESCRIPTION OF THE METHOD

A signal rms value of which is to be accurately determined
is converted into dc form by a relatively inaccurate rms
converter, such as e.g. AD 637 (Counts and Kitchin, 1984) and
after that converted into digital form by an analogue/digital
converter (ADC, e.g. high accuracy digital dc voltmeter). The
result (E_{x1}) is both stored in memory data register (MDR) of
microcomputer and converted into dc voltage (E'_{d1}) by a
digital/analogue converter (DAC). The last signal is given as
input to the rms converter (fig.1). The dc output of the rms
converter is measured again (E_{d1}) and stored in MDR. Quite
apart from the results E_{x1} and E_{d1}, three next voltages are
measured in cycles and stored in MDR:
1) E_o – zero voltage from rms converter when its
input is connected to ground,
2) E_{oD} – zero voltage from rms converter when its
input is connected to the output of DAC
(set to 0),
3) E_{fs} – full scale output voltage from DAC.
The rms value of the original input can be computed from:

$$E_x = \frac{E_{x1} - E_o}{E_{d1} - E_{oD}} \, E_{x1} \qquad (1)$$

The number of discrete states of DAC m_{x1}, equal to E_{x1}, is set according to:

$$m_{x1} = \frac{(2^n - 1) \, E_{x1}}{E_{fs}} \qquad (2)$$

where n - word length of DAC. Then:

$$E'_{d1} = E_{x1} \qquad (3)$$

Let's assume that systematic error of the rms converter is a linear function of the input signal, and systematic error of the ADC is neglected. Thus the output signal can be expressed as:

$$E_{x1} = E_{xr} \left(1 + \delta_m \right) + E_o, \qquad (4)$$

where E_{xr} represents the value of the original unknown rms input and δ_m represents the multiplicative component of the systematic error. If the resolution and linearity of DAC are good enough to meet the accuracy requirement of equality (3), the second conversion gives:

$$E_{d1} = E'_{d1} \left(1 + \delta_m \right) + E_{oD} =$$

$$= \left[E_{xr} \left(1 + \delta_m \right) + E_o \right] \left(1 + \delta_m \right) + E_{oD} \qquad (5)$$

Taking into consideration formulas (1), (4) and (5) we obtain:

$$\underline{E_x = E_{xr}} \qquad (6)$$

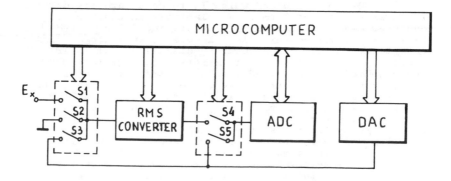

Fig. 1   The rms iterative measurement system

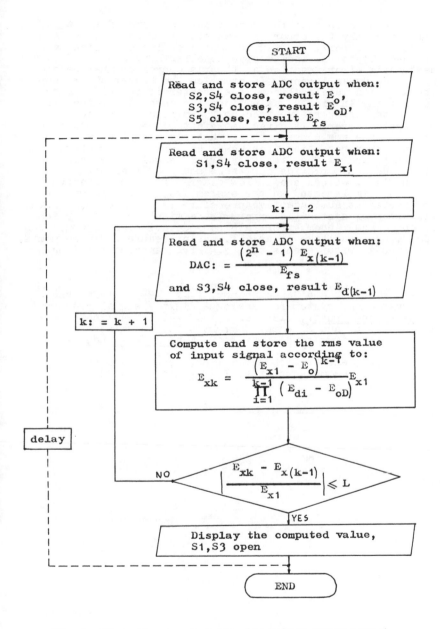

Fig.2 Flow diagram for the iterative measurement
process

239

One can prove that the result obtained from formula /1/ will have only the error of measurement of dc voltage /ADC systematic error/. The DAC is tested by the ADC which makes it possible to use inexpensive monolithic version of DAC. The systematic error function of the rms converter must be smooth over wide dynamic range. In practice the nearer can we approach the accepted idealizations, the greater is the efficiency of correction. In model examinations the authors obtained in the first step of iterative process 20-50-fold reduction of conversion errors. It gives satisfactory results, but when choosing the rms converter one must pay a particular attention to the error resulting from ac-dc conversion characteristics diversity and to the frequency error as well. The correction method does not provide for those sources of error.

In Fig.2. we can see a flow diagram for the iterative procedure implemented in the system from Fig.1. It may be considered as auto-adaptive procedure in accordance with the criterion of systematic error. There might be other options if one takes random error into consideration. Procedure from Fig.2. does not take it at all. The procedure could be modified for special requirements too, e.g. frequency corrections.

REFERENCES

1. Alijev, T.M, and Schekichanov, A.M.: 1983, Iterative measurement system for analogue/digital signal processing Mess. and Pruef., 19: No.12, 757 + 60, 764 /in German/
2. Counts, L, and Kitchin, Ch.: 1984, Second-generation monolithic rms-to-dc converter, Analog-Dialogue, 18, No 1, 11-13
3. Goyal,R., and Brodie, B.T.: 1984, Recent advances in precision ac measurements, IEEE Trans. IM - 33: No 3, 164-167.

THE AUTODIAGNOSIS SYSTEM OF AN INTELLIGENT MEASURING INSTRUMENT

F. Ferraris, M. Parvis

Dipartimento di Automatica e Informatica
Politecnico di Torino
Torino,Italy

INTRODUCTION

In this paper we describe the autodiagnosis system of a particular high performance measuring instrument (see Ferraris et al.[1] for the description of the whole apparatus), but the methodological aspects and the design features can be extended to most of intelligent measuring systems.

Before everything, it is worthwhile to state that we intend the word "autodiagnosis" in a quite wide sense: in our meaning, the results of the autodiagnosis procedures are not only "pass-fail" responses, but also statements such as "the instrument is to be used in degraded accuracy (or with feature loss) condition"; in fact, such condition cannot be usually detected by inspecting only measuring results, but require a more complete tests to be executed.

The autodiagnosis is obviously carried out on both digital and analogue circuits, but, for the sake of brevity, in this paper we point out only the analogue section, rather well known techniques having been used for the digital section.

THE AUTODIAGNOSIS SYSTEM

The complete autodiagnosis system can be logically splitted into two major tasks conceived for separate aims:
- full-test off-line autodiagnosis;
- on-line fast test.

The full-test autodiagnosis

This task performs a complete test inside the whole instrument and checks over the metrological and functional performances.

Test running is incompatible with normal working, so it is carried out off-line, at power up, on user request or when the on-line fast test detects hard problems.

For a good test a hardware improvement is usually required, but this is not normally a severe problem, because of the slight modifications needed. A simple example, clearifying the concept, is sketched in fig. 1, where a simplified high-performance instrument input stage is presented: a powerful test can be easily carried out whith only two access points

241

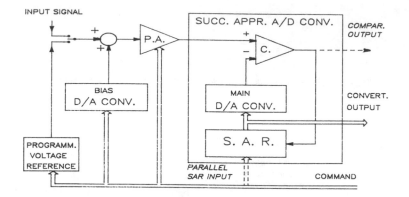

INPUT SIGNAL

SUCC. APPR. A/D CONV.

COMPAR. OUTPUT

P.A.

C.

BIAS D/A CONV.

MAIN D/A CONV.

CONVERT. OUTPUT

S. A. R.

PROGRAMM. VOLTAGE REFERENCE

PARALLEL SAR INPUT

COMMAND

Fig. 1   A simplified high-performance instrument input stage
Dashed lines indicate the new access points
P.A. :   programmable amplifier
C.   :   comparator
S.A.R.:   successive approximation register

(highlighted in the figure with dashed lines and slanted characters)[1].
     The analogue full-test autodiagnosis can be viewed as a superset of
the self-calibrating procedures and carried out in two steps.
A)   The first step is just a static calibration cycle performed by means
     of a calibration standard. The scaling and correction parameters
     extracted in this step are used to calibrate the measuring channels,
     but are also stored and compared with the old parameters recorded in a
     permanent memory (environmental and historical memory); in such a way
     a calibration history is obtained, allowing the prediction of the
     short-time behaviour by means of a previously identified degradation
     model.
B)   In the second step the instrument behaviour is checked under dynamic
     conditions by means of "analogue stimuli" provided inside   the
     apparatus. We have used three types of stimuli:
     - Slow triangular waves, to monitor all the A/D converter states,
       revealing missing codes.

Note 1. A possible test procedure can be sketched as follows:
             - Digital tests are performed by loading, via parallel input,
               suitable patterns to the Successive Approximation Register and
               by monitoring the converter output.
             - Functional tests are executed on comparator and bias D/A
               converter, using the latter D/A to force comparator switching.
             - Functional and static characteristic tests on D/A converters
               are obtained by monitoring the comparator output while both D/A
               converters are handled.
             - Autocalibration of the A/D converter is performed, using the
               programmable voltage reference.
             - Dynamic tests on A/D converter are executed using the bias D/A
               converter as stimulator and presetting the main D/A converter
               by SAR parallel input.

242

- Programmable fast triangular waves, to test the measuring channels at high speed sampling rate, revealing slew rate limitations.
- Microsteps superimposed on a programmable constant voltage (see also Souders[2]), to monitor the fast dynamic response of both converter and signal conditioning systems, revealing parasitic oscillations.

The information obtained from the tests is processed according to the block diagram sketched in fig. 2. Three possible responses arise:
a) "The instrument is usable", if all tests are passed.
b) "The instrument must be used in degraded conditions", if for any reason the rated accuracy cannot be achieved or if any not basic hardware element failed the test.
c) "The instrument is not usable", if any basic element failed the test.

If the instrument is usable (case a or case b), after processing the test results and the content of the historical memory, a short term performance prevision is emitted. The test results will be also used in normal working to correct the measurements carried out and to improve the uncertainties estimation (see Ferraris et al.[1]).

If any problem arises during tests (case b or case c), a failure analysis to obtain reparation aids is performed and hypotheses about failures aggravation are presented.

The on-line fast test

While operating, it is important to warrant the user that the instrument still meets all the specifications.

The requirement is checked in two ways:
- The critical elements of the hardware (for example, the A/D converter) are continuously tested to reveal potentially dangerous fails while still in asymptomatic stage; test results are then processed in order to obtain a fail-safe behaviour. The monitor process can be:
  - direct, i.e. applied to the potentially dangerous quantities (for example, voltages or currents in selected points are measured and checked with bound values);

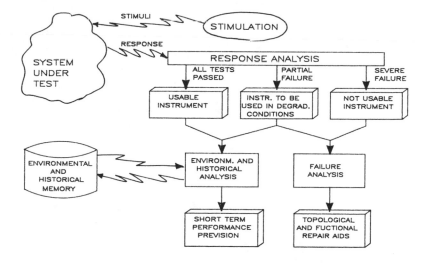

Fig. 2   Full-test autodiagnosis block diagram

- indirect, i.e. applied to quantities related to normal work (for example, the components temperatures).
- As frequently as possible (at the end of the measuring interval or during the acquisition process, if the sampling rate allows it), a known standard signal is measured and the measurement results are compared with the known values.

If a fail or accuracy loss is recognised, the instrument automatically stops the normal working and starts the full-test procedures.

REFERENCES

1. F. Ferraris, I. Gorini and M. Parvis, Intelligent multi-input in-strument for electrical measurements, IMEKO Symp. "Intelligent Measurement - Inquamess 86", Jena (1986).
2. T. M. Souders, A dynamic test method for high-resolution A/D converters, IEEE Trans. on I. M., n. 1 (March 1982).

A MICROCOMPUTER-BASED METHOD FOR CALIBRATION

Heinz Holub and Bernd Michaelis

Technical University Otto von Guericke
3010 Magdeburg
German Democratic Republic

INTRODUCTION

A typical measurement problem are nonlinearities caused by the sensor. Additionally the characteristic is affected by drift and time-dependent deformation. Further inaccuracies are produced by the amplifier, the analog-digital converter and other elements (Seifart, 1986).

Precise determination of the physical quantity requires the calibration of the measuring device. In intelligent devices this is performed with the aid of microcomputers. Considering repeated calibration or selfcalibration, problems arise if nonlinearities are to be compensated on the basis of a few calibrating points only.

GENERAL CALIBRATION-ALGORITHM

The paper deals with a general method requiring only a few calibration points and using a-priori-information. The general calibration-algorithm is presented in Fig. 1. At first the functions $f_i(x)$ for a series expansion are estimated using a-priori-information (a representative number of measured characteristics $y_k(x)$). This calculation needs to be performed only once with the aid of a suitable computer. It results in a series expansion for the real characteristic of the measuring device in the form:

$$\widetilde{y}(x) = \sum_{i=1}^{m} a_i \, f_i(x).$$
(1)

In eqn. 1 x there are the measured values, $\widetilde{y}$ are the calculated (linearized) values and $f_i(x)$ represent the functions of the series expansion used.

The coefficients a_i must be calculated for the desired nonlinear characteristic of the measuring device. This calibration is fulfilled with the aid of the microcomputer of the measuring device.

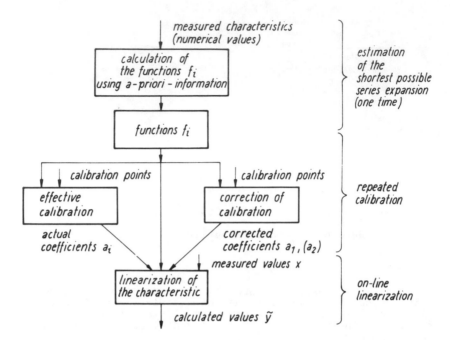

Fig. 1. Presentation of the used calibration method

Using l calibration points x_ν, y_ν, the coefficients a_i follow from the well known condition (least mean square):

$$\sum_{\nu=1}^{l}\left[y_\nu - (a_1 f_1(x_\nu) + a_2 f_2(x_\nu) + \dots)\right]^2 \Rightarrow \text{Min!} \quad (2)$$

This calibration is necessary for every measuring device and must be repeated if the characteristic of the device is changed.

If the variation of the characteristic is very small, only one or two coefficients of the series expansion must be corrected. For this the criteria eqn. 2 can be used in a slightly modified form.

The on-line linearization during the measurement uses the characteristic eqn. 1 with the actual coefficients. For the reduction of computing time the characteristic can be stored pointlike or other methods may be used (Holub, 1985).

Only few calibrating points are desired in real measurement. This demands a very short series expansion (small m) in eqn. 1. Using a representative number of measured characteristics as a-priori-information, such a series expansion can be estimated (Michaelis, 1983).

ESTIMATION OF THE SERIES EXPANSION

The well known Karhunen-Loéve-expansion (Chien and Fu,

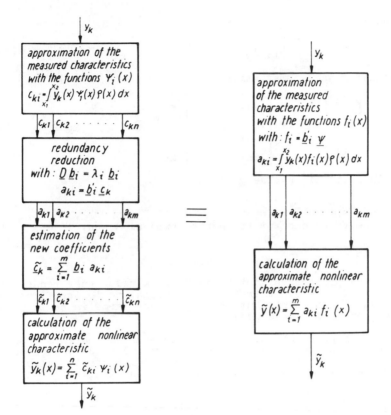

Fig. 2   Information reduction for the generation of
the shortest possible series expansion

1967; Peipmann, 1975) is applied to describe the nonlinear
characteristic of the measuring device resulting in a very
short series expansion.

The measured characteristics should be approximated in
the form:

$$y_k(x) = \sum_{i=1}^{n} c_{ki} \, \psi_i(x) \qquad k = 1,2, \ldots ,p. \qquad (3)$$

The functions $\psi_i(x)$ are identical for all characteristics.
Now the shortest possible series expansion

$$\tilde{y}_k(x) = \sum_{i=1}^{m} a_{ki} \, f_i(x) \qquad (1a)$$

must be estimated. In eqn. 1a the functions $f_i(x)$ are ortho-
normalized. A useful criterion is

$$R_m = \frac{1}{p} \sum_{k=1}^{p} \int_{x_1}^{x_2} \left[y_k - \sum_{i=1}^{m} a_{ki} \, f_i(x) \right]^2 dx \Rightarrow \text{Min!} \qquad (4)$$

247

Assuming

$$f_i(x) = \sum_{j=1}^{n} b_{ij}\, \psi_j(x) \tag{5}$$

and using the abbreviations

$$\underline{\psi} = \begin{bmatrix} \psi_1(x) \\ \cdot \\ \cdot \\ \cdot \\ \psi_n(x) \end{bmatrix} ; \quad \underline{b}_i = \begin{bmatrix} b_{i1} \\ \cdot \\ \cdot \\ b_{in} \end{bmatrix} ; \quad \underline{c}_k = \begin{bmatrix} c_{k1} \\ \cdot \\ \cdot \\ c_{kn} \end{bmatrix} ; \quad \underline{D} = \frac{1}{p}\sum_{k=1}^{p} \underline{c}_k\, \underline{c}_k' \tag{6}$$

follows the eigenvalue problem

$$\underline{D}\,\underline{b}_i = \lambda_i\,\underline{b}_i \tag{7}$$

and the functions for the series expansion eqn. 1

$$f_i(x) = \underline{b}_i'\,\underline{\psi} \; . \tag{5a}$$

The proposed algorithm performs a redundancy reduction resulting in a concentration of information in the first terms of eqn. 1. The contribution of the last terms is negligible and a small m may be used.

In Fig. 2 the redundancy reduction and the interconnection of $f_i(x)$ and the original functions $\psi_i(x)$ is demonstrated. The weighting function $\rho(x)$ introduced generalizes the algorithm for unequal errors in the range of measurement.

MINIMIZATION OF THE RELATIVE DEVIATION

Above the functions $f_i(x)$ are estimated on the assumption of the least mean square of linear deviation between real characteristics and approximations. This assumption is used for the calculation of the actual coefficients a_i, too (calibration).

Some problems demand the minimization of relative deviation in the measuring range.

Supposing the real case of small approximation errors and little time-dependent deformation of characteristics, a simple modification of the algorithm derived above can be used.

The measured characteristics $y_k(x)$ must be normalized with the aid of the middle characteristic $f_N(x) \approx \bar{y}$ as

$$\eta_k(x) = \frac{y_k(x)}{f_N(x)} \; . \tag{8}$$

From eqn. 8 follows for the deviation of $\eta_k(x)$

$$\Delta\eta_k(x) = \frac{\Delta y_k(x)}{f_N(x)} \; . \tag{9}$$

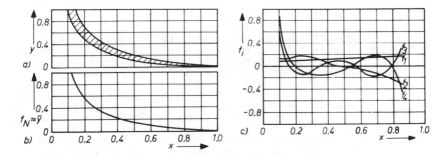

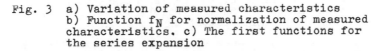

Fig. 3 a) Variation of measured characteristics
b) Function f_N for normalization of measured
characteristics. c) The first functions for
the series expansion

Assuming only small variation of the measured characteristics,
we obtain approximately

$$y_k(x) \approx f_N(x). \qquad (10)$$

Therefore $\Delta\eta_k(x)$ is approximately the relative deviation
of $y_k(x)$

$$\Delta\eta_k \approx \frac{\Delta y_k}{y_k} \qquad (11)$$

and the algorithm suggested can be used for the normalized
characteristics $\eta_k(x)$ minimizing the relative deviation.

More details are described by Michaelis and Holub (1982).

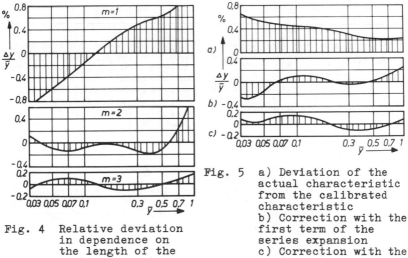

Fig. 4 Relative deviation
in dependence on
the length of the
series expansion

Fig. 5 a) Deviation of the
actual characteristic
from the calibrated
characteristic
b) Correction with the
first term of the
series expansion
c) Correction with the
first two terms of the
series expansion

NUMERICAL EXAMPLE

The power of the derived method is demonstrated with a numerical example. Fig. 3 shows the typical nonlinear characteristic and the range of deviation. The first functions of the series expansion are illustrated on the right.

The relative difference between the real characteristic and its approximation is shown in Fig. 4 for a typical example with the number of terms m of the series expansion as parameter. If the characteristic is not considerably deformed after calibration a correction must be applied only. Fig. 5 demonstrates this situation. For correction only the estimation of one or two coefficients of the series expansion and some few calibration points are needed (Holub, 1985). This makes the programming of the microcomputer very simple.

CONCLUSIONS

The wide use of microcomputers in measurement gives the possibility for simple application of the suggested method for calibration. The sensor or device must permit the input of calibration points and the a-priori-information for calculation of the functions of the series expansion should be available.

The functions $f_i(x)$ of the series expansion eqn. 1 need to be calculated only once on a suitable computer using a number of measured nonlinear characteristics (a-priori-information).

In an intelligent device the method suggested needs additionally only a small programme for calibration and a memory for the functions of the series expansion.

REFERENCES

Chien, Y. T., and Fu, K. S., 1967, On the generalized Karhunen-Loéve-expansion, IEEE Trans.Inform.Theory, IT-13:518.
Holub, H., 1985, Mikrorechnergestützte Kennlinienapproximation mit problemangepaßten Entwicklungsfunktionen zur effektiven Kalibrierung von Meßfühlern, Wiss.Z.Techn.Hochsch. Magdeburg, 29:7:97.
Michaelis, B., and Holub, H., 1982, Ein Verfahren zur Linearisierung nichtlinearer Kennlinien und effektiven Kalibrierung unter Nutzung von A-priori-Information, msr, 25:365.
Michaelis, B., 1983, Die Einführung zusammengesetzter Meßgrößen - ein Konzept zur Meßdatenreduktion, msr, 26:300.
Peipmann, R., 1975, Grundlagen der technischen Erkennung, VEB Verlag Technik, Berlin.
Seifart, M., 1986, Stand und Trend der Meßwerterfassung mit Mikrorechner, Materialien der 4. Fachtagung "Anwendung von Mikrorechnern in der Meß- und Automatisierungstechnik", Magdeburg, pp. 65-79.

ADAPTIVE CONTROL OF FREQUENCY AND PHASE

IN VECTOR ANALYZER

Mart Min, Ants Ronk and Hanno Sillamaa

Department of Automation
Tallinn Technical University
Tallinn, Estonia, USSR

INTRODUCTION

Vector analyzers are the instruments for measuring vector coordinates of electrical signals' harmonic components. Our analyzer Model 8687 QUADRA enables to perform precise measurements within the frequency range 1 Hz to 1 MHz in case of additive disturbances exceeding the signal up to 100,000 times. The analyzer is based on quadrature (two-phase) synchronous demodulation[1], performed as hardware multiplication of analog signal $S_A(t)$ to be analyzed by digital coordinate signals[2] $\widetilde{\sin}$ and $\widetilde{\cos}$ (see Fig. 1). The coordinate signals are phase locked onto external reference signal $S_R(t)$ with phasing error $\phi_e < 0.1°$. The signal channel contains a two-phase synchronous detector SD, based on two multiplier-type digital-to-analog converters[4] Mult1 and Mult2 driven by the input signal $S_A(t)$. The converters transform $\widetilde{\sin}$ and $\widetilde{\cos}$ into the SD output signals $\hat{I}$ and $\hat{Q}$. These signals enable to estimate inphase and quadrature coordinates $\hat{I}$ and $\hat{Q}$ of the vector $\overset{\star}{S}_A$. The signals I_0, Q_0 or M_0, Φ_0 at the output of signal channel are products of digital processing of $\hat{I}$ and $\hat{Q}$ in the signal processor. Due to the phasing error ϕ_e of the coordinate system, the values I_0, Q_0 and Φ_0 will not coincide with the true values I, Q and Φ (see Fig. 2) of the vector $\overset{\star}{S}_A$.

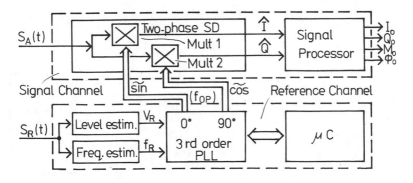

Fig. 1 Block diagram of the vector analyzer

251

The actual Cartesian coordinate system, used in the analyzer is determined by signals $\widetilde{\sin}$ and $\widetilde{\cos}$ as (positive) inphase and quadrature coordinate vectors (see Fig. 2). These signals are formed as the special digital codes by the phase-locked loop (PLL)[2,3] in the reference channel (see Fig. 1).

The next requirements are placed upon the PLL to be satisfied over the whole frequency range. It is necessary

(1) to keep $\widetilde{\sin}$ and $\widetilde{\cos}$ mutually quadrature and normalized,
(2) to obtain and keep $\widetilde{\sin}$ locked onto the reference signal $S_R(t)$, accompanied by noise or disturbances not exceeding its level, with phasing error $\phi_e < 0.1°$ (see Fig. 2),
(3) to attain phase-lock with the minimal transient time,
(4) to obtain accurate frequency and phase acquisition in case of the drifting and sweeping reference frequency f_R.

We propose to use the time optimal third order PLL, adaptive control of which is realized by microcomputer µC (see Fig. 1). Estimators of the reference signal level V_R and frequency f_R are used to provide the µC with expert information about $S_R(t)$.

THE PLL AND ITS PHASING CAPABILITY

The PLL (see Fig. 3) contains a phase detector PD, similar to Mult1 and Mult2 in the signal channel (Fig. 1), a low-pass filter LPF and a voltage controlled oscillator VCO with quadrature outputs for $\widetilde{\sin}$ and $\widetilde{\cos}$.

The PD may be considered as a phase comparator for signals $\widetilde{\sin}$ and $S_R(t)$ (or its harmonic component). Then, however, the PD output signal S_e is not a pure error signal. The LPF is needed to suppress different significant noises and disturbances, partly caused by properties of the PD (for example nonlinearity). As $\sin = \sin(2\pi f_{op}t + \phi_e)$, exact phasing can be achieved due to slight driving of the operating frequency f_{op} by the LPF output signal S_c. In case $\Delta f = f_R - f_{op} \neq 0$ being sufficiently small, a linear model of the PLL is applicable and the PLL is able to lock $\widetilde{\sin}$ onto $S_R(t)$ (i.e. to obtain $f_{op} = f_R$ and $\phi_e \approx 0$) in acceptable time.

For a narrow band of f_R a suitable LPF with transfer function

$$F(s) = \frac{\tau_2 s + 1}{(\tau_3 s + 1) \cdot T_F s} \tag{1}$$

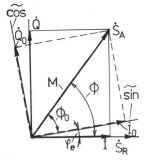

Fig. 2 Vector diagram

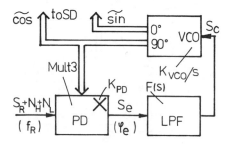

Fig. 3 Block diagram of the PLL

can be found. Transfer function of the remainder part of the PLL

$$L(s) = \frac{V_R}{2} K_{PD} \cdot \frac{K_{vco}}{s} = \frac{1}{T_L s} \qquad (2)$$

together with $F(s)$ determine the open-loop transfer function of the linearized PLL[3]:

$$G(s) = F(s) \cdot L(s) = \frac{\tau_2 s + 1}{T_F T_L s^2 (\tau_3 s + 1)} \qquad (3)$$

It has been shown[3], that the PLL is optimal, if

$$\nu = \frac{\tau_2}{\tau_3} = 6 \ldots 7 \qquad (4)$$

and, for each value of the parameter ν

$$\text{opt } \chi^2 = \frac{\tau_2}{\sqrt{T_F \cdot T_L}} = \frac{2\nu^2}{9\nu - 27} \qquad (5)$$

Then, in case $|\Delta f / f_{op}| < 0.05$, the signals $\widetilde{\sin}$ and $\widetilde{\cos}$ will be locked onto $S_R(t)$ so, that $\phi_e < 0.1^\circ$ is achieved with minimal transient time.

It is evident that for keeping the PLL optimal and able to track varying $f_R(t)$, adaptation of the PLL to the reference signal is required. On the other hand, some means are needed to measure the frequency f_R with relative error not exceeding a few per cent and to get some information about the reference signal level V_R.

THE ADAPTIVE PLL CONFIGURATION

An appropriate configuration of the adaptive PLL is shown in Fig. 4. Here the LPF consists of analog filter AF and digital integrator DI based on reversible (up/down) counter RC and digital-to-analog converter DAC. The VCO is composed of voltage controlled clock VCC, binary frequency--divider FD and coordinate code former. The AF output signal V_D acts on the VCC (via the summing input) introducing so a zero into the LPF transfer function (1).

The reference frequency estimator consists of adaptive pulseformer APF (based on peak detectors) and binary counter BC. Synchronous detector SD (Mult4), similar to Mult1 and the others, enables to estimate the reference signal level.

The microcomputer based on 8-bit microprocessor realizes

(1) control of the time constants of the analog filter and the digital integrator (code A0-A7, N and n),
(2) control of the dividing factor 2^N of the frequency divider (A0-A4, N) and the current frequency subrange of the voltage controlled clock (A5-A7, n),
(3) control of current frequency of the VCC by changing the reversible counter output code (B0-B7, k),
(4) control of the reference frequency estimator operating mode (L0-L2).

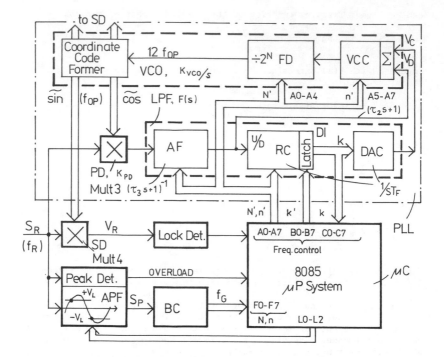

Fig. 4 Block diagram of the reference channel

REFERENCE FREQUENCY MEASUREMENT

Sufficiently precise estimation of reference frequency f_R in conditions of low-frequency and high-frequency disturbances N_L and N_H is possible due to the APF. The pulse-forming process in the APF is shown in Fig. 5. Triggering levels V_{Li} of the APF depend on peak values V_{pi} of the signal $S_R(t) + N_L(t) + N_H(t)$, and on factor q corresponding to frequency estimation mode, chosen by the µC. When the signal $S_R(t) + N_L(t) + N_H(t)$ crosses the level V_{Li-1} for the first time, the value of V_{Li} is fixed according to $V_{Li} = qV_{pi-2}$ and peak detector is initiated to determine V_{pi-1}. If q=0.5, the use of such adaptive hysteresis band guarantees the needed accuracy in case $V_R/(N_L \ N_H) > 3$ and even in many situation where $V_R/(N_L \ N_H) < 3$. For example, if $N_H \gg N_L$, and $f_H \gg f_R$ then q=0.5 makes $V_R/(N_L \ N_H) > 1/3$ acceptable.

CONTROL OF OPERATING FREQUENCY f_{op}

Varying the digital integrator's DI output signal k=k(t) the PLL attains, both, $f_{op}=f_R/t)$ and $\phi_e \approx 0$, and realizes so the frequency fracking within the window determined by n and N (see Fig. 4), both given by the µC. k(t) may be presented in the form

$$k(t) = k_G + k_T(t) + k_F(t) \qquad (6)$$

Table 1 Window-shifting modes M

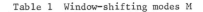

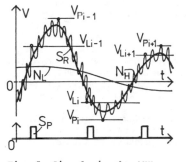

Fig. 5 Signals in the APF

M	Q	r	δ_D	δ_I
2	k > 215	-2	0	12.8%
	215 > k > 120	-1		
0	k > 215	-1	-3.4%	3.5%
	k < 33	+1		
-2	33 < k < 120	+1	11.4%	0%
	k < 33	+2		

where k_G is the initial value of $k(t)$, which correspnds to the approximate value f_G of the reference frequency f_R, measured by the frequency estimator or given via external control by the aid of front panel or GPIB. $k(t)$ is a slowly varying component of frequency control characterizing the trend of f_{op} towards the f_R, and $k_F(t)$ may be considered as rapidly varying nonstabilities of frequency and phase.

The PLL is built up so, that, not taking in account the small analog component V_D of feedback signal (see Fig. 4), the VCO operating frequency is

$$f_{op} = (2^{20} + 10^3 \cdot (k - 120)) \cdot 2^{-(N+n/8)} \qquad (7)$$

where $N = 0, 1, \ldots, 20$, $n = 0, 1, \ldots, 7$ and $k = 0, 1, \ldots, 255$ determine the frequency dividing factor 2^N, the frequency subrange of the VCC, and the frequency within the subrange, respectively.

In order to hold (a) $f_{op} = f_R(t)$ observable, and (b) the PLL able to perform its functions normally (to keep $0 \le k \le 255$) during the next period T_M, the microcomputer μC replaces, if it is needed, the current frequency window by another one according to the next window shifting algorithm.

WINDOW SHIFTING ALGORITHM

The window-shifting is realized by the microcomputer which performes with a time interval T_M the next operations:

(1) reading and storing a current value of $k(t) = k(T_M \cdot i) = k_i$,
(2) choosing according to the behaviour of $k(t)$ and previous window-shifting mode M, the most suitable new mode M and proper value of shift-step r,
    Comment: Table 1 presents the main characteristics of three possible modes M. The value of r is determined by chosen M and by true (in the case of given k_1) inequality Q, if such exists. Otherwise r=0. δ_I and δ_D are the maximal relative increase and decrease of f_{op} per T_M, which in some cases cannot be realized by the PLL.
(3) if r≠0, computation and following simultaneous output of new values for k, n and N according to

$$k' = -A + (A + k) \cdot 2^{r/8}, \qquad (8)$$

where $A = 2^{20} \cdot 10^{-3} - 120$, and

255

$$n' = \begin{cases} n+r+8 \\ n+r \\ n+r-8 \end{cases} \qquad N' = \begin{cases} N-1 & \text{if } n+r < 0 \text{ and } N > 0 \\ N & \text{if } 0 < n+r < 8 \\ N+1 & \text{if } n+r > 8 \text{ and } N < 20. \end{cases} \qquad (9)$$

FINAL REMARKS

Detailed analysis of frequency acquisition dynamics will be presented elsewhere. Fig. 6 illustrates possibilities opening up to the PLL due to the microcomputer-control. It's evident, that one ought to avoid too frequent shifting the windows as result of varying of $k_F(t)$. This can be done by a proper choice of (1) size, overlapping, and the number of windows, covering the frequency range, and (2) window-shifting algorithm. As the problem is similar to that, which arise in connection with gain control , so we prefer to consider it in a wider context in future.

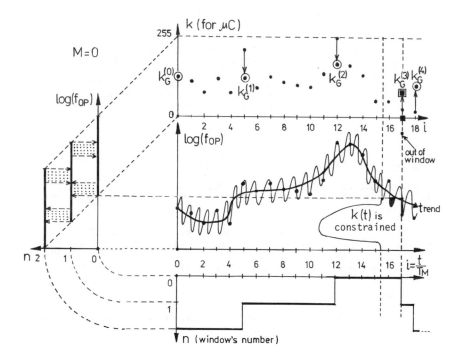

Fig. 6 Variations of f_{op}, n and k

REFERENCES

1. M. L. Meade, "Lock-in Amplifiers: Principles and Applications", Peregrinus, London (1983).
2. M. Min, Phase lock of coordinate signals in vector measuring devices (in Russian) in: "Proc. Tallinn Technical University", No. 592 (1985).
3. M. Min, A method for design of the time optimal third order phase locked loop, in: "Proc. 7-th European Conference on Circuit Theory and Design (ECCTD'85)", Part 1, Prague (1985).
4. M. Min, T. Parve, Kh. Khyarm, and T. Pungas, Quadrature stepwise frequency converter, US Patent No. 4,409,555. Oct. 11, 1983.

MICROCOMPUTER CONTROLLED RADIO FREQUENCY SIGNAL ANALYZER BASED ON POLARITY CORRELATION

Eckart Schröter, Reiner Thomä, Peter Peyerl

Dept. for Communications Engineering and Theory
of Electrical Engineering
Technische Hochschule Ilmenau
6300 Ilmenau, GDR

INTRODUCTION

Digital signal analysis is a well-established method for signal and system analysis e. g. in the field of acoustics, speech and vibration studies. But for higher frequencies it becomes increasingly expensive to digitize and record the incoming analog waveforms. In the case of periodic (or at least repeatable transient) signals, undersampling is a convenient way to reduce the rate of digitization. This, however, requires some triggering or synchronisation with regard to the signal. It fails, if the signal period is unknown or if the signal to be investigated is random.
In the following a data acquisition scheme using RF sampling and polarity correlation is described which becomes the basis for 2-channel signal analysis in the RF range. It is preferably employed if the signals explored are stochastic in nature.

POLARITY CORRELATION

Eq.(1) describes an estimate of the polarity correlation function (PCF), provided the stochastic process is ergodic and has zero mean:

$$\Upsilon_{P(xy)}\ (k\Delta\tau) \approx \frac{1}{N} \sum_{i=0}^{N-1}\ \mathrm{sgn}\left[x(t_i)\right] \cdot \mathrm{sgn}\left[y(t_i + k\Delta\tau)\right].$$

$$(1)$$

If the process has furthermore a uniform or Gaussian probability density function, the scaled common correlation function Υ_N results from Eq.(2). (For unscaling the mean power values $\Upsilon_{xx}(0)$, $\Upsilon_{yy}(0)$ have to be determined[1].)

$$\Upsilon_{N(xy)}(k\Delta\tau) = \frac{1}{\sqrt{\Upsilon_{xx}(0)\ \Upsilon_{yy}(0)}}\ \Upsilon_{xy}(k\Delta\tau) = \sin\left[\frac{\pi}{2}\ \Upsilon_{P(xy)}(k\Delta\tau)\right]$$

$$(2)$$

If the probability distribution function of $x(t)$, $y(t)$ is unknown, auxiliary signals $a(t)$, $b(t)$ are added to $x(t)$, $y(t)$:

$$x'(t) = x(t) + a(t)$$
$$y'(t) = y(t) + b(t).$$

The signals $a(t)$, $b(t)$ must be uncorrelated with respect to each
other and to the signals to be analysed. Further demands are:

- zero mean $(\overline{a(t)} = \overline{b(t)} = 0)$
- uniform probability density
  $p(a) = 1/2A$ on $[-A, A]$, $p(b) = 1/2B$ on $[-B, B]$
- $|x(t)|_{max} \leq A$; $|y(t)|_{max} \leq B$.

For convenience $a(t)$, $b(t)$ should be simple periodic signals
without special requirements to frequency and frequency stability. If the PCF is determined in this way, the (common) correlation function follows directly from:

$$\Upsilon_{xy} (k\Delta\tau) = A \cdot B \cdot \Upsilon_{P(x'y')} (k\Delta\tau). \qquad (3)$$

In order to determine $\Upsilon_{xy}(k\Delta\tau)$ from Eq.(2) the mean power values $\Upsilon_{xx}(0)$, $\Upsilon_{yy}(0)$ are estimated directly from the probability distribution functions $P(x)$, $P(y)$ according to Eq.(4):

$$\Upsilon_{xx}(0) \approx \sum_{i=1}^{N-1} x_i^2 \, \Delta \, P(x_i). \qquad (4)$$

Starting from the correlation function $\Upsilon_{xy}(k\Delta\tau)$ a variety of signal characteristics can be derived. One major goal is the po-
wer spectrum as indicated in Fig. 1. It is evident that this method prefers the indirect way of spectral estimation[2]. The well-
known alternative direct periodogram estimation requires spectral averaging and multiple execution of the FFT-procedure. In
contrast the indirect way is often avoided because of the time-
consuming calculation of the correlation sum. But polarity correlation offers an effective way of (i) hardware realization of
Eqs.(1), (4) and (ii) signal sampling to reduce data rates as
will be pointed out later.

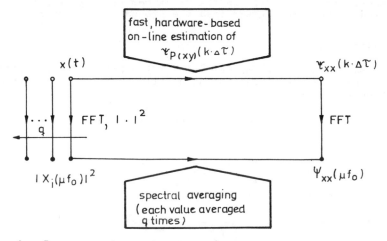

Fig. 1   Power spectra estimation (indirect way using polarity
          correlation)

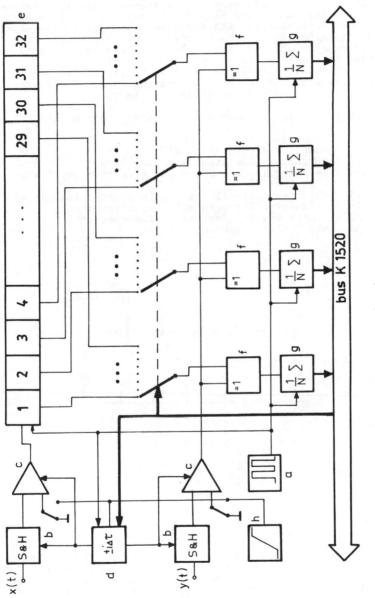

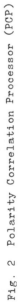

Fig. 2 Polarity Correlation Processor (PCP)

259

RF POLARITY CORRELATION PROCESSOR (PCP) HARDWARE DESIGN

According to Eq.(1) the main functions to be performed by the PCP are these: determination of signal polarity, signal delay, multiplication and mean value estimation. Since only the sign is to be determined and processed hardware implementation using standard TTL circuits becomes relatively simple (two circuit boards 215 x 170 mm, including the sampling unit) as compared to a multilevel correlation concept.
In this connection should be noted that all functions given are hardware implemented for high processing speed. For further time saving parallel-serial processing is used.

The circuit in Fig. 2 shows the following parts of the PCP[3]:

    (a) - clock generator
    (b) - sample and hold circuit
    (c) - comparator for sign detection
    (d) - sampling impulse delay for effective small step
        signal delay
    (e) - shift register for signal (sign function) delay
    (f) - EXOR-gates for sign multiplication
    (g) - binary counter for mean value estimation
    (h) - bias voltage source.

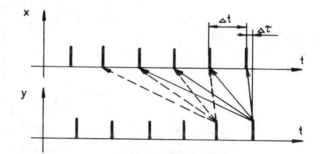

$$\tau_1 = \Delta\tau$$
$$\tau_2 = \Delta\tau + \Delta t$$
$$\tau_3 = \Delta\tau + 2\Delta t$$
$$\tau_4 = \Delta\tau + 3\Delta t$$

Fig. 3 Sampling time scheme

As indicated in Fig. 3 four samples of the PCF are evaluated in parallel. Because of the immediate (on-line) estimation of the polarity correlation function it is merely the delay interval $\Delta\tau$ that is determined by the sampling theorem, whereas the signal sampling interval Δt could be substantially greater as shown in the timing scheme (Fig. 3). Note that the signal sample sequences $x(t_i)$, $y(t_i)$ are not recorded at all. Hence, a

maximum τ of 63 µs and $\Delta\tau = 0,25 \dots 500$ ns can be chosen so
that a frequency range from 0,01 to 2000 MHz is covered for PCF
estimation.

Step by step increasing of the bias voltage source (h) over the
entire input voltage range finally provides a means to determine
the signal probability distribution $P(x_i)$ (in order to evaluate
Eq.(3)).

SOFTWARE ENVIRONMENT FOR SIGNAL ANALYSIS

After recording $\tau_{P(xy)}(k\Delta\tau)$ and $\tau_{xx}(0)$, $\tau_{yy}(0)$ from the
PCP output the elementary system software allows the calculation
of $\tau_{xy}(k\Delta\tau)$ by Eq.(2).
Now the integrated signal processing software permits the determination of the following characteristics:

- auto- and cross spectra
- coherence
- system transfer function and impulse response
- complex time domain functions (signal envelope and
  instantaneous frequency)
- cepstrum .

These methods for signal analysis are essentially based on a FFT-
algorithm.
For further extension it should be possible to use the directly
measured correlation function for autoregressive spectral estimation (e. g. to enhance spectral resolution).

DEVICE CONFIGURATION

The hardware described fits into the MFA-100-System (option
MFA 103)[4]. It is configurated to acquire PC-level (including
graphical output of results). The operating system is floppy
disc based.
An arithmetic board is used to speed up the signal processing
software.
In addition a PN-generator board for system excitation is available.

REFERENCES

1. D. Kreß, "Theoretische Grundlagen der Signal- und In-
   formationsübertragung", Akademie-Verlag, Berlin (1977)

2. S. M. Kay and S. L. Marple, "Spectrum Analysis - a
   Modern Perspective". Proc. IEEE, New York, Vol. 69,
   No. 11, 1981 pp. 1380-1419

3. E. Schröter, "Entwurf von Polaritätskorrelatoren für die
   HF-Meßtechnik", in: "29. Int. Wiss. Koll.", TH Ilmenau
   (1984)

4. W. Lüdge et al., "Modularer Fourieranalysator MFA 100".
   radio fernsehen elektronik, Berlin, Vol. 34, No. 9
   (1985), pp. 583-587).

MULTIPROCESSOR SYSTEM FOR MEASUREMENT SIGNAL ANALYSIS

Gabriel Ionescu, Radu Dobrescu, Radu Vărbănescu

Department of Control and Computers
Polytechnical Institute of Bucharest,
Romania

INTRODUCTION

The paper presents the results of the third stage research-
es made by the authors in the field of real time measurement
signal analysis. At this stage, a dedicated complex analyser
was built, which satisfies almost all the demands of signal
processing.

This analyser, actually a multiprocessor system, includes
a performant dedicated real time FFT processxor /1/ and some
specific structures for data handling between a master and
several slave processors, according to the requirement of
the IEEE 488 interface.

THE FUNCTIONS OF THE SYSTEM

The system allows data processing both in time and frequency domain and performs the following functions:

- data aquisition from transducers or measurement equipments,
  in analog or digital form, the maximum number of entries
  being $N_i = 16$;
- sampling, holding and analog to digital conversion of
  signal and storage of data in digital form of 16 bit
  words;
- computing the time domain characteristics: auto- and
  inter-correllation;
- computing the frequency domain characteristics; power
  spectra, Fourier spectra, transfer functions, coherence
  functions;
- graphical and numerical display.

These functions are accomplished by different tasks. A
task concentrates several mathematical operations or a complex
algorith; in this way the data processing flow is separated
in independent stages which can be processed in parallel.

The aim is to speed the computation, according to the real time requirements.

There are four essential tasks: FFT computing task /TASK 1/, arithmetical task /TASK 2/, data handling task /TASK 3/ and displaying task /TASK 4/. The independence of tasks implies a special hardware architecture.

THE HARDWARE STRUCTURE

_ The first two tasks are performed by dedicated processors - a FFT processor and an arithmetic processor, havint a similar structure implemented around a I 3000 /uP. Notice that the FFT processor allows the simultaneous computation of two Fourier transforms /direct or inverse/, being in fact a dual FFT processor. These processors are slaves in relation with a master microcomputer FELIX M 118 /a standard Roumanian general purpose microcomputer /3/, which is responsible with the general data flow and with the display of results /TASK 3 and TASK 4/. The Fig.1. depicts the block-diagram of the system.

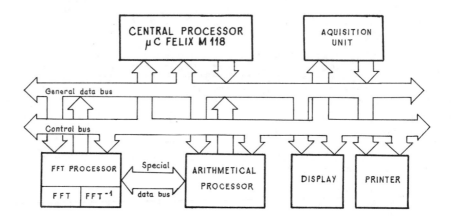

Fig.1.

The speciality of the structure is the use of two data buses. The general data bus, controlled by the master /uC which handle data from the aquisition unit, amplitude and phase values generated by the FFT processor, the data stored in the /uC - RAM necessary to the arithmetical processor for computing different functions /correlation, coherence etc/. The special data bus is devoted to link the two slave processors, allowing a rapid transfer of data in different stages of computing a Fourier spectrum /e.g. for computing $\sqrt{a^2 + b^2}$ arctg a/b/. For these reasons, the arithmetical processor operates in an interrupt mode, whenever the Fourier processor operates in a synchronous mode assisted by the master.

A short example is offered with the aim to explain the flow of data for a particular program : to compute coherence function. $\mathscr{C}(f) = G_{yx}^2(f)/G_{yy}(f)G_{xx}(f)$, where G_{xx} and G_{yy} are the power spectra of a system input and output signals and G_{yx} is the cross spectrum.

Step 1 - The data (N samples) of the input and of the output are converted and stored in the µC memory.

Step 2 - The data are transmitted to the FFT processor which compute $F_x(f)$ and $F_y(f)$ simultaneous (F_x, F_y - Fourier spectra of input and output).

Step 3 - The blocks of data $F_x(f)$ and $F_y(f)$ are transferred via the special data bus in the arithmetical processor input memory.

Step 4 - The arithmetical processor computes successively $G_{xx}(f) = F_x(f) \cdot F_x^*(f)$, $G_{yy}(f) = F_y(f) \cdot F_y^*(f)$, $G_{yx}(f) = F_y(f) \cdot F_x(f)$ and finally $\mathscr{C}(f)$. In this step the FFT processor can operate on other data, furnished by the master from the entry.

Step 5 - The blocks of data G_{xx}, G_{yy}, G_{yx}, $\mathscr{C}$ are transferred into µC memory; the special data bus is available for another transfer.

Step 6 - The µC prepare the display of $\mathscr{C}$ function; G_{xx}, G_{yy}, G_{yx} can be displayed too.

CONCLUDING REMARKS

The main feature of the system is the possibility to perform almost all computation algorithms involved by basic signal analysis, in a quite reasonable computing time. Essential in this respect is the intelligent use of the hardware and software resources : a performant independent FFT processor, a dedicated bus for communication and an appropriate tasks establishment.

REFERENCES

1. Ionescu,G.; Dobrescu,R.; Vârbănescu,R. - FFT Algorithm and computing block (in Romanian), Bult.of the Politechnical Inst. of Bucharest, vol.XLIII, nr.1 1981
2. Ionescu,G. - The use of the IEEE 488 standard interface in a measurement system (in Romanian), Proceedings of the 4-th CSCS Conference, Bucharest 1981
3. Petrescu,A. - Microcomputers FELIX M 18, M 118 (in Romanian), Ed.Tech., Bucharest 1984.

DESIGN OF INVARIANT SENSORS FOR ON-THE-SPOT CONTROL OF
TWO-COMPONENT FLOW PARAMETERS IN PIPELINE

B.V. Lunkin, A.P. Gridasov, A.V. Ivanov

Institute of Control Sciences,

Moscow, USSR

INTRODUCTION

In measuring the parameters of two-component flows in pipelines the errors mainly occur due to the changes of controlled media properties. The error is particularly great in such cases when an absolute value of a measured parameter is small, for example, when measuring moisture content of fuel oil. In this case the value of an output parameter changes by media properties change may several times exeed the change of an output parameter at the account of the controlled parameter change[1]. The traditional approach of the error reduction is based on the obtaining of preliminary information containing media properties, for example, by taking samples. However it essentially complicates the measurement process and makes an accurate on-line control impossible.

THEORETICAL BASIS OF THE METHOD

The proposed method is based on a possibility of designing a sensor, invariant to distrubances, to control the parameter of a two-component flow by utilizing two channels with asymmetric output characteristics followed by their subsequent calculation [2]. Channels asymetry is achieved by introduction of a standard body into the channels, for example, an ampule filled with water when emulsion moisture content is controlled. In fact, effective dielectric permeability of emulsions according to Wiener[1] is

$$\mathcal{E}_e = \mathcal{E}_\rho (1 + 3w)$$

where $\mathcal{E}_\rho$ is dielectric permeability of the initial substance, w relative moisture content of emulsion.

Upon introduction of a standard body, which is a spherical ampule filled with water, into a controlled sector of a pipeline, effective dielectric permeability becomes

$$\mathcal{E}_T = \mathcal{E}_\rho (1 + 3w)$$

where W_T is a relative volume of the ampule.

Using the dependence between a resonance frequency of an empty primary converter and the one filled-up with media having dielectrical permeability $\mathcal{E}_e, \mathcal{E}_T$:

$$f_1 = f_o\, \mathcal{E}^{1/2} \quad , \quad f_2 = f_o\, \mathcal{E}_T^{1/2}$$

where f_o is a natural resonance frequency of an empty primary converter, we obtain the desired invariant algorithm in the form

$$W = W_T\, f_2^2 / (f_1^2 - f_2^2) - 1/3$$

Thus, introduction of a standard body allows us to obtain the asymmetrical output characteristics of primary converters which enable the realization of the algorithm invariant to the media properties.

IMPLEMENTATION OF THE METHOD

A construction consisting out of two radio resonance primary converters designed, for example, in the form of a long line section can be installed in series in a controlled pipeline, with the standard body placed in one of their localization fields. This construction is one of the variants of the sensor, realizing the proposed method. Under excitation of natural electromagnetic oscillations in primary converters, the oscillation frequencies will vary depending on the variations of effective dielectric permeability of a controlled medium. The measured values of resonance frequencies are processed by the above algorithm, the result of which is a desired parameter. Provision of a sufficient accuracy in measurement requires high accuracy of measuring the resonance frequencies of primary converters which is provided by application of a special circuit of frequency tracking. Inertiality of the tracking circuit (in the order of milliseconds), stipulated by a high accuracy of frequency measurement, determines the overall inergiality of a device. In the considered construction the primary conveeters are connected in series and it provides for a sufficient measurement accuracy in many important cases. However in a number of cases when media properties variations within the sector between primary converters cannot be neglected it is necessary to carry out two simultaneous measurements on one sector. In this case, realization of two asymmetrical channels may be carried through by means of entering a standard body of an elongated shape and generating two types, of oscilations in one primary converter so that the vectors of the fields strength would be directed along or across the standard body. In this case asymmetry of characteristics also provided by the shape factor.

CONCLUSION

The standard body used in construction of invariant sensors can be also applied to measurements of two components-flow parameters such as gas and dust content. In this case the main problem is a proper choice of a standard bady and a corresponding structure of a primary converter providing for channel asymmetry which exceeds the noise level.

REFERENCES

1. E.C. Kritchevski et al., Theory and application of
   express moisture control for solids and fluids.
   Publ. Energia, 1980, p.240.
2. B.N. Petrov, et al., Principle of invariance in
   measurement technology. Publ. Nauka, 1976, p.244.

NEW CAPABILITIES RESULTING FROM INCORPORATION OF MICROPROCESSOR

IN INSTRUMENTS FOR PHYSICO-CHEMICAL MEASUREMENTS

Zbigniew Moron, Zbigniew Rucki, Zdzislaw Szczepanik

Technical University of Wroclaw
Institute of Electrical Metrology
Wroclaw, Poland

INTRODUCTION

The introduction of microprocessor "intelligence" into simple instrument for elektrochemical analysis such as ion- and conductivity meter can
considerably improve their metrological properties. It is a characteristic of
these instruments that: (i) they need calibration (in ionmetry a very frequent one), (ii) measurement results may depend on many quantities not being
the subject of measurement but influencing both sensor and object, (iii) measurement results are often complicated. Individual calculating means enable
to avoid most of this advantages resulting from the above characteristics as
well as creating new possibilities: (i) converasation between user and instrument, (ii) calculations, particulary these performed in the way yet impossible for realization in a single instrument e.g. iterative determining of
parameters from analytically insolvable relationships, numerical integration, etc. That, in turn enables practical application of methods hitherto
difficult to realize, e.g. ionmetric incremental methods, or elaboration of
methods not applied yet, e.g. conductometric incremental methods.

The role and capabilities of such "intelligence" measuring instruments
are presented below with examples of a microprocessor ion-meter and conductivity meter elaborated and designed by a team including the authors.

ION-METER. CLASSIC INSTRUMENT LIMITATIONS

Ion-metr, an instrument meant for ion concentration measurement, is
commonly used in laboratory and industrial conditions due to simple design,
operation, and relatively low cost of sensors. Ion-selective electrodes,
theoretically described by Nernst equation, in practice show great production
and exploitation changes which results in a shift and changes of an electrode
characteristic slope with regard to a theoretical one. The instrument preparation to operation - calibration - is done by placing electrodes in turn in
in two standard solutions (buffers) and by such switches adjusting enabling
shift and turn of characteristic to obtain repeatability of readings in each
standard solution. These activities have to be done many times because the
shift and turn are not independent. The measurement accuracy depend among
others, on a stable setting mainly of potentiometers.

In the case of hydrogen ion concentration measurement, a temperature
variation of the measured solutions is accepted, which affects the change of
the electrode characteristic slope. The consideration of these changes

271

requires setting of an isopotential point potential and making a hardware temperature correction. In the case of concentration measurement of ions X other than H^+ a logarythmic scale pX is not usually used. Such measurement at the range of two decades the most, is possible only with the use of analog meter non-linear scale. Greater problems are caused by incremental method - most frequently auxiliary tables and diagrams have to be used. An error of the measurement of very low concentration results from the presence of a small number of ions determined in the solvent. The consideration of these ion concentration in so called "blank correction" manner is impossible in hardware way.

MICROPROCESSOR ION-METER

A fundamental feature of a microprocessor ion-meter is the transfer of all (except of the gain) operations being performed on the electric signal to the operations being done on the information[1].This enables the controls to remove with all consequences of that. Than, the calibration requires only single dipping of the electrodes into each standard solution and a measurement result for the H^+ ions is calculated directly from the equation:

$$pH_x = -\left[\frac{1}{S}\frac{273.2 + t_w}{273.2 + t_x}(E_x - E_{iso}) - (E_w - E_{iso})\right] + pH_a \qquad /1/$$

where: S - electrode characteristic slope
pH_b, pH_a - buffer solutions values
E_b, E_a - SEM values corresponding to buffer solutions
E_{iso} - SEM of isopotential point
t_w, t_x - calibration and measurement temperatures, respectively

Direct calculation of result performed by the processor enables so called "blank correction" considerably reducing error of small concentration measurement. Except for that ion-meter microprocessor does not influence the measurement accuracy increase directly but indirectly by (i) an easy and quick calibration (calculations instead of manual adjusting) which can be done frequently, (ii) obtaining direct readings despite of measuring procedure.

CONDUCTIVITY METER. CLASSIC INSTRUMENT LIMITATIONS

A conductivity meter may be applied for measurement of the concentration of a one-component solution or concentration change of one of the components of a multicomponent solution. Conductometric method is very attractive because of its great simplicity, durability and reliability of sensor used and instruments cooperating with them, as well as their relatively low cost. Concentration measuremnt by using a classic conductivity meter requires analytical description of a surface: conductivity - concentration - temperature $(\varkappa, c, t)$. This description is performed in the form of two analytical functions $c = F(\varkappa)$ and $\varkappa = f(t)$, being the cross-sections of the surface $\varkappa, c, t$. Application of the method is limited by such factors as: (i) non-linearity and ambiguity of the relationship conductivity - concentration, (ii) strong, often non-linear and changing itself with the concentration relationship of conductivity - temperature, as well as non-selectivity of the method[2,3].

MICROPROCESSOR CONDUCTIVITY METER

Microprocessor application for a conductivity meter results in not only easier operation but improves measurement accuracy and extends measuring and methode application ranges as well.

Calibration improvement

In a microprocessor conductivity meter it is possible to store conductivity values of the most frequently used standard solutions e.g. KCl (depending on concentration and temperature). The user informs an instrument about a standard solution and places a cell in it and the instrument calculates the cell constant when measuring conductance. Also, due to an easy realization of a conductance ratio measurement it is possible to determine a cell constant with regard to standard one.

Direct concentration measurement. Temperature correction of readings

In the microprocessor instrument it is possible to store with a assumed resolution a full $\varkappa, c, t$ surface and obtain concentration recalling from the memory value of c corresponding to the given pair of $\varkappa, t$. The classic problem of temperature compensation, especially temperature coefficient variation, does not exist any longer. This manner meets actually some limitations resulting from the lack of sufficient data describing the surface $\varkappa, c, t$ of any measured solutions. Another possibility is to store only some sections of the surface $\varkappa, c, t$ characterizing a given solution and to make calculations by means of interpolation. These calculation may have iterative character, which allows for temperature coefficient correction in relation to a concentration and temperature, for example[3].

Expansion of conductometric cell measuring range

Errors of conductivity measurement by two-electrode cell result among others from interfacial phenomena. This error can be minimized by the measurement of a frequency characteristic of electrolyte resistance R(f) and by extrapolation of these results to the value $R(f=\infty)$. Microprocessor device makes possible both measuremnent and extrapolation of this characteristic, what according to the authors enables up to tenfold reduction of the error resulting from interfacial phenomena when using an ordinary two-electrode cell.

Incremental methods in conductivity measurement

A microprocessor introduction to a conductivity meter makes possible a realization of procedures not used before - conductometric incremental methods (analogously to the ionmetric incremental methods). For example a method analogous to the analate addition one based on adding small and known volume of a solution of unknown concentration to known volume of a standard solution seems to be especially usefull. This method can be used for the measurement of high concentration solution. It is ipmportant that conductometric cell can be placed all the time in a solution of a much lower concentration. Assuming a linear relation of conductivity - concentration (which can be accepted in the range of small changes of concentration resulting from adding examined solution sample to a standard) the concentration of an examined medium can be determined from the following equation:

$$c_x = c_n \left[\underline{\quad\quad} (1 + p) - 1 \right] \qquad\qquad /2/$$

where: c_x - examined solution concentration

c_n – standard solution concentration
p – ratio of examined solution volume to standard solution volume
$\varkappa_n$ – standard solution conductivity
$\varkappa_x$ – solution conductivity after adding sample to standard

Conductivity ratio in Eq. /2/ can be substituted by a conductance ratio due to invariability of a cell constant. Eq. /2/ becomes:

$$c_x = c_n \left[\frac{}{} (1 + p) - 1 \right] \qquad\qquad /3/$$

where: G_x, G_n – conductances corresponding to solutions before and after adding of sample, respectively

Analogously, a measurement can be conducted by means of addition of a standard solution to a sample. The advantages of these methods are: the possibility of measuremnt without known cell constant value, very wide measuring range with the use of one cell only, redundancy of temperature compensation, multiplicative errors minimizing by measurement of conductance ratio.

REFERENCES

1. Z. Moron, Z. Rucki, "Metrological aspects of the application of microprocessor in an ino-meter" (in Polish), Pomiary, Automatyka, Kontrola, 1984, No 10
2. Z. Moron, "Problem of compensation of temperature variations influence on measurement results", (in Polish), Pomiary, Automatyka, Kontrola, 1983, No 1
3. Z. Moron, Z. Rucki, "Extending the capabilities of conductometric measurement of electrolytic solution concentration by application of microprocessor", (in Polish), Pomiary, Automatyka, Kontrola, 1986, No 8

MICROPROCESSOR METER OF RADIATION ATTENUATION

COEFFICIENT IN SOLUTIONS

Janusz Mroczka

Institute of Electrical Metrology
Technical University of Wrocław
50-370 Wrocław, ul. Wybrzeże Wyspiańskiego 27, Poland

INTRODUCTION

The development of the scattering media optics and its application create new metrological problems connected with the measurements of these quantities describing relationship between electromagnetic radiation and the matter which determine its properties. These problems appear at the optics of sea and atmosphere, astrophysics, biophysics, physical chemistry, physics and chemistry of polymers as well as at biology and medicine. The quantity described is the attenuation coefficient being the sum of absorption and scattering coefficients,which has often been demonstrated (Van de Hulst, 1957; Jerlov, 1976). It belongs to the inherent optical properties as shown by Preisendorfer (1976). The main feature of the inherent optical properties is the fact that they do not depend on the external lighting conditions, but only on the nature of the investigated solution i.e. its chemical composition and physical structure.

Hitherto existing way of the attenuation coefficient evaluation needs comment because of the diversity of readings, given in the literature, which results mainly from the method errors. With the assumed model of the investigated phenomenon the systematic errors of the measurement being the result of the diversity between theoretical assumptions of the attenuation coefficient measurement and real conditions, result from the identification of the radiation scattered at a small angle coming to a detector with the radiation transmitted through the examined medium.

So far a transmission measurement of cylindrically limited light beam has been done and known expotential equation analogous to Lambert absorption law has been used to evaluate attenuation coefficient. To make this measurement correct one has to eliminate totally radiation to the area surrounding thin and parallel beam, scattering of the second and higher orders and which is the most important registering of the radiation scattered at a small angle by a detector. In this case measurement inaccuracy is caused by the fact that the radiation detector of a finite angle of sight receives not only radiation transmitted through the medium but radiation scattered at a small angle as well. The amount of error depends on a detector's angle of sight, measuring system geometry as well as on the optical properties of the investigated medium. The ways of this error elimination discussed in Mroczka (1984) use theoretically evaluated quantities of the volume scattering function, by means of which the scattered part of radiation reaching detector has been determined. Theoretically evaluated quantities of the volume scattering function at a small angle can differ very much from the real volume scattering function quantities of the examined medium.

The idea of the measurement by means of the proposed meter is based on introducing an additional measurement of the volume scattering function at a small angle to the existing measurement of the attenuation coefficient. The measured volume scattering function for small angles and known real angle of sight of a detector as well as measuring system geometry enable the determination of this part of radiation which reaches detector and its consideration for the final result of the attenuation coefficient measurement.

ANALYSIS OF ATTENUATION COEFFICIENT MEASUREMENT

The basis for the determination of the attenuation coefficient c is the equation of the radiation energy transmission which is a steady state for the medium with the constant refraction coefficient n and without internal light sources on the path r:

$$\frac{dL(\varepsilon,\eta)}{dr} = c\,L(\varepsilon,\eta) + L_x(\varepsilon,\eta) \tag{1}$$

where: ε,η - direction angles,
  L - radiance transmitted in the direction (ε,η),
  L_x - path function expressed by the equation

$$L_x(\varepsilon,\eta) = \int_0^{\pi}\int_0^{2\pi} L(\varepsilon',\eta')\beta(\varepsilon',\eta',\varepsilon,\eta)\sin\varepsilon'\,d\varepsilon'\,d\eta \tag{2}$$

where: $\beta(\varepsilon',\eta',\varepsilon,\eta)$ - volume scattering function.
After integration of Eq. 1 along the path r under the assumption that the path function L_x equals zero and that the radiance L is measured exclusively from the direction (ε,η) we obtain:

$$c = -\frac{1}{r}\ln\frac{L_r}{L_o} \tag{3}$$

where: the radiances L_r/L_o ratio is the transmission of the radiation beam.

Due to incorrect geometry of the spectrophotometer optical system or its unadjustment to the attenuation coefficient measurement in a scattering medium a part of radiation scattered forward at a solid angle is also received by a radiation detector. The attenuation coefficient determined according to Eq. 3 will be smaller of a quantity

$$\Delta c = \int_0^{\gamma}\int_0^{2\pi} \beta(\varepsilon',\eta',\varepsilon,\eta)\sin\varepsilon'\,d\varepsilon'\,d\eta \tag{4}$$

where: γ - angle of sight a detector, than its real value.
Presented here the attenuation coefficient meter measures its value according to Eq. 3 considering c value according to Eq. 4.

276

INSTRUMENT DESCRIPTION

A scheme of an instrument for the attenuation coefficient measurement is given in Fig.1. The instrument consists of optical elements and an electronic system. The He Ne laser is a light source. Radiation beam falls at a measuring cell containing examined medium which is placed in a lens focus. The scattered radiation falls at a focusing lens and then at a head with a system of opto-electronic radiation detectors. The head has a cylinder shape with hollow openings at a determined distance from the cylinder axis of symmetry and parallel to it. At the ends of openings there are the radiation photodetectors. The openings diameter corresponds with that of photodetectors photosensitive surface. The head dimensions are arranged in such a way that the detectors' field of view covers the whole scattering medium illuminated by a laser beam. Photodetectors sygnal is transmitted to the ampliphier by means of analogue switches controlled by a microcomputer, then ampliphied and processed into a digital sygnal it is stored in a microcomputer memory. The value measured by radiation detectors placed in the head is the energy luminance of the scattered radiation. On the basis of the measured value of energy luminance at the angle range of 0°-9° and that of radiation falling at a measuring cell the volume scattering function for a given angle is estimated.

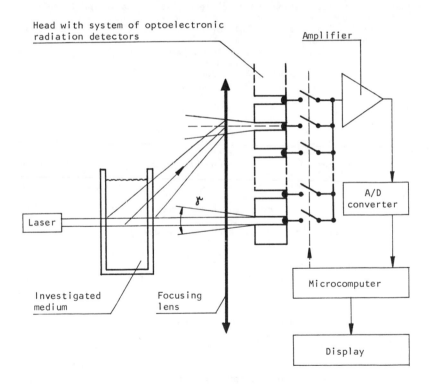

Fig.1 Instrument scheme

A function of volume scattering function alternation is evaluated by means of a microcomputer and then by interpolation the value of the volume scattering function for the angle of 0° is determined, which allows us to estimate the error of attenuation coefficient measurement according to the Eq.4. The used microcomputer functions as a couting device as well as a measurement controlling one.

METROLOGICAL PROBLEMS

The required accuracy and reliability of measurement results demand a detailed analysis of errors occurring at all stages begining with the medium preparation through measurement itself to the results processing. Radiation attenuation coefficient measurement errors result from an instrument design, quality of measurement, properties of separate sub-assemblies, and changes of technical parameters during exploitation. The occurring errors are systematic and random. Due to an appropriate choice of a measuring method and careful measurement organization the influence of random errors could be neglected. The problems considered when designing a device measuring attenuation coefficient by means of the proposed method have been:
- the dependence of the volume scattering function on the scattering angle,
- the affect of radiation refraction appearing on the measuring cell walls,
- the affect of radiation reflection from the measuring cell walls,
- the affect of attenuation and multiple radiation scattering in an examined medium,
- the determination of a real angle of measurement of the measured volume scattering function.
The affect of particular factors on the attenuation coefficient measurement at a determined geometry of measuring device has been considered by correction coefficients estimated by a microcomputer.

CONCLUSIONS

On the example of the presented here microprocessor meter of the radiation attenuation coefficient of solutions it can be said that microprocessor introduction to a measuring device allowed:
- greater measuring possibilities of device by the evaluation of unmeasurable quantities in a direct way,
- the improvement of metrological parameters of the instrument due to the elimination of systematic errors,
- the improvement of instrument reliability due to the reduction of the integrated circuits number and assembly simplification,
- the simplification of the instrument service due to the elimination of many operating and controlling activities.

The presented example of the radiation attenuation coefficient measurement of solutions allows us to state that besides design, technology, and application works it is necessary to conduct some theoretical studies whose aim would be:
- the creation of new models of the measured objects and phenomena and the analysis of usefulness of hitherto existing definitions for the sake of microprocessor measuring technique,
- the revision of measurement algorythms for their usefulness to new applications.

A rapid development of microprocessor devices and connected with it an increasing stream of measuring information create new approach to standardization and unification of measuring devices and calibration methods.

REFERENCES

Jerlov, N. G., 1976, "Marine optics," Elsevier Scientific Publishing
Company, Amsterdam-Oxford-New York.
Mroczka, J., 1984, Method error of the total attenuation coefficient
measurement in the scattering solutions, XVII Intercollegiate
Conference of Metrologists, Poznań, pp. 213-219 (in Polish)
Preisendorfer, R. W., 1976, "Hydrologic optics," U.S. Department of
Commerce, NOAA, ERL, Honolulu, Hawaii.
Van de Hulst, H. C., 1957, "Light scattering by small particles,"
John Wiley and Sons, New York.

INTELLIGENT VENTILATED PSYCHROMETER

Jerzy Bolikowski, Emil Michta
Andrzej Pieczyński, Wiesław Miczulski
Higher School of Engineering

PL - 65246 Zielona Góra, POLAND

The presented psychrometer is one from the modules of the measurement system designed for meteorology.

The psychrometer module consist of the four-channels analog multiplexer and the resistance to standard voltage signal converter. The two inputs of the analog multiplexer are joined with the temperature sensor. The third input is conected with the sensor which measured temperature inside the instrument. The fourth input is joined with standard resistance 100Ω (for t=20°C) with known temperature coefficient α_T.

The humidity measurement is preceded by the autocalibration procedure of the measurement circuit. This procedure has a three cycles. During the first cycle, the temperature (t_x) inside the instrument is measured. Microprocessor compute the value of the standard resistance in the current temperature

$$R_r = 100 \ (1 + \alpha_T \ (t_x - 20)) \ ; \tag{1}$$

During the second cycle, input of the R/U converter is shorted. The value of the offset voltage (n_0) is writtn into the RAM memory. During the third cycle the standard resistor is joined and the result of conversion (n_1) is remember in RAM memory. Next, microprocessor compute the sensitivity of the measurement circuit in the current temperature of work:

$$S = \frac{n_1 - n_0}{R_r} \tag{2}$$

The relative humidity measurement is taken place in two cycles. In turn are converted: temperature of the dry-(t_d) and the wet-bulb thermometer (t_w). The conversion results are stored in the RAM memory. Microprocessor computes the relative humidity based on following formulas:

$$R_d = \frac{n_d - n_0}{S} \ , \qquad R_w = \frac{n_w - n_0}{S} \ ; \tag{3}$$

$$t_{d(w)} = a_2 R_{d(w)}^2 + a_1 R_{t(w)} + a_0 \; ; \tag{4}$$

$$P_{d(w)} = c_4 t_{d(w)}^4 + c_3 t_{d(w)}^3 + c_2 t_{d(w)}^2 + c_1 t_{d(w)} + c_0 \; ; \tag{5}$$

$$RH = \frac{P_w - A(t_d - t_w) \cdot p}{P_d} \; . \tag{6}$$

where:

a_0, a_1, a_2	– coefficient polynomial approximation of the temperature sensors characteristics,
P_d, P_w	– saturated vapour pressure in the air at the dry and wet temperature,
p	– atmospheric pressure $[h\,Pa]$; if p is not measured then p = 1000,
A	– constant of the psychrometer head,
$c_0 \dots c_4$	– coefficients polynomial approximation of saturated vapour pressure in the temperature function as shown by Jeżewski (1957).

The A and $c_0 \dots c_4$ values are chosen by microprocessor from two tables. The first contain coefficients for $t_d \geq 0.01°C$ and the secound contain coefficients for $t_d < 0.01°C$.

Thanks to use of the autocalibration procedure the measurement circuit error is very little in relation to calculation error of the psychrometer head constant.

REFERENCE

M. Jeżewski, J. Kalisz - Tables of physical quantities, Warszawa, PWN (1957).

3
OPTICAL MEASUREMENTS—
MEASUREMENT OF MECHANICAL QUANTITIES

INTELLIGENT OPTOELECTRONIC MEASUREMENT

L. Benetazzo, C. Narduzzi, C. Offelli

Università di Padova
Istituto di Elettrotecnica e di Elettronica
Padova, Italy

INTRODUCTION

Solid state optoelectronic transducers are gaining increasing consideration in a number of applications ranging from research instrumentation to industrial process control. Integrating them in a data acquisition and processing system offers interesting capabilities in the analysis of optical signals, particularly in spectroscopy, recognition of object shapes and remote measurement of geometric dimensions.
A solid state sensor can be considered as an array of independent transducers (either photodiodes or CCD's), called "pixels", that turn a light signal into an electrical one; analogue outputs from each pixel are sent over a serial output line to an analogue to digital converter. The set of data supplied by the sensor during a scan is called a "frame"; the data acquisition system must process control and timing signals from the sensor to obtain information concerning pixel position within the array, and associate them to each frame. It should be reminded that transducer output is proprotional to the time integral of the incident light energy, thus making output amplitude also depend on the time interval of exposure, which we shall refer to as the observation interval. Such interval must be set by the user on the basis of sensor performance and the features of the analysed light source.

When a measurement system is required to combine real-time performance with a high degree of adaptability to different set-up conditions and measurement procedures, adequate characteristics may be obtained by realizing instrumentation with built-in intelligent features. To achieve this, emphasis should be placed on aspects connected with the organization of computing resources in the system, with regards to its hardware architecture and to a suitable structuring of its software[1].
The paper relates about those aspects more strictly connected with intelligent operation by discussing their application in a microprocessor based optoelectronic measurement system we realized, and illustrates how these features contribute to improved measurement accuracy, optimised system performance and ease of use.

REAL-TIME OPERATION AND POSTPROCESSING

If the system is to operate in real time, the time needed for data

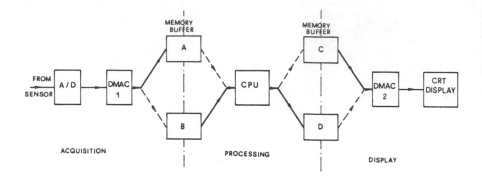

Fig.1 block diagram of real-time structure

acquisition and processing must not exceed the length of the observation interval. Resulting constraints prompted us to adopt a system structure whereby the microprocessor can perform data processing for a major part of its time, while data acquisition and the output of results are dealt with by two DMA controllers, respectively.

Three processes are thus concurrently executing in the system (fig. 1). A frame is acquired from a 12 bit A/D converter through DMA controller no. 1 and placed in a memory buffer area, either A or B. At the same time the Z80 CPU reads the preceding frame from the other area, processes it, then stores it either in output buffer area C or D, depending on which one is accessible. The other output buffer area contains the previous frame, that is being read by DMA controller no. 2 for display on a CRT. The use of twin buffer memory areas improves throughput.

The whole structure calls for some form of supervision by the Z80 CPU, namely, concurrent processes are synchronized and the correct sequence of operations is checked. This also results in a preliminary control on the consistence of acquired data, since the CPU detects improper activities and immediately halts data acquisition.

The sensor, as already mentioned, is of the integrating variety, that means it is not possible to synchronize it with the processing unit. It is the latter, therefore, that has to adapt itself to sensor peculiarities. To achieve this, a very simple support circuitry has been introduced, and data processing software has been organised accordingly. As soon as the whole system is powered, the sensor begins to operate, asynchronously with the processor; however, as long as the CPU has not completed system set up and preliminary operations, data acquisition is not allowed. When the CPU is ready a control signal is activated and support hardware enables acquisition, ensuring it begins from the start of the next frame. Synchronization is thus achieved and samples can be unambiguously associated with pixels.

The instrument executes all essential data processing in real time, so that displayed frames provide reliable indications: we designed it to regularly operate on a 50 ms observation interval with a 512 pixel sensor, such a specification deriving from its intended main field of application. It may be interesting to observe that real time performance is much helpful in cutting down set up time for the optics that precede the sensor, the effect of any modification being displayed immediately.

The system features postprocessing facilities that allow users to perform

286

further calculations. These operate off-line on a set of "frame registers" within system memory, and comprise elementary arithmetic operations, averaging, variance calculations, smoothing, finite integration with variable extremes, etc.[2].

SELF-CONFIGURATION

In a complex measurement system it is essential that the user be helped to properly set up the system and issue correct commands.
In the design and realization of our system steps were taken to relieve users from the need to take care of most details in the set up procedure. In fact a user is only requested to set the length of the observation interval as suitable: at power on the system measures this interval and determines how many pixels there are in the sensor to which it is connected; in this way, a variety of solid state sensors can be supported and a wide range of measurement needs be satisfied.
Knowledge of these parameters is instrumental in the optimal use of system resources. The size of memory buffers (fig. 1), and of frame registers as well, depends on the number of sensor pixels; since a fixed number of buffers is needed, available frame registers vary with pixel number.
The amount and the complexity of real-time processing the system can perform are influenced by both parameters. Therefore, after completing self-configuration the system itself determines its maximum real-time performance, in terms of processing activities, for the parameter values it has measured. Subsequently, whenever the user requires features that exceed such capabilities in a given context a warning message is issued.
It is readily appearent that the length of the observation interval and the number of sensor pixels influence most activities; therefore, system software is organized so that they are accepted as external parameters.

The instrument can execute a number of checking procedures that ensure optimum set up of the sensor and of its interface with the processor; they are concerned with the extent of systematic and random measurement error, and with a check of sensor to processor synchronization. To carry out such procedures the operator is required to perform some very simple actions, for which he is instructed by messages the instrument writes on the screen.
Measurement errors may originate either in the sensor itself or in the interface with the processor. To determine overall systematic error the system acquires a number of frames after the sensor has been obscured, and computes their average. From this information and from knowledge about the kind of sensor employed, it is easy to determine which sources contribute most to systematic error and, if necessary, to vary system set up in order to reduce such terms. The resulting value, plus the related measurement uncertainty, may be stored at this stage, so that similar checks can be made during normal operation to detect system misadjustments.
Once this value is known, subtracting it from frames acquired with the sensor obscured yields the contribution of random errors; from this it is easy to determine whether the influence of random errors is acceptable. The instrument also displays the resulting "error frame", to provide quick visual reference.
A periodic check can also be made of the synchronization between sensor and processor interface. Total loss of synchronization is easily detected, since data processing yields completely meaningless results; however, more subtle problems may arise from a condition of near synchronization, whereby sample value from pixel N is arbitrarily attributed to pixel N+n owing to interface malfunctions, n being usually a small integer. In this case the user is requested to provide for very sharp transitions of light intensity on the sensor; one way to proceed is darkening only part of it, but sometimes it may be easier to make use of a light source whose spectrum features strong lines. The difference between two consecutive frames is zero or nearly so if

the system works correctly, but if some loss of synchronization has taken place it becomes large where transitions are located.

For display purposes the abscissa on the instrument screen, that is proportional to pixel sequence numbers, can be turned into a more meaningful output parameter, such as wavelength or distance. To do this, the user should make use of a reference light source, whose features are input to the system.
A self-test procedure has been implemented and partly devoted to verifying interconnections among system elements.

ACCURACY ENHANCEMENT

Concepts of intelligent instrumentation may prove particularly useful in adapting a measurement system to a wide range of applications, at the same time enhancing accuracy and resolution.
In the implemented system a programmable gain amplifier, whose gain can be varied in powers of two under CPU control, is placed before the A/D converter; the microprocessor increases gain whenever the maximum sample value is less than half full scale, or decreases it if pixels reach saturation. The instrument displays each frame as a diagram on the screen and the self-scaling feature is applied accordingly to an entire frame.
It should be noticed, however, that very often useful information are found in just a part of the frame: for instance, in the neighbourhood of a single spectral line. The system allows users to restrict processing and visualization to a selected part of the frame by positioning a variable size observation window.
This results in improved real time performance, since pixels outside the window are assumed to contain redundant information and are not dealt with, achieving faster processing times. Moreover, self-scaling may be selectively applied to the same pixel subset, giving the user optimised amplitude resolution within the observation window: such a feature becomes particularly useful when the range of amplitudes in a frame is wide, and a traditional self-scaling procedure would hide most details of low amplitude components. A common situation when this feature proves worthy is the inspection of low amplitude spectral lines in the spectroscopic analysis of light emissions.
To evidence the fact that remaining pixels yield unreliable indications, their position is appropriately marked on the screen to avoid confusion.

Present manufacturing techniques do not always yield sensors with uniform pixel characteristics. This results in sensor non linearity, that has to be corrected by the system to improve accuracy. Sensor support circuitry allows for some form of compensation that however doesn't completely eliminate the problem; moreover it may be necessary to vary compensation with the application being considered.
The instrument, therefore, should be able to adapt itself to any sensor non linearity, also taking into account the effect of hardware compensation. In the implemented calibration procedure the user must supply a light source that uniformly excites all sensor pixels. The system acquires a number of frames and averages them to suppress noise contribution, then it computes the mean output level. A correction factor is automatically calculated and applied only when a pixel is found to deviate from the mean value greater than a predetermined threshold. The extent of the correction was found to depend mainly on sensor quality and performance.
It should be stressed that the averaging facility mentioned above is generally available and selectable by the user to improve accuracy by reducing the effect of random errors in the system.
Usually, the light source to be analysed is superimposed on background lighting; thus, the sensor actually detects the combination of such signals.

Suppression of background effects may be carried out, in the hypothesis that related energy is stationary, by first acquiring a frame when no light comes from the source under study. Such frame is subtracted from all following acquisitions.

TRANSIENT ANALYSIS

As already mentioned, an optoelectronic sensor cannot be started by an external command. If a transient light signal is to be analysed two possibilities arise, either to use the instrument to provide the transient start trigger, or to implement an acquisition procedure that allows recording of asynchronous transients.
In the latter case, continuous acquisition of frames is immediately initiated, pending the start of the transient. In this stage a frame is only stored in memory until acquisition of the following one is completed. When an external signal indicates the transient has begun, the stored frame is kept and N+2 more are acquired, where N is automatically computed from the estimated transient duration. This yields a total of N+3 acquired frames, of which the first only contains background noise and provides a reference to check actual transient length. This is done by comparing it with the last frame, that should differ negligibly. N+1 data frames are needed, as usual, to take into account the uncertainty due to asynchronous interaction between phenomena to be analysed and the system (fig. 2). The user is left the choice on the kind of processing to be performed. It must be noticed that a systematic error, albeit small, is introduced whenever the transient exceeds a duration corresponding to the acquisition time of a single frame.
Transient length supplied by the user is first checked by the system, to determine if enough memory space is available; when the acquisition is completed, a comparison between the first and last frame allows to verify the correctness of duration estimate; should their difference be noticeable, the instrument assumes that memory has been saturated by an exceedingly long transient and warns the user about data unreliability, at the same time preventing them from being processed. Such feature is typical of this mode of operation and was introduced with the aim to ensure full validation of data also in this particular case.

FINAL REMARKS

The aspects discussed in this paper emphasize how an intelligent system does not merely perform measurement. It can at the same time check its correct operation, and set itself up to ensure the best configuration with accurate results; a power-on self-test, like the one implemented, is also a useful asset.
A wide choice of output data presentation modes is provided, and combines with the availability of screen cursors to allow immediate and detailed interpretation of displayed results.

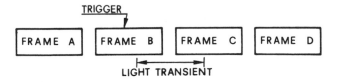

Fig. 2. An example of transient acquisition where N=1, that is, light transient duration is less than or equal to integration time: A-D=0, (B-A)+(C-D)=light transient without background noise

The measurement system contains a host computer interface, mainly used to save and retrieve measurement data.

Instrument use is remarkably easy, since not only does the system provide a user friendly interface, but it also relieves the user from the need to consider its internal structure, requested inputs only concerning measurement.

REFERENCES

1. C. Offelli and A. Licciardello, Acquisizione continua di segnali analogici con strumentazione a microprocessore, in: "Il microprocessore nelle Misure Elettriche", CLUP, Milan, Italy (1983).
2. C. Offelli and A. Licciardello, Sistema a microprocessore per misure spettroscopiche con array lineare di fotodiodi, in: "Rendiconti della LXXXIV Riunione Annuale AEI", Cagliari, Italy (1983).

IMPACT OF INFORMATION TECHNOLOGY ON MEASUREMENT AND QUALITY ASSURANCE

H.J. Warnecke, M. Rueff, K.W. Melchior

Fraunhofer-Institut für Produktionstechnik
und Automatisierung (IPA), Nobelstr. 12,
7000 Stuttgart 80, Federal Republic of Germany

Abstract

Examples of application of information technology in measurement and quality assurance are given. The examples are taken from the enlarging field of application of image analysis systems in industry. Some statements concerning image analysis software are given.

1. Introduction

As a consequence of the rapid development of semiconductor technology in the past thirty years computers have found acceptance in most fields in which numerical data must be handled. Beside those computers which are applied in science to solve complicated numerical problems another class of computers are applied by which physical signals coming from different external sources and changed into numerical values by appropriate hardware components are directly processed. These computers play an important role in measurement as well as in quality assurance. This firstly is due to their capability processing an immense amount of data in very short times and to extract essential features out of the physical signals. Secondly this is due to their stability in analysing the signals over long time periods. Parallel to the development of the semiconductor technology the theory of information also have had an explosive extension starting from such famous works as e.g. of Wiener (see e.g. /1/) and of Shannon and Weaver (s.e.g. /2/). Both, semiconductor technology and information theory form the basis of information technology which is more and more applied also in measurement and quality assurance.

In this arcticle examples of such complex measurement systems are reported. Among the many different areas of information technology a restriction is done to the technology of image analysis.

The organization of the article is as follows. Chapter two deals with some applications of different vision systems whereas in chapter three some statements concerning image analysis software are contained. Chapter four gives some outlook.

2. Image analysis systems/Some industrial oriented examples

In this chapter we shortly demonstrate three image analysis systems for industrial purposes. The first system shown in Fig. 1 was developed at IPA, the Fraunhofer-Institute for Manufacturing Engineering and Automation, for industrial measurement practice /3/.

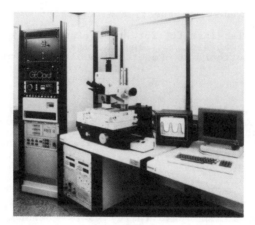

Fig. 1 "GEOPRÜF" A complex vision
system for geometrical
measurement purposes

Most of the geometrical measurement tasks arising in industry can be solved with this system with high accuracy. The essential part of the concept of this system is the combination of a greyvalue processing system and a system for fast binary image analysis. With the greyvalue system preprocessing of images can be done very effectively. The binary images to be further processed are therefore of high quality. By way of automatical focusing measurements of geometrical quantities in several parallel planes can be done.

The second system is shown in Fig. 2. IPA developed a complex programmable unit for the assembly enginge parts. As an example the assembly of a waterpump on an engine-block was demonstrated at the Hannover-Fair in 1983.

Fig. 2   Programmable assembly unit
         with vision system for quality
         control

An essential subsystem of the unit is the vision system. During assembly the vision system has to control the
quality of the gasket on the waterpump:

Fig. 3   Greyvalue picture of the
         waterpump with attached
         gasket

The position of the gasket on the waterpump as well as defects of the gasket (s. Fig. 3) can be recognized by the greyvalue processing system in very short times (less than 3 sec.).

The third system is shown in Fig. 4. As with the second system a quality assurance problem is solved. By a robot guided camera different positions on deep drawing parts from the Automobile Industry can be inspected /4/. In particular checks of holes (Fig. 4), contours

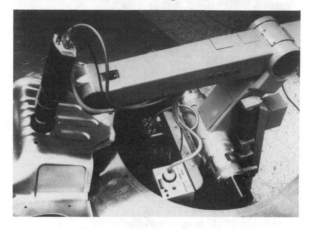

Fig. 4   Robot guided camera

tears (s. Fig. 5) etc. can be done. The vision system used is capable for greyvalue processing.

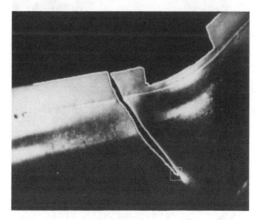

Fig. 5   Deep drawing part; recognized tear

Such flexible systems with robot guided cameras can find many application in different industrial areas.

3. Software for vision systems

To be able to perform the great variety of tasks by which we are daily confronted with from industry we have built up a softwarelibrary especially for the automation of visual inspection problems. Some of the above shown examples partly had been realized with algorithms from this library.

The structure and some of the content of this software library called BILDLIB are already reported (s.e.g. /5/,) and we don't want to go in detail here. We currently enlarge BILDLIB with specific algorithms that may have furture applications in measurement as well as in quality asurance. We have done a comparsion with some other software libraries for image analysis purposes, which partly are available on the market.

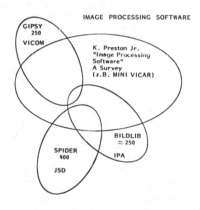

Fig. 6   Image processing Software.
Schematic represenation of the
respective overlap of sublibraries

We have found that there is a great overlap in most of the respective sublibraries. This is schematically displayed in Fig. 6. But there are some packages that are only contained in BILDLIB. Particularly this is for scale invariant object recognition, for special dimensionality reduction algorithms as well as for algorithms with which different fractal dimensions (e.g. Hausdorff dimension) can be calculated.

Up to now there are no unified mathematical methods in image analysis as well as in pattern recognition. Software libraries as a tool to apply these methods therefore must be very flexible and adapative to new results in the basic research of the above noted sciences.

4. Conclusion

We have shown how modern information technology can be used to solve special problems in measurement and quality assurance. Some of the discussed examples are plants and had been shown at different fairs. We mean that they exhibit to some extend the potentialities that are given by this special information technology, the image analysis.

One cannot claim the field of applications of image analysis in industry but one knows that this field will be very great in future. Visual inspection problems for example are given in nearly all industry regions and a lot of these tasks can automatically done by vision systems.

An essential research is structuring problems of measurement as well as quality assurance with a view to well known methods of image analysis and pattern recognition. This is done for example at IPA and this work defines also the future development of the software library BILDLIB.

Experiences gained in applications more and more will show the most useful among these methods. Unfortunately a very essential criteria os the processing time that can be reached with specific image analysis systems. Contrary to other fields of image analysis application, as e.g. medecin, the often required short processing times are barriers to introduce this new technology in wide areas of industry. This circumstance will change in the next few years. Better and cheaper hardware components as well as more efficient information theoretical methods will help to destroy most of these barriers. Especially for quality assurance a new aspect is coming into discussion. It is the coupling of vision systems with computers of the AI-Generation. AI-computers can powerful support the image analysis systems because experience gained during training periods will influence the recognition processes. Even such combinations can enlarge the field of application of vision systems in industry in the next five years.

REFERENCES

1. Wiener, Norbet, "Cybernetics". 2nd Ed., MIT Press and John Wiley and Sons, Inc. New York, 1961.
2. C.E.Shannon and W. Weaver, "The Mathematical Theory of Communication". University of Illinois Press, Urbana, 1949.
3. Warnecke, H.-J.; Keferstein, C., "Optoelektronisches Sensorsystem zur Geometrieprüfung mit automatisierter Bildverarbeitung". TM 52, Heft 9, 1985.
4. E.J.Schmidberger; R.-J. Ahlers, "Quality Control with a Robot-guided electro-optical Sensor". Proc. of IV Conf. on Robot Vision and Sensory Control, Okt. 84, London, U.K.
5. M. Rueff, K.W. Melchior, "BILDLIB the Image Analysis Software at IPA". Journal of Robotic Systems, 2, 1985, John Wiley and Sons, Inc, New. York.

COMPUTERIZED SCATTERING MEASUREMENT FOR ROUGHNESS EVALUATION OF SMOOTH SURFACES

Horst Truckenbrodt, Martin Weiss, Claus-Peter Darr
Department of Technology for Scientific Instruments
Friedrich-Schiller-University of Jena
6900 Jena, German Democratic Republic

INTRODUCTION

Now and in the near future the evaluation of surface properties of smooth materials is met with more and more interest. For example, in optics and microelectronics, the influence of roughness in the range of nanometers in vertical direction and parts of micrometers in horizontal direction of surfaces becomes more and more important. Reasons for these developments are increasing integration of electronic elements in smaller areas, the increasing demands on qualities of optical systems in the field of lithography and picture analysis. On the other hand the demands on in-process testing without physical and chemical contact also increase together with the better material machining.

MAIN DIRECTIONS IN CALCULATING THE ROUGHNESS PARAMETERS WITH THE AID OF SCATTERING MEASUREMENTS

In our department the utilization of light scattering for the inspection of surfaces of several materials is of great interest. There are two kinds of measurements, the integrated measurement above a certain scattering angle range (e.g. the totally integrated scattering measurement (TIS) [1]) and the other one, the angular resolved scattering measurement (ARS)[2]. The latter measurement permits the detection of more details of surface roughness. The possible scattering ranges of the ARS- measurement are larger than the angular ranges of integrated measurements in most of the devices. The ARS-measured relative scattering intensities (indicatrix) can also be integrated

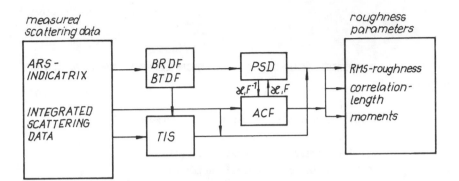

Fig.1 Main directions in calculating the of roughness parameters with the aid of measured scattering data

into a computer, if we have samples with isotropic roughness. For the determination of roughness parameters by substitution of measured scattering data we used coherent light sources, e.g. He- Ne- laser in a range of RMS < λ. In case of reflecting scattering measurements the measured indicatrix allows the determinition of the bidirectional distribution function of reflection (BRDF) and the BTDF by transmission. With the aid of these values it is possible to calculate the power spectral density (PSD) [3] by utilization of scattering theories [3],[4],[5]. The autocovariance function (ACF) with RMS = δ (the value on the position of the point zero) and the correlationlength T as horizontal roughness parameter is given by the inverse Fouriertransformation of PSD (in the case of isotropic surfaces by Hankeltransformation). Under the assumption of special formulas of PSD or ACF (e.g. gaussian or exponential) there are several procedures to calculate roughness parameters [3]. These procedures contain certain bandwidth problems and are only valid for special materials. TIS- measurements permit a simple calculation of the RMS for isotropic surfaces if the reflectivity of sample is known. The technical utilization of recently published straightforward ACF- measuring [6] by changing of laserspot contains for our opinion too many experimental problems. Many sensors of stray light work by integration of a certain angular range and these integrated scattering values are often usual for the surface inspection after calibrating.

OUR DEVELOPED AUTOMATED ARRANGEMENT OF LIGHT SCATTERING

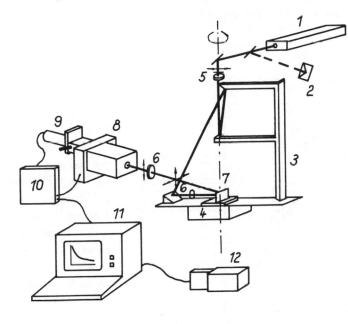

1 laser
2 calorimeter
3 rotating sam-
  ple arrange-
  ment
4 stepping mo-
  tor
5 $\lambda/4$-plate
6 polarization-
  plate
7 sample
8 system of
  attenuators
9 photomulti-
  plier
10 electronic
   for 8,9
11 microcomputer
12 interface

Fig.2 Our computerized arrangement of scattering [7]

In our arrangement we use an He- Ne- laser with λ = 632,8 nm, I=40 mW and an Ar- Kr- laser with 4 wavelengthes (with intensities between 50 and 250mW) as light sources (1). Only the sample arrangement (7,6,3) is moved by a computer controlled stepping motor. The axis of rotation is a tangent to the reflecting surface. The scattering intensity is measured in one azimuth. Besides the sample can be rotated by fixed scattering angle θ_s around the axis wich is a normal to the reflecting surface. A system of light attenuators (controlled by a microcomputer (11,8)) allows to find the measured intensities in the linear range of the characteristic of the photomultiplier (9). The microcomputer (11) controls the rotations of the sample by stepping motors in angular steps of 0,006 degrees. The computer controlled measurement permits measuring regimes to calibrate attenuators, the noise power and the measurement of an θ_s- indicatrix between -90° and 90° in some minutes independent of the value of angular steps. It is possible to obtain the graph during the measurement. With the aid of an interface it is possible to calculate roughness parameters with an calculator. We developed FORTRAN- software to realize the system as shown in Fig.1. We calculate the

ACF by numerical Hankeltransformation of the the PSD and therefore we do not
assume particular preconditions for the ACF or PSD.

RESULTS

Isotropic media make it possible to get some more exact information of
surface rough-
ness. A typical
indicatrix (Fig.
3a,3b) demands
the measuring of
10^{-9} to 10^{-11} as
a part of the
incident inten-
sity. We investi-
gated the pro-
blem of bandwidth
limitation (BWL).
This problem con-
tains the depen-
dence of measured

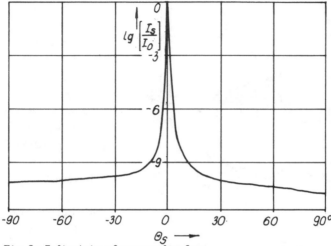

Fig.3a Indicatrix of measured wafer

roughness parameters on a suitable scatter range of the measuring
arrangement with the finite wavelengths λ. This problem can be demonstrated
by a model of the profil of surfaces. The roughness of surfaces can be
interpreted with the aid of the Fourieranalysis as an heterodyning of
sinusiodal gratings with different amplitudes, space wavelengths and
direction. The gratings with the short wavelengths diffract the light in the
wide angle range and those with the long wavelengths do this in the near
angle range. This relationship explains the so-called short scale (SC) and
long scale (LC) problem. By decrease of the wide angle range the registered
gratings with the short wavelengths are limited and increase of the near
angle leads to a limitation of the gratings with the long wavelengths. The
BWL- problem complicates both the comparison of roughness parameters
measured with other devices and the interpretation of these parameters. The
RMS and ACF with T highly depend on the measured minimal angle 0_m and are
comparable with typical LC-data in [3] (Fig.4,5). The correlationlength T
will be longer with a better measurement of the long scale range of the
little scattering angle θ_m(Fig. 5) and it will be shorter with a better

302

measurement of the short scale range of the great scattering angle θ_M

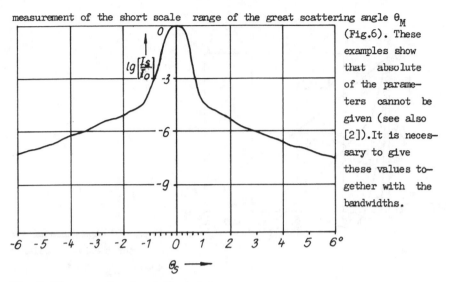

(Fig.6). These examples show that absolute of the parameters cannot be given (see also [2]).It is neces- sary to give these values to- gether with the bandwidths.

Fig.3b LC range of wafer indicatrix

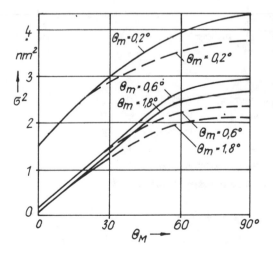

Fig.4 Bandwidth dependence of RMS (glass BALK) --vector theory[5], —— scalar theory [4] θ_m minimal scat- tering angle, θ_M — maximal scattering angle

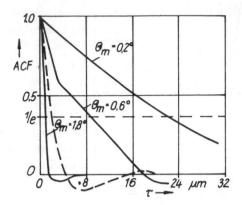

Fig.5 LC dependence of ACF
$--$ vector theory, $-$ scalar
theory, θ_M= 12,8°

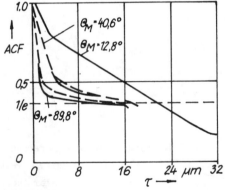

Fig.6 SC dependence of ACF

REFERENCES

1. H.E. Bennett: Optical Engeneering 17(1978)5 pp.480
2. J.M. Bennett: Optical Engeneering 24(1985)3 pp.380
3. R.J. Noll, P. Glenn: Applied Optics 21(1982)10 pp.1824
4. P. Beckmann, A. Spizzichino: New York: Pergamon Press 1963
5. H.E. Ponath: Annalen der Physik 36(1979) pp.438
6. H.E. Ponath, H.G. Walther: Annalen der Physik 39(1982)
   pp. 233
7. C.Bickel: Applied Optics 18(1979) pp.1707

SURFACE ROUGHNESS MEASUREMENT BY SPECKLE CORRELATION METHODS

Bernd Ruffing

Institut für Mess- und Regelungstechnik
Universität (TH) Karlsruhe
D-7500 Karlsruhe 1
Federal Republic of Germany

INTRODUCTION

The common method to measure the roughness of technical surfaces is the stylus method. Stylus instruments are highly automated and the standardized roughness parameters are derived from this special technique. However, the usefulness of stylus instruments is limited for applications such as on-line measurement, non-destructive testing and measurement during production processes. In these fields, optical roughness measuring techniques have a great potential. The coherent-optical speckle correlation methods have proved their applicability to surface roughness measurement and will be described in the following.

PRINCIPLE OF SPECKLE CORRELATION

A speckle pattern is a random intensity distribution, which is generated when coherent light is reflected from an optically rough surface. The rough surface is a stochastic process which modulates the phase of the incident laser beam. The properties of the speckle pattern are also determined by the light source and the optical system which is used to produce the pattern in the observation plane.

When we deal with a Fourier transform optical system, we speak of an objective or far-field speckle pattern, whereas an imaging optical system leads to a subjective or image speckle pattern.

The rough surface $h(x,y)$ is assumed to be an isotropic zero-mean Gaussian stochastic process, which is described by its standard deviation σ_h. σ_h is equivalent to the rms-roughness R_q. In the case of Gaussian height statistics, there exists a direct relation to the arithmetic average R_a,

$$R_q = \sqrt{\pi/2}\, R_a \quad , \tag{1}$$

which is presently the most important standardized roughness parameter. Furthermore, the surface shall be rough compared to the illuminating wavelength ($\sigma_h > \lambda$). We then deal with a so-called fully developed speckle pattern, which alone does not carry information on surface roughness. We therefore consider two speckle patterns which are generated under slightly different conditions by varying either the illuminating wavelength or the angle of incidence (Fig. 1). The two patterns are compared by defining the degree of speckle intensity correlation,

$$\gamma_{12} = \frac{\langle i_1 i_2 \rangle - \langle i_1 \rangle \langle i_2 \rangle}{[(\langle i_1^2 \rangle - \langle i_1 \rangle^2)(\langle i_2^2 \rangle - \langle i_2 \rangle^2)]^{1/2}} \tag{2}$$

where i_n with $n = (1,2)$ are the speckle intensities and $\langle ... \rangle$ denotes ensemble averaging. The spectral-speckle-correlation (SSC)-method characterized by a wavelength change is generally distinguished from the angular-speckle-correlation (ASC)-method which is characterized by a change of the angle of incidence.

We assume relatively small differences in wavelength or angle of incidence

$$\Delta\lambda \ll \lambda_n \quad , \quad \Delta\alpha \ll \alpha_n \quad n = (1,2) \ . \tag{3}$$

Using scalar diffraction theory and observing the speckle patterns on their respective optical axis, evaluation of Eq. (2) yields for the ASC-method[1]

$$\gamma_{12}^{ASC} = \exp[-16\pi^2 \sin^2\alpha \ \sigma_h^2 \ \Delta\alpha^2/\lambda^2] \tag{4}$$

and for the SSC-method[3-6]

$$\gamma_{12}^{SSC} = \exp[-16\pi^2 \cos^2\alpha \ \sigma_h^2 (\Delta\lambda^2/\lambda_1\lambda_2)/\lambda_1\lambda_2] \ . \tag{5}$$

The principal dependence of γ_{12} on the roughness parameter σ_h following Eqs. (4) and (5) is shown in Fig. 2.

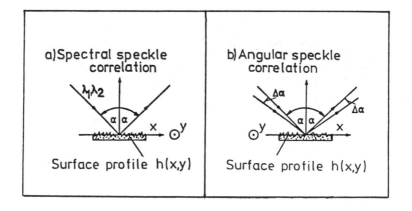

Fig. 1   Different speckle correlation methods

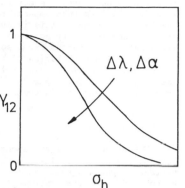

Fig. 2 γ_{12} as a function of σ_h

APPLICATION TO SURFACE ROUGHNESS MEASUREMENT

The measuring range of the speckle correlation methods can easily be varied by choosing suitable values for $\Delta\lambda$ or $\Delta\alpha$ (Fig. 2). It has been shown[6] that the minimum roughness which can be measured is approximately $\lambda/2$. The highest roughness value which has been measured[2], is $\sigma_h = 32\mu m$.

For surfaces with a Gaussian height distribution, (produced by grinding, sand blasting or spark erosion), absolute roughness measurements are possible. The quantity γ_{12} is, of course, also characteristic for non-Gaussian surfaces, however, calibration with a stylus instrument is then necessary.

A special experimental arrangement (Fig. 3) is now considered which has similarly been used by many authors[3-6]. It applies the SSC-method in a far-field plane geometry to surface roughness measurement. A dichromatic laser source illuminates a surface sample and the scattered intensity is observed through a pinhole in the observation plane. The pinhole resolves one single speckle grain and the lens L produces the image of the pinhole onto the photo-detectors PD1 and PD2. Colour separation is performed by interference filters IF. The surface sample is rotated and we obtain the time-dependent intensity signals $i_1(t)$ and $i_2(t)$. The signals are sampled with an A/D-converter and a microcomputer calculates γ_{12} and σ_h according to Eq. (5). Some experimental results[4] are shown in Fig. 4, which have been obtained from grinded surfaces.

The motion of the surface is in many applications not desirable. Therefore, a TV-camera positioned in the observation plane can principally be used to record a larger area of the intensity patterns. Eq. (2) can then be evaluated by using digital image processing techniques and the ensemble averages are substituted by spatial averages over the recorded area.

Highly coherent gas lasers have generally been used as light sources. It has been shown[5] that speckle correlation measurements can be performed with semiconductor lasers which allow a more compact and economically reasonable sensor design.

Although theoretical and experimental investigations of speckle correlation methods with respect to surface roughness measurement showed promising results, no instrumental realization for industrial applications is
available up to now. This is due to the fact that all speckle correlation
methods require a very accurate adjustment of the optical set-up. Misalignment of the system light source, rough surface, detector unit leads to
an additional decorrelation of the speckle patterns which deteriorates the
measurement. The main task of future work in this area of research will
therefore be the development of a stable speckle correlation set-up, which
in a certain range is insensitive to misalignment.

CONCLUSION

The principles of the spectral and angular speckle correlation methods
have been described. These techniques have proved their applicability to
surface roughness measurement, which was discussed by means of a special,
often used, SSC-arrangement. Although speckle correlation methods have
many advantages, the main drawback, namely their high sensitivity to misalignment, has to be overcome in the future in order to make possible an
instrumental realization.

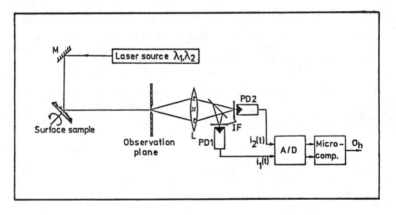

Fig. 3 Experimental set-up for the SSC-method

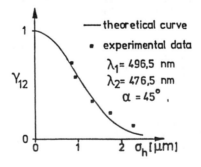

Fig. 4 Experimental results with the SSC-method[4]

308

REFERENCES

1. D. Lèger, E. Mathieu, and J.C. Perrin, Optical surface roughness determination using speckle correlation techniques, Appl. Opt. 14, 872:877,(1975).
2. D. Lèger and J.C. Perrin, Real-time measurement of surface roughness by correlation of speckle patterns, J. Opt. Soc. Am. 66, 1210:1217, (1976).
3. M. Giglio, S. Musazzi and U. Perini, Surface roughness measurements by means of speckle wavelength decorrelation, Opt. Commun. 28, 166:170, (1979).
4. G. Bitz, Verfahren zur Bestimmung von Rauheitskenngrößen durch Specklekorrelation, Doctoral dissertation, Universität (TH) Karlsruhe, (1982).
5. B. Ruffing, Optical surface roughness measurement by speckle correlation with semiconductor lasers, 10th IMEKO World Congress, Prague, (1985).
6. B. Ruffing and J. Fleischer, Spectral correlation of partially or fully developed speckle patterns generated by rough surfaces, J. Opt. Soc. Am. A, Vol. 2, No. 10, 1637:1643, (1985).

DETERMINATION, MEASURING, VALIDITY AND APPLICABILITY OF THE SURFACE ROUGHNESS PARAMETERS

Ilona Pap

Department of Production Engineering
Technical University for Heavy Industry
H-3515 Miskolc-Egyetemváros, Hungary

INTRODUCTION

In order to judge the quality of product and manufacturing, among other things, the scientific investigation of surface roughness is necessary, too.

For characterizing surface roughness many kinds of roughness parameters are used in research and practice, interpretation and determination of which are not unified. In the technical practice only R_a, R_z and R_{max} are generally prescribed by the designers.

Accuracy of the surface roughness features measured depends on the present level of the instrument technique. Different publications and my own experiences prove that the values of roughness features measured by different types of instruments correspond to one onother only within a certain error percent-age.

In consequence of the international exchange of documentation and goods, as well as at the purchasing of licences we can often meet foreign instructions and markings which have to be interpreted or maybe adapted. A significant part of the instruments and equipment required for examination of surface roughness rise from abroad on which the marks and interpretations correspond to that of the producing country. All these have necessitated an analytical summarizing the surface roughness features prescribed by the Hungarian Standards (MSZ) and used also in the international literature (In detail see:[6]).

ACCURACY OF ROUGHNESS FEATURES MEASURED

Geometrical features of surface roughness can be determined on the basis of the observed roughness profile. The observed profile is an approximation of the real one to the accuracy limit practicable from the point of view of instrument technique. From the previously mentioned things it follows

that the different measuring procedures conduce to different
measuring results. The measured value can be influenced by several factors [3,6]. From these I should like to show only
the influence of filters - denoted by M and RC - on the basis
of a surface roughness diagram taken up from a few reference
standards machined. The filters can change the form of the
roughness profile, in the first line the roughness profiles generated by tools with regular edge geometry.

The roughness profilograms taken up by filters M and RC
from a reference standard machined by milling demonstrate this
fact well. (see: Fig.1.). Because the basis of evaluation of
the roughness features is the roughness profile which can be
taken up with filters M and RC, consequently the surface roughness features measured by the alternative filters also differ
from one another.

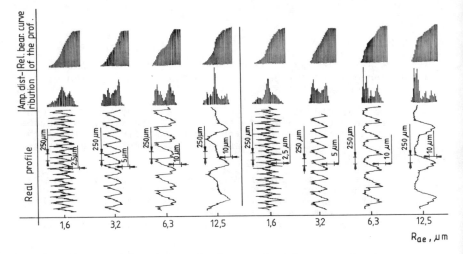

Fig.1   Measured roughness features of reference standars
         machined by HSS face milling cutter (F-RO-C)

From the roughness features measured, I compose two ratios
for R_a and R_p using the measured values of the alternative filters both at the standards machined by grinding and at the standards machined by milling:

$$k_1 = \frac{R_{aRC}}{R_{aM}} \; ; \qquad\qquad k_2 = \frac{R_{pRC}}{R_{pM}}$$

The values got in this way are shown in Fig. 2.

On the basis of Fig.2. it can be verified as follows:

- To R_{ae} = 1.6 µm, independently of the machining method,
  $k_1 \approx 1$ and k_2 changes from k_2= 0.82 to k_2=1.2; accordingly

312

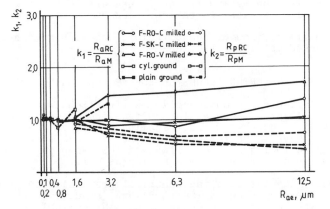

Fig.2   Ratios of roughness features measured with filters RC
and M in case of standards machined by grinding and
milling respectively

the roughness features measured by the two different filters are approximately the same ones.

- In the range of R_{ae}=3.2 - 12.5 µms the features measured
on distinct reference standards machined by milling have
to be separated. The maximum difference occured on the
standard denoted by F-RO-V where k_1 = 1.42 - 1.64 and
k_2 = 0.83 - 0.42. The changes on the standard F-SK-C
are as follows: k_1=0.88 - 1.02 and k_2 = 0.72 - 0.5;
- At last, on the basis of the above mentioned things it can
be ascertained that the filters, M and RC influence the
different roughness features in not equivalent way.

From the discussion written previously and considering the
works[2], [5] and [6] it can be established that the values of
roughness features measured are influenced by several factors.

CONDITIONS REQUIRED FOR IDENTITY OF ROUGHNESS FEATURES MEASURED

On the basis of own measuring results and literary data it
can be given that in case of qualification by roughness-measu-
ring instruments supplied with contact (scanning) head what kind
of instructions have to be included in the text of the transport
engagement to avoid the unsettled questions occuring by chance:

- identification (marking) of the surface(s) to be measured;
- marking the places to be scanned;
- the measuring length;
- the filter to be used (M or RC);
- the wavelength of the filter to be used;
- the specification of roughness-examining equipment accepted
by all the two signatories;
- the type of contact (scanning) head;
- the feature(s) to be measured;

- it have to be marked if the values are to be taken out of
  a diagram or   are to be indicated by the instrument.

Even if only one of these factors was missing the identity
of the marked value was limited.

PROSPECTIVE DEVELOPING TRENDS OF SURFACE ROUGHNESS SPECIFICATION

Lately the correspondence to the functional requirements
has come into prominence at the surface roughness prescriptions
[1, 2, 4, 5]. Hence the specification method of the roughness
parameters have to be further-developed to such an extent that
it should be indicatable and clear in the technical documentation and in the process of machining as well.

One of the tasks is the numerical presciption of roughness
features corresponding to the functional conditions which needs
wide-ranging research. The paper shows the suitable prescription
of the roughness parameters for a few surfaces functioning under
different circumstances.

For surfaces to be coated:

- R_{max} - maximum roughness/uneveness
- S_{max} - number of tops
- S_i - wavelength of local protuberances of profile
- T_2 - depth number
- K_p - profile-emptyness ratio

For surfaces before cold plastic deformation:

- R_{max} - maximum roughness
- K_p - profile-emptyness ratio
- S_{ni}^p - wavelength of roughnesses

For hard-wearing surfaces:

- R_z - roughness height
- K_p - profile-emptyness ratio
- t_p - bearing length curve
- l_p - length of profile
- S^p - number of tops
- S_k - skewness factor.

Having determined those roughness parameters by which a
given surface can be characterized the best of all from the
point of view of functioning, the mechining method, by means
of which the surface can be produced, has to be selected, and,
at last, the surface has to be checked.

REFERENCES

1. Berkes,R.,Gribovszki,L.,Pap,Mrs.(1981), Possibilities of
    Endurance-increasing at Laminated Springs of Vehicles
    Scientific Publications of Technical Academy for Traffic
    and Telecommunication, 3rd Sc.Congress,Section for Traffic-
    machinery, Győr, Hungary (in Hungarian).

2. Duthaler,V.(1962), Oberflächengüte und Rauheitsprobleme in Theorie und Praxis, Tehnische Rundschau, Vol.54, No.47, pp.33-37.

3. Frdődy,I., Pap,Mrs.(1981), A Few Features of Connection of Surface Roughness and Machinings, Gépgyártástechnológia,Vol. XXI.(in Hungarian)

4. Pap,J.(1980), Investigation of Technologies of Finishing and Surface Rolling at Hydraulic Work-cylinders, Engagement No.: 96-XXV.-6/76, Miskolc (in Hungarian).

5. Pap, Mrs.(1981), Prospective Trends of Specification of Surface Roughness Features, Borsod Technical Weeks (oral presentation, in Hungarian).

6. Pap, Mrs (1981): Determination, Measuring, Reliability and Applicability of Surface Roughness Features, Dr.techn.Dissertation, Technical University for Heavy Industry,Miskolc.

COMPUTER CONTROLLED REFLECTION AND TRANSMISSION

MEASUREMENTS WITH THE SPECTROPHOTOMETER SPECORD 75 IR

Ra Tsol Su, A. F. Rudolph, K. H. Herrmann

Bereich Experimentelle Halbleiterphysik
Sektion Physik der Humboldt-Universität zu Berlin

DDR-1040 Berlin, German Democratic Republic

INTRODUCTION

The evaluation of interferences is a well-known method
for determining film thickness if the refractive index n is
known. In semiconductors, on the other hand, the refractive
index exhibits strong dispersion (Opyd, 1973) resulting in
non-equidistant interferences, if observed over a wide
wavenumber region $\Delta\tilde{\nu}$ (Fig. 1).
It was the aim of this paper to develop a method for determining the dispersion of refractive index of semiconductor
thin films deposited on isolating as well as semiconducting
substrates by evaluating interferences in transmitted and/or
reflected light in a wide spectral region.

Use was made of a high-resolution grating infrared spectrophotometer Specord 75 IR (VEB Carl Zeiss Jena, G.D.R.) and
of a desk-top computer EMG 666 (EMG, Plants for Electronic
Measuring Equipment, Budapest, Hungary).

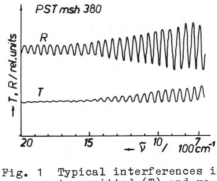

Fig. 1   Typical interferences in
         transmitted (T) and re-
         flected light (R) -
         PbSnTe film on BaF$_2$

Following problems had to be taken into account:
- wavelength-dependent refractive index of the semiconductors investigated (small-gap lead chalcogenides),
- useful wavenumber range limited by absorption losses both in the interband and the free-carrier absorption regions,
- changing amplitude of the interferences and changing medium transmission due to wavelength-dependent absorption.

It was the aim to exhaust the full optical resolution of the spectrometer as well as data under additional parameter variation (of temperature, e.g.).

Because the memory of the EMG 666 is too small for storing the whole spectrum and for evaluating it subsequently, a compromise had to be chosen between computer control and control by the operator.

HARD-WARE IMPLEMENTATION

The measuring configuration is built-up by following units (Fig. 2):
- EMG 666 with 8 kByte memory, tape recorder and peripheric units (plotter, printer and/or type-writer),
- Specord 75 IR, equipped with non-standard reflection units and thermostat,
- digital voltmeter for digitizing transmission and reflection data,
- wavenumber decoding unit.

The interface parts A, B, C, and D have the following functions:
part A: interface for realization of IEC-bus for measuring system,
part B: data transmission (bit-parallel, byte-parallel into bit-parallel, byte-serial),
part C: motor drive control,
part D: wavenumber input to the computer (see below).

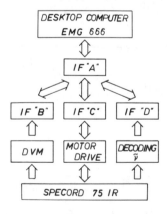

Fig. 2 Architecture of the measuring system

SOFT-WARE

The soft-ware (Fig. 3) has following functions:
- evaluation of transmission/reflection maxima and minima, resp.,
- evaluation of a first approximation of thickness or refractive index from the interference pattern at medium wavelength, where $n(\tilde{\nu})$ const.,
- corrections due to nonvertical incidence in reflection measurement,
- interpolation of the actual index of refraction of a given sample according to material composition within the given mixed-crystal system, for instance $n(x)$ for $PbSn_{1-x}Te_x$ $(0 \leq x \leq 1)$,
- a last square fit of $n(\tilde{\nu})$ to theoretical wavelength dependence in the limiting cases of interband and free-carrier contribution.

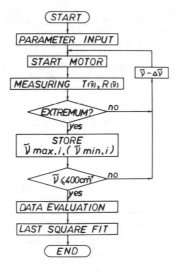

Fig. 3   Program structure

RESULTS AND DISCUSSION

In the present state usually first a spectrum is being registrated without computer control, and then it is decided by the operator,
- what is the actual wavenumber range to be measured with computer control, and
- to what special dependence $n(\tilde{\nu})$ the results are to be fitted.
After that, a computer-controlled precision measurement in the selected wavenumber range is being carried out.

The system described has the advantage, that the interference spectrum of a sample can be measured with 0.5 cm^{-1} resolution. At the same time the system has the disadvantage, that such a precise measurement takes more than one hour.

In order to overcome this disadvantage the structure of interface D was slightly changed.
For evaluating thickness from interference spectra only the wavenumbers of extrema are needed while information from intermediate wavenumber regions is redundant. From this fact the following idea evidences: only transmitted/reflected signals above (below) a certain selected level must be evaluated if the position of a maximum (minimum) is being searched for.

This mode of operation becomes possible with a modified structure of interface D as shown in Fig. 4. The interface additionally contains a five-digit register, an ADC unit with ICC520 D, and a comparator (with exclusive OR).

The function is as follows: After the interface from the computer has got information concerning transmittivity/ reflectivity limits for searching extrema, it starts spectrometer tuning. Let be transmission initially below the selected value. The transmittivity signal is digitized and fed into the comparator. The comparator ejects a STOP (initiating a real measurement by means of the digital voltmeter) only if the measured value equals the selected value. In this way the spectrometer is being tuned quickly without really measuring as long as the amplitude is outside the selected region. Following a STOP real measurements are being done and evaluated by the computer as long as the amplitude is within the selected interval.

The measuring time can be further decreased by means of program techniques.

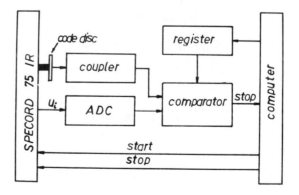

Fig. 4 Modified interface D

CONCLUSION

    As compared with manual data evaluation based on the
transmission and/or reflection plots delivered by the Specord,
following improvements have been obtained:
- enhanced accuracy due to automatic detection of interferences maxima and minima,
- increased accuracy due to spectra accumulation of noisy
  interferences in the case of strongly absorbing layers.

REFERENCE
        Opyd, W. G., U. S. Nat. Tech. Inform. Serv. AD Rep.
            767677/8 GA (1973)

APPLICATION OF CCD-SENSOR TECHNIQUE FOR QUALITY CONTROL OF GLASS-WARE

Bernhard Bock, Dagmar Hülsenberg, Gerd Kleemann

Institute of Technology,
DDR-6300 Ilmenau, Ehrenberg
German Democratic Republic

INTRODUCTION

The specific properties of glass material as well as the techniques which are typical for the manufacturing of glass products often require the quality control of each single product. This applies to all branches of glass industry such as the production of flat glass, glass tubes and container glass, or the manufacturing of chemico-technical hollow glass ware. The purely visual test made by man which was typical of this kind of control became more and more difficult in the last years and is now almost impossible. The large variety of parameters to be tested (product geometry, distribution of wall thickness, bubbles, cracks, tiny stones, splinters and the like), on the one hand, and the rate of the production lines which has been considerably increased, on the other hand, are considered to be the reasons for this fact.

Another aspect appears in the analysis of many possible applications of industrial robots in glass industry which reveals that the worker occupied with one manual operation additionally realizes a control function. Thus, such working places can only be automized in a suitable and economically favourable way if the operation process and the test function are both automized.

In order to meet the increased requirements, attempts have been made for some years to automate the test process. In doing so,
- mechanical contacting elements (geometry test),
- light barriers (geometry test),
- single phototransistors or diodes (detection of cracks),
- vidicon cameras (errors in geometry and glass defects), and
- CCD-cameras (errors in geometry and glass defects)
were being used in this order.

Only since optical CCD-sensors, line or matrix arrays, have been available an efficient sensor could be designed. After a relatively short time, there were devices on the base of hardware evaluating the visual information and of processing the measured values on the basis of simple algorithms which is then not an "intelligent" way of data processing.

The design of "intelligent" measuring systems by connecting sensor technique with microcomputer technology is a solution of these problems.

INFORMATION PROCESSING FOR INTELLIGENT MEASURING SYSTEMS

If it is desirable to test the whole surface of a typical chemico-technical hollow glass ware such as a beaker glass with a diameter of 10 cm for defects up to a magnitude of 0.1 mm and to classify these defects, about 3100 single photos must be taken during the turning of the product in case of the CCD line sensor.
Such products are manufactured with rotary blowing machines whose productivity is considerably below that of the in-series machine. In this case, piece numbers of 60/min are typical. The quality demand, however, are higher than in case of container glass. Since manipulation processes are also necessary for the test process besides the attaching of samples, an effective time of about 0.5 seconds remains for the actual pick-up. Thus, the image frequency will amount to 6.2 kHz. In order to obtain a sufficient resolution of the product height as well, the application of a CCD line camera with 1024 pixels is at least required.

This results in a data set with $3.2.10^6$ image points for each product and in an image point frequency of 6.3 MHz. These data sets cannot be processed without a reduction of information in real-time processing.

The first essential way for obtaining amounts of information which can be processed is an optical reduction of image points. By choosing appropriate methods of illumination the image is reduced to essential elements or, by increasing the contrast, the detection of defects and of the edges of products is then simplified or even becomes possible.
Figures 1 - 3 show glass products with different illumination.

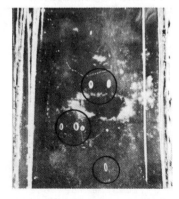

Fig. 1: bubbles(dark field)   Fig. 2: stone(light field)

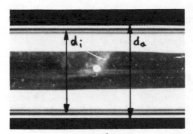

Fig. 3: geometry (shadow stripe)
of a glass tube

In addition to the optical image reduction, a hardware-
based information preprocessing is necessary which can be
realized by a runlength coding of the incoming visual information. Here, not all the recorded series of image points
is stored. In fact, only the black-white transitions occuring are acquired and made available for the image evaluation.

STRUCTURE OF AN INTELLIGENT MEASURING DEVICE FOR CHEMICO-
TECHNICAL HOLLOW GLASS WARE

For designing a test equipment for the solution of the
tasks mentioned, a test apparatus was constructed at the
department of glass/ceramics technology of the Ilmenau Institute of Technology which is being tested at present.
The basic structure is shown in Fig. 4. The run-length coding is realized by means of an additional circuit card,
which was developed at this department. Simultaneously, this
card allows to mask out any section of the image on the
basis of the hardware. Inspite of the optical hardware-
based pre-reduction of information, the information processing had to be split up into image processing, detection of
defects, and sequence control. This resulted in the hierarchical multicomputer system shown in Fig. 4.

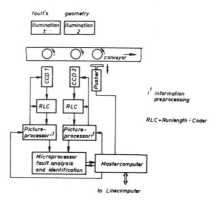

Fig. 4: structure of the measuring device

Being used as an essential element in a control unit of
hierarchical structure for production lines in the glass
forming process, a system like that offers good possibilities for further developments to be made in the future.
Then, the data concerning quality supplied by the test
system are used for directly influencing the forming machine and other processing machines in the forming process
with redard to control technique. By means of such a system,
the transition is made from the passive control of quality
to the active line control aimed at a maximum quantity at
high quality.

INTELLIGENT MEASUREMENT SPEEDS UP PATTERN RECOGNITION

APPLIED TO INDUSTRIAL ROBOTS

Gabriel Ionescu, Dan Popescu

Department of control and Computers
Polytechnical Institute of Bucharest
Bucharest, Romania

Abstract. The paper presents an intelligent measurement
system which is able to determine the most important features
and position of industrial parts commonly encountered in assembling robot applications. It is a programmable system built up
around a Z-80 microprocessor which controls a set of hardware
modules (image processors). These modules are dedicated for
primary processing of visual data and also for determining
various features, position and angular orientation.

1.INTRODUCTION

In order to fit the requirements imposed by on-line pattern
recognition for industrial robot applications, such as high
speed processing and high flexibility, both hardware and software resources must be utilised. In this respect, the general
structure of a visual pattern recognition system is proposed
in fig.1. The blocks represented in fig.1 diagram are rather
functional units denoting the specific tasks which are to be
performed. The complexity of these tasks together with a great
variety of requirements imposed by the applications, environmental conditions,etc. implies a certain degree of intelligence
for solving, in a resonable period of time, such problems as:
- to determine the limits of light variation and, consequently,
to establish the decision levels for the image A/D convertion;
- to minimize the number of features, from a fixed set, satisfactory enough for classes separation with an estimated error;
- to carry out the classification by using the above mentioned
various numbers of features;
- to allow convenient adjustments of some parameters involved
in the measurement and classification as: the classification
thresholds, the number of classes, the reference positions,etc.
For this reason two stages of operation are provided: firstly,
a learning one and, secondly, for measurement and recognition
one. At the first stage, the system acquires the information

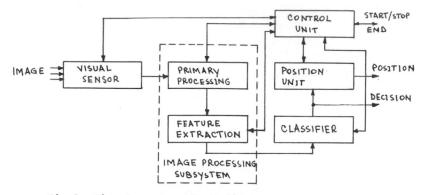

Fig.1   Visual system block diagram

by measurements on the classes prototypes and answers to the
question if the recognition is posible by using the established
features. At the next stage the system performs the measurements
on the parts actually involved in the robot application and, by
processing the data, effects their recognition in a time short
enough to consider it an on-line operation according to the
usual velocity of the parts conveyor.

2. THE HARDWARE MODULES

     Following the above mentioned head-lines, in fig.2 the
hardware structure for on-line image data processing is presented. This structure includes specialised hardware, high-speed
operating modules which are performing the following tasks:
a) The primary data processing of visual sensor signal in order
to achive the noise filtration, the adaptive binarisation and
the contour extraction, all of them being performed on-line
during one frame image acquisition. The synthesis of the specific modules is based on algorithms processing a 3x3 neighbourhood of the under consideration pixel. For instance, in fig.3
the median filter module is detailed, $F(i,j)$ and Me $F(i,j)$
being the brightness of the (i,j) point, respectively, the
median of the brightness on a 3x3 vecinity of that point.
Similarly, the logical smoothing and the contour extraction are
obtained by appropriate logic functions operating on the bina-
rised image signals.
b) The extraction of two important features, the area A and the
perimeter P follows directly by counting the "1" at the output
of the logical smoothing module and,correspondingly, at the
output of the contour module (some corrections for lines having
a $\pm$ 45º slope are also included). In the same manner, the
products Ai_G and Aj_G are determined by repeated additions of

the row, respectively the column order number at the rate of
the logical smoothing module output "1". These products are
useful for an easier computation of the weighting center coordo-
nates i_G, j_G.

c) The angular orientation module, a more complex one is, in
fact, the hardware implementation of the algorithm describing

328

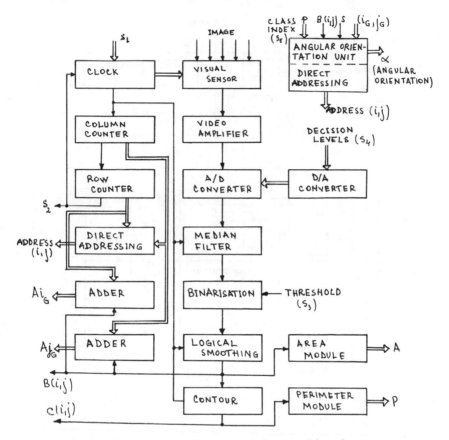

Fig.2 The hardware structure for on-line image
processing

the method known as the circular scanning of polar coded image
[1], [3]. In comparison to the software procedure, the speed is
much higher, for this reason the complexity being acceptable.

3. THE MICROPROCESSOR SYSTEM

The structure of the microprocessor system is sketched in
fig.4. The main tasks to be accomplished by the microprocessor
system are:
- the control and the syncronisation of the hardware modules
operation both at the learning stage and the recognition one,
- the settlement of the overall system parameters at the learning stage,
- the computation of the (i_G, j_G) coordinates and other supplementary features if necessary (min or/and max radius, number
of holes, etc.),
- the classification achievement by using minimum distance criterion,
- the communication with superior hierarchical level,

329

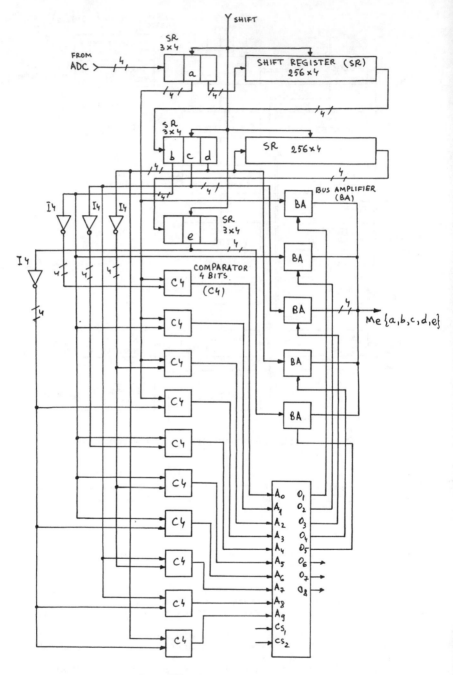

Fig.3   The median filter structure

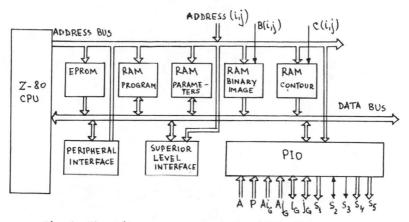

Fig.4   The microprocessor system structure

- the communication with the operator allowing external adjustment of the parameters and programs.

4. CONCLUDING REMARKS

The proposed system, under a reasonable hardware complexity,
performs image data processing in a very short time, so as for
256x256 image dimension, the measurement and the recognition
using the A and P features are performed in only 30 ms. The
duration for the angular orientation determination is depending on the required resolution; for the 2° resolution 200 ms
are necessary [1] .

REFERENCES

1. Popescu,D., Sprânceană,N.,ʺ Sistem vizual destinat recunoaş-
   terii formelor plane cu aplicaţii la roboţi industriali în
   linii de montaj", Proc. of the 6-th International Conference
   on Control and Computers, Bucureşti, Romania, 22-25 may 1985.
2. Ionescu,G., Popescu,D., "Prelucrarea în timp real a semnale-
   lor video în cadrul sistemelor de recunoaştere ale roboţilor
   industriali", National Conference "Teoria şi tehnica măsură-
   rii", Timişoara,Romania,5-7 dec,1985.
3. Martini,P.,Nehr,G.,"Recognition of Angular Orientation of
   Objects with the help of Optical Sensor", Industrial Robot,
   No.2/1979.

THE SAMPLING CONTROL OF VIDEO SIGNAL USING PERSONAL COMPUTER

Tsunehiko Nakanishi, Komyo Kariya, and Hiroshi Takano

Department of Electricty, Faculty of Science and Engineering,
Ritsumeikan University
56 Tojiin-kitamachi, Kita-ku, Kyoto, Japan

1. INTRODUCTION

As the video pictures have many informations, the many applications on measurement using video system has been reported. Generally, the method is achieved on a computer by information processing of quantized video signal which is sampled video signal from a camera. This method requires high speed A-D converter and large capacity computer to have high resolution. As the informations of the measuring object exist on plane position and time position, these informations are converted as time series signal by plane scanning in a camera. On this case, also much long time is required to process for information selection because many data must be treated in spite of the case that requested informations are not exist in whole part of video pictures. This fact occurs time delay until to be got optimal output from video signal acquisition. In the case that requested informations exist in only finite part of video pictures, information management is possible on a small size computer and simple system construction by appropriate pre-selection of video signal.

The basic detecting method of the informations which are brightness on the video pictures is achieved by the optical detector like as a photo transistor. This method was applied to automobile traffic flow measurement to ease monitor of measuring circumstances and simple construction.[1] But the method is difficult to aim the optical detector at the correct points on a video screen. An improvement of this difficulty is achieved by the proposed method which employs the graphic display function of personal computer and superimpose function of video picture. As the usual video pictures are constructed by the scanning lines which are the sets of the brighting point and brightness information is time series signal, the information collection of the finite parts on video picture is done by the time series sampling, that is, sampling control for the video signal.[2]

This construction was considered upon the concept of the harmony of system functions. The harmony that is allocation of the function of system components is very important to achieve appropriate measurement especially computer aided measurement.

2. VIDEO SIGNAL

There are two methods to get the position informations on video measurements. One is a scanning method by one photo element on a picture, and other is parallel output method by two dimensional array of many photo elements. The TV broadcasting system and also usual video system for the measurement are one of examples of the former method.

The scanning method of usual system in Japan is standardized to NTSC system which standard is 60 fields per second, linear scanning with interlacing, and 525 scanning lines in a picture frame. The scanning lines are converted to time series signal which has video information of brightness and position. The synchronizing pulses for horizontal scanning and vertical scanning are composed to the time series video signal as reference of position information. On this composit video signal, the position information of a video picture is included in time delay degree from the synchronizing pulses. Fig. 1 shows the correspondence between position on a video picture and time delay degree on video signal, and these relations are shown as following.

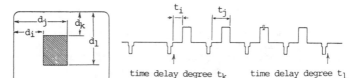

(a) length on a video screen,   (b) schematic video signal,
Fig. 1 An example of correspondence of picture size on video
screen and its video signal

The time delay degree t_k and t_l those are time from a vertical synchronizing pulse, t_i and t_j those are time from a horizontal synchronizing pulse are written as

$$t_k = \frac{d_k}{(\text{screen height})} \cdot \frac{1}{60} [\text{sec.}], \quad t_l = \frac{d_l}{(\text{screen height})} \cdot \frac{1}{60} [\text{sec.}],$$

$$t_i = \frac{d_i}{(\text{screen width})} \cdot \frac{1}{60 \cdot 2 \cdot 525} [\text{sec.}], \quad t_j = \frac{d_j}{(\text{screen width})} \cdot \frac{1}{60 \cdot 2 \cdot 525} [\text{sec.}]$$

(1)

The information acquisition from a finite part of video picture is possible by appropriate sampling for composit video signal. The sampled video signal $S(t)$ which corresponds to a hatched part in Fig. 1(a) is done by the gating for the time series video signal $V(t)$ as

$$S(t) = G(t) \cdot V(t) \tag{2}$$

where $G(t)$ is the gate operator. The gate operator is shown as

$$G(t) = G_h(t) \cdot G_v(t) \tag{3}$$

$$G_h(t) = 1 ,(m \cdot t_h + t_i < t < m \cdot t_h + t_j), \text{ or } = 0 ,(\text{ else }) \tag{4}$$

$$G_v(t) = 1 ,(n \cdot t_v + t_k < t < n \cdot t_v + t_i), \text{ or } = 0 ,(\text{ else }) \tag{5}$$

t_v ; vertical scanning period,
t_h ; horizontal scanning period,
m, n; natural numbers.

On this method, the time delay degrees should be decided before putting into practice though the signal processors of next stage are replaced by the simple elements and also time delay of output is made shorter. But it is not easy to decide accurate time delay degrees because the correspondence between information position on a video picture and time position on video signal is not equal as fly-back time of scanning is not zero and not so stabilized. Furthermore, information position has

334

unstableness as the destiny of usual relocatable sensing equipment. It is necessary to confirm the position of pre-selection and measuring objects on a video display screen by some man-machine interface to decide proper sampling position with high accuracy.

3. SAMPLING CONTROL SYSTEM

A figure on graphic display unit of personal computer is constructed by the sets of many brighting points which are controlled on each as one picture cell, and figure pattern is generated in a micro processor according to the software, and these informations are transmitted to a display unit on time series signals as information of illuminating position which has been corresponded to time delay. The time delay degree is determined similarly as in Fig. 1. This video signal P(t) is shown as follow,

$$P(t) = C(t) \cdot B(t) \tag{6}$$

where B(t) is brightness information and C(t) is the convertion operators between position and time degree as same as in eq. 3. Since P(t) accords with C(t) under a condition of constant brightness, P(t) are able to replace with C(t) in eq. 2 as the gate operator for appropriate sampling. Fig. 2 shows the schematic system construction for sampling of video signal and for monitoring of the sampling region used convertion operator. The fundamental elements are an analogue gate on the video signal path to the following signal processor, gate control unit, and sampling position indicator. The gate control unit is consisted of a personal computer which generates the convertion operator for gate control, signal synchronizer which adjusts timing of convertion operator with composit video signal, and gate driver. The sampling position indicator is consisted of signal adder to add convertion operator signal to composit video signal, and display unit of video pictures. The gate control signal is displayed on the display unit as position information and also video pictures of the measuring object are superimposed on one display unit to identify the relation of both position information. This function is important and useful to confirm the pre-selection region on a video display screen without special knowledge about the structures of video signals, and to setting and changing of sampling regions are achieved on computer software easily.

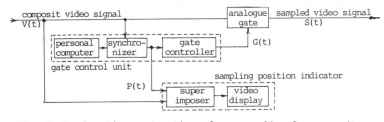

Fig. 2　A schematic construction of pre-sampling for composit
video signal

The plural sampling regions are able to set independently in one graphic display system by use of color pattern generator system. On a color display system, three parallel signals which correspond to primary colors are transmitted on simultaneously from micro processor to display unit. The three individual sampling regions are set easily because the color pattern signals are consisted of three primary color signals, and further number of sampling regions are set by illuminating with neutral color and encoding of fundamental three color signals. These easiness are important to generalization of measurement using video system for practical application.

4. APPLICATION

A combination of a video camera system and the pre-selection system was applied to measure speed distribution of automobile traffic flow.[3] On this case, as the measuring object is traffic flow on a road, the relocatable video camera was used for the 1st order detection. And as required information is the time of existence of automobile on a road by brightness change on the video picture, the two detecting points were settled on the video display screen for the 2nd order detection which two detecting points were used to measure passing time of an automobile in unit distance to know speed of each vehicle. As accuracy of measured speed are depended on setting of detecting points, accurate setting and accurate sampling of detecting points are required. Fig. 3 shows the system construction to achieve appropriate signal sampling and accurate aiming of the detecting points.

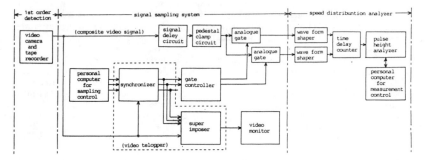

Fig. 3 The system construction for measurement of speed distribution on automobile traffic flow using video camera system

By use of the sampling method, the following merits were obtained.

(1) Sampling positions were confirmed with original measuring picture on a video display unit as a man-machine interface.
(2) Positioning stability was good with high accuracy.
(3) Change of detecting positions was easy on software of a personal computer.
(4) System construction for sampling was not so complicated.
(5) Signal processor after sampling was very simple.
(6) Short time delay was required to output final information after acquisition of primitive information.

These are achieved by proper and effective allocation of the functions of each constructing components.

REFERENCES
1. T. Nakanishi, H. Hasegawa, K. Kariya, Measurement Method of Transition of Amplitude Probability Density Distribution, Proc. IMEKO Symp. on Measurement and Estimation, 77 (1984).
2. T. Nakanishi, S. Hayashida, K. Kariya, A Study for Sampling Control Method of Video Signal using Personal Computer(Japanese), Memoirs of Res. Inst. of Sci. and Eng. of Ritsumeikan Univ., 43:157 (1984).
3. Y. Makigami, T. Nakanishi, K. Seill, Simulation Model Applied to Japanese Expressway, Jour. of Trans. Eng., 111:9 (1984).

PATTERN ANALYSIS OF PERIODIC TIME-VARIANT SCENES

G. Sommer (1), D. Gottschild (2),
(2)
F. Kubenk (2), A. Brandstaet (3)

(1) Department of Technology, (2) Hospital for Radiology and (3) Div. of Electronic Data Processing, Medical Department, Friedrich-Schiller-University Jena, DDR-6900 Jena, German Democratic Republic

INTRODUCTION

One of the problems of dynamic scene analysis is detection, localization and discrimination of changes in temporal image sequences [1]. The most frequently studied dynamic scenes are such, where movements take place in the plane of the image. Our concern contrary to this are scenes, which are characterized by constraint movement in place and by a fixed period for any object of the scene. Examples are top views of piston pumps or membranes which cause periodic contracting volumes.

Vision control of such periodic time-variant scenes relates to cognize the amount of gray level variations and distortions of the synchronism of movement.

Human cognition only hardly can be adapted to follow such dynamic processes. Therefore we modified a method for semantic segmentation of an image, which is used in a similar manner for multispectral analysis of remote sensed data. This method is clustering in a multidimensional feature space. The base is compression of data by kinetic and semantic modelling of the movements.

MODEL DRIVEN COMPRESSION OF TEMPORAL IMAGE SEQUENCES

Temporal image sequences can be compressed, if it is possible to model the dynamics of the basic process by any kinetic equation. This method is state of the art in nuclear medical functional image processing. This parametic imaging is applied to elastic objects like the heart chambers and liquid motion like blood flow in vessels.

If we have any image sequence with T slices of dimension N^2 (see fig. 1) we have to fit N^2 measurement vectors x (t) of dimension T with the kinetic equation $y(t,\underline{P})$.

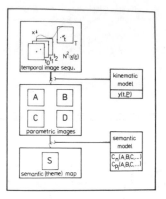

Fig. 1 Model driven compression of temporal image sequences

This is realized with the help of the parameter vector $\underline{P} = [P_1, P_2, \ldots, P_R]$. A set of parametric images $\{A, B, \ldots, M\}$ is computed on the base of this parameter vector, which semselves is not suitable to reflect the human imagination of the kinetic model.

These parameters are highly correlated. This is an essential source for semantic classifying the actual patterns of parameter images. But the power for visual recognition of this correlation will diminish the more parameter images are to be analysed. Extraction of parametic images by fitting a kinetic equation of motion is the first step of data reduction. The second step of compression is to automate the semantic segmentation of parametric images. This segmention is based on experiences in regional classification of scenes of interest. The result is a semantic map or a theme map S, which assigns each pixel in the image plane or in the region of interest to one of a set of classes.

Classification is done in the m-dimensional parameter space or m-dimensional histogram of parameters by partitioning this space in mutually exclusive regions [2]. Each region will correspond to a particular pattern class in the set of parameter images. So image segmentation in the image domain is replaced by clustering in the parameter domain. The following steps must be done (see fig. 2).

1. Imaging of the selected region in the set of parameter
images $\{A, B, \ldots, M\}$ in to the parameter space R^M.

2. Clustering and labeling the selected clusters with signs of membership to one of the semantic modelled classes or with the sign of rejection.

3. Mapping back the signed clusters to the image domain, where the maximal connected components of the labeled clusters constitute the image segments.

In our case we have to define also semantic classes, which are defined by their shape at the same position. This results from the correlation of parameters among one another. So step two of the classification procedure will be enlarged

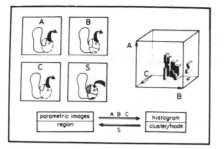

Fig. 2 Regional segmentation in the domain of parametric
        images A, B, C by histogram mode clustering in the
        parameter space R^3_{ABC}

by shape analysis of selected modes with the help of
first and second order statistical features. These features
are well known from texture analysis using cooccurrence
matrices[3]. In the space of these secondary features semantic
classes are to be separated.
        Nonstationary nonlinear smoothing within the semantic
defined populations will result in higher correlated
clusters with better statistical significance.

THE KINETIC MODEL OF THE HUMAN HEART

        One important problem in Nuclear Cardiology is to
study the kinetics of the human heart wall, particulaly
that of the left ventricle. As vehicle the change of volume
of the left ventricle is applied by labeling the blood with
a special radioactive nuclide and measuring an image
sequence, consisting of 30 frames in the period of one
heart cycle. The images are of the dimension 64 x 64 pixels.
The blood volume over this period follows approximately a
sinusoid.
        The kinetic model of this dynamics, restricted to the
position (x,y) at time t, $0 \leq t \leq T$, is approximately described by a Fourier series, limited to only first order

coefficients .

$$V_{xy}(t) = A_o + A \cdot \cos \left(\frac{2\pi t - P}{T}\right) + \dots,$$

with the mean volume A_o, the amplitude A and phase P.

$$A = \left(a_1^2 + b_1^2\right)^{1/2}$$

$$P = \arctan \left(\frac{a_1}{b_1}\right)$$

a_1 and b_1 are the Fourier coefficients of first order,
which are the parameters of the model. These amplitude and
phase values are computed for each pixel of the image
matrix.

339

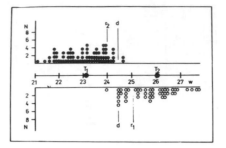

Fig. 3 Result of discriminant analysis, black circles: P:N,
blank circles: P:D, non elementar discriminant
feature w

THE SEMANTIC MODEL OF HEART WALL MOTION

The amplitude image reflects the amount of volume
changes. The phase image represents the degree of synchronism of this changes within the period.

For normal hearts we get in the region of the left
ventricle a homogeneous distribution of phases, which is
restricted to only 2 or 3 gray levels in a scale of 16.
Amplitudes are in this case intensively expressed with a
maximum at the most mobile parts of the ventricle.

Some diseases of the heart result in local or global
pathologic movements of the ventricles wall which are indicated by reduced amplitudes and/or spreaded phases.

The common semantic model relates the following classes
to sets of actual parameter constellations:

1. amplitude: normal (N), hypokinetic reduced (H),
              akinetic (A)
2. phase: normal (N), dyskinetic spreaded and shifted (D),
              paradox as strong dyskinetic shift (P),
              undefined (U)

CLASSIFICATION OF PARAMETRIC IMAGES

Supervised training of the classificator was done
interactively with the help of AP-histograms.

The training set comprised 128 patients including 76
with normal and 52 with dyskinetic phases respectively 52
with normal and 76 with hypokinetic amplitudes.

With a semiautomatic contour following method the
region of interest was selected. The pixels belonging to
this region were imaged in a three dimensional parameter
space R^3_{API}, which was spanned by amplitude, phase and a
stationary intensity at the phase of maximal filling of the
ventricle. The quantisation of each parameter was 16 levels.
Only the projection R^2_{AP} , the two dimensional AP-histogram,
was used for classification. Automatic histogram seeking
relating phase gave up to 5 modes. For these modes a set

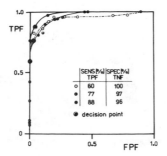

Fig. 4 ROC-analysis, blank circles: univariate discriminant
    function, black circles: multivariate discriminant
    function, TPF: true positive fraction, FPF: false
    positive fraction

of secondary features, describing cluster shape, on the
base of 1. order and 2. order statistics was computed. These
features are in effect moments until the fourth order in
one and two dimensions.
    For classification a parallel epiped method in the
space of secondary features was used. Only to discriminate
between normal and dyskinetic phase we used a 7-dimensional
discriminant function.
Classification was done in three steps:

1. Cleaning of the histogram regarding artefacts

2. P-mode-classification: Identification of the classes
                A:N, A:H, A:A, P:N, P:D

3. Subsegmentation of P-modes: Dyskinetic classes are sub-
   segmented on the base of local structure features into
   components P:N, P:D and P:P

    Univariate variance analysis confirmed, that the
variance of phase is the best discriminating single feature
concerning phase. But this feature was not further included
in the optimal set of features which are used by the linear
discriminant function.
    Instead of this the discriminant function based on
higher univariate moments and some bivariate features like
entropy and energy of the modal AP-distributation. Fig. 3
shows result of discriminant function applied to the
training set.
    Results of ROC-analysis (fig. 4) indicate, that with
classification of phase on the base of the discriminant
function sensitivity increased from 60 % to 88 % against
some decrease of specifity from 100 % to 96 %.
    An aneurysm shows fig. 5. It causes paradoxal movements
and large akinetic regions and results in an enlargement
of the left ventricle to compensate the reduced pumping
function. The strong dyskinesia results in hypokinesia.
Phases of akinetic regions are undefinded.

CONCLUSIONS

    Until now we did not test the power of the semantic

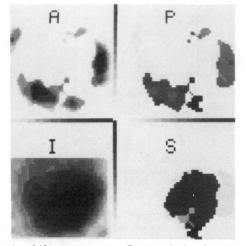

Fig. 5 Patient with aneurysm. Parametric images and theme
       map

segmentation method with an extensive set of unclassified
patients. But we think, it will have some general meaning
not only for functional imaging in medicine but also for
analysis of in place movements in the technical field.
     We only have demonstrated the procedure of semantic
segmentation. This map is the input for a scene analysis,
which is not the matter of this paper.
     The training procedure confirmed our conjecture that
the unsharp linguistic definition of semantic classes
causes problems in a sharp labeling procedure. Therefore
we conclude that classification in fuzzy sets would increase the classification accuracy.
     As another problem classification results only maxi-
mally can be as good as the world model. We used a model
which is composed by a kinetic equation and a sementic part.
Recently two new models have been published, which are based
on the eigenvalues of the image sequence.
     Perhaps these methods are better approaches to reflect
the ground truths of Nuclear Cardiology.

REFERENCES

1. H.H. Nagel, Overview on Image Sequence Analysis, in:
   T.S. Huang (ed.), "Image Sequence Processing and Dynamic
   Scene Analysis", NATO ASI Series F: Computer and Systems
   Sciences No. 2, Springer-Verlag, Berlin, Heidelberg,
   New York, Tokyo, 1983, 2-39

2. R.M. Haralick and L.G. Shapiro, Survey: Image Segmentation Techniques, Comp. Vis., Graph. and Image Process.
   29:100 (1985)

3. R.M. Haralick, K. Shanmugan and I. Dinstein, Textural
   Features for Image Classification, IEEE Trans. Syst.,
   Man and Cyb. SMC-3:610 (1973)

COMPUTER AIDED LASER SCAN SENSOR FOR OPTICAL MICRODEFECT DETECTION AND CLASSIFICATION

Claus-Peter Darr, Thomas Seifert,
Horst Truckenbrodt, Martin Weiss

Department of Technology for Scientific Instruments
Friedrich-Schiller-University of Jena
DDR-6900 Jena, German Democratic Republic

INTRODUCTION

Highly finished surfaces as used in microelectronics, optics, magnetic and optic storage techniques require a surface defect inspection for quality assurance. The inspection is directed on identification of the magnitude and the kind of the defects.
Laser scan methods based on the stray light indication concept are suitable for automatic microdefect detection.

STRAY LIGHT INDICATION CONCEPT

Stray light caused by light diffraction on a surface defect.
On principle, the stray light thus contains all information about magnitude and kind of defects. Defect determination may be carried out by direct or indirect stray light indication.
In direct indication a suitable designed optoelectronic detector is used for direct stray light collection.
The registration of the reduced direct reflex is used in the indirect detection concept. The cause of reducing of the direct reflex may be the light deflection after scattering, or it may be light absorbed by the surface defect (Fig. 1).

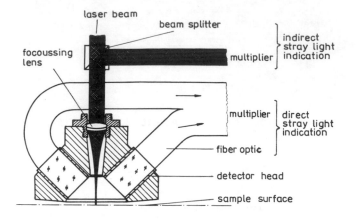

Fig. 1 Stray light defect detection concept and sensor head configuration

PATTERN ANALYSIS SENSOR PRINCIPLE

Foundation of all the following illustrations is the stray light concept in connection with laser scanners.

Defect Area Proportional Scatter Method (DAPSM)

DAPSM results from the fact that a surface defect which is smaller than the laser spot diameter produces a light scatter pulse during the scan which is proportional to the defect-spot-overlap area. Thus the amount of the optoelectronic detector output pulse is proportional to the defect magnitude. A threshold switch allows the adjustment of the minimum detecting defect magnitude.

Fundamental research showed that there is a good agreement between theory and experiment (Fig. 2).

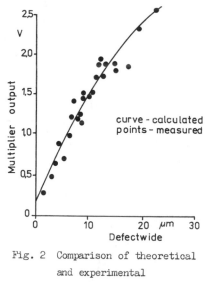

curve - calculated
points - measured

Fig. 2 Comparison of theoretical and experimental DAPSM-values

Pattern Analysis Method (PAM)

PAM may be used if geometrical conditions of the defects are the basis of identification. For them it is necessary to subdivide the sample surface in small pixels (circa 60 x 20 um^2). In our device we use a spiral scan and a polar subdivision. This is advantageous because our samples are circular. The angle element subdivision is electronically produced by an incremental rotating pulser (IRP) rotating synchronous with the sample.

During the scan the angle element adress (14 bit) of each defect affected element is stored in a microcomputer together with a two-bit-code for the track radius. The storing is triggered by the DAPSM-generated threshold switch output signal (Fig. 3).

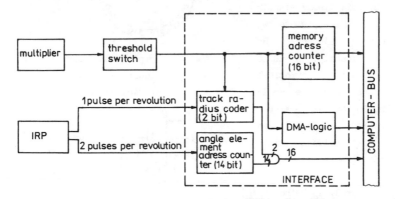

Fig. 3 Block diagram of computer aided pattern analysis sensor

After the scan the microcomputer starts a pattern recognition programme fo analysing the stored position adresses. The programme, for instance, is able to distinguish between point-, line-, and area-defects.

LOGIC LEVEL ANALYSIS SENSOR PRINCIPLE

The spatial distribution of scattered light contains a lot of information about the defect.

Scatter Light Geometry Analysis Method (SLIGAM)

A special sensor head with different angle detection ranges can give information on magnitude and geometric destribution of stray light (Fig. 4).

Of special interest is the distinction between surface defects and contaminations. Contaminations produce a comparatively high stray light level tangential to the sample surface. With SLIGAM contaminations and defects can be differentiated. Special designing of one of the angle detection ranges allows a distinction between point- and line-defects. Line defects produce an unisotropical linear scatter with orientation perpendicular to the scratch. This linear scatter can only strike one ore two of the triple elements A, B, C simultanieously. A point defect with isotropical scatter strikes the whole triple. In this way a simple electronic device can distinguish between point- and line-defects.

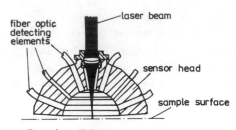

Fig. 4a SLIGAM-sensor configuration

Fig. 4b Triple subdivision of the middle detector element of Fig. 4a

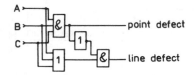

Fig. 4c Electronic distinction of point- and line-defects

Frequency Analysis Method (FAM)

Special kinds of defects like surface ripple can be detected by their periodical scatter signal. Presupposition is a constant scan velocity. Then it is possible to identify the periodical receiver signal with the aid of frequency filters. The filters produce a logic signal which indicates the kind of the periodic defect.

Logic Level Analysis Method (LOLAM)

LOLAM connects all methods illustrated till now with the exception of PAM. All the methods realise logic output signals. The sum of all signals may realise a special bit pattern for each kind and magnitude of defects which can be interpretated by a microcomputer.

A PHOTORECEIVER - BASED OPTICAL SENSOR

O.K. Arobelidze

Institute of Control Sciences

65, Profsoyuznaya, Moscow 117342, USSR

Further automatization of industrial process, wide application of industrial robots cause the need of increasing the technical flexibility of robots at the account of robots control advancement by means of sensitization and adaptation. One of the important problems which is necessary to solve is to provide a robot with environment perception. The main communication channel of a robot with the environment which carries the main information flow, needed for its rational control, is the visual (optical) channel. Visual information is received and processed by an optical sensor.

The choice of a sensitive element for the optical sensor is determined by the set of factors connected with the necessity to satisfy the requirements of the given task and with the tendency to minimize technical means in the process of this task solving. Great effectiveness of photoreceivers with a radial electric field is ensured by their ability to perform some definite information conversions of images in the process of their being converted into an electrical signal[1].

The operation principle of such devices is based on dragging electric field generation within a semiconductor plate which is obtained by means of a metallic circular and point electrode. Under the effect of this field minority nonequiligrium current carriers are formed as a result of semiconductor illumination. The carriers are moving along the radii from the centre toward a circular electrode. When an object image is projected to a sensitive surface of a photoreceiver, the concentration of nonequilibrium current carriers is distributed within the surface. This distribution reflects the light flow intensity distribution conditioned by an analized object image. Hereat, an output signal of photoreceivers with a radial electric field characterizes a total illumination of all points along the scanning radius. At a definite moment of the time a photoreceiver signal describes an image integrally along radius, i.e. it corresponds to a modified integral radial - circular (IRC) description of an image[2].

In realizing the IRC-description implemented by photo-receivers with a radial electric field an image is characterized by a periodic signal which does not change its character with the change in illumination or orientation of processed images. With an image illumination change and a change in orientation leads to a change in a signal phase with respect to a scanning beginning while an image scanning period remains constant with a signal form unchanged.

In order to get an output (information) signal from a videosensor, adequately describing an analized object, first of all one should centre the image, projected to a sensitive photoreceiver surface with respect to its geometrical centre, being the beginning of a coordinate system of the IRC-scanning, i.e. aligning the energy centre of an object image with the geometrical photoreceiver centre. Fulfilment of this procedure is necessary for invariant recognition in the IRC-description and it allows the norming of an image with respect to its transfer.

The IRC-description, on the whole, is a swept contour of an image whose characteristic points are given as maximums (peaks). With application of the analysis revealing the particularities of the modified IRC image description it is recomended to use as information features the features found from amplitude and time characteristics of a photoreceiver signal. To obtain an amplitude feature an information signal of a photoreceiver is compared with some threshold voltage which allows us to isolate maximums in a photoreceiver signal for further processing (counting of their quantity). Time features can be maximal or minimal pulse duration, maximal or minimal interval between pulses, total pulses duration etc. Formation of time characteristics may be realized by discrete counting method, widely applied in measurement technique.

The increase of image recognition selectivity is achieved by forming the features set along several channels with different values of threshold voltage to which the features are relatively formed. The number of channels (the number of threshold voltage) is defined by a class of objects to be recognized and requirements to recognition validity. The operation of the channels is characterized by the value of threshold voltage with which an output signal of a photoreceiver is compared. The totality of amplitude and time features obtained from a photoreceiver signal is transformed in a more compact kind - code, characterizing the analized image.

Subsequent processing of information is reduced to comparing the obtained codes with the reference codes stored in a memory unit, i.e. to implementing the analysis of features closeness and in the case of satisfying the closeness conditions - to forming a signal of image recognition in totality in all the channels.

The above data, obtained by means of the present videosensor are supplied to the control unit of actuators robot's memory and that is where a corresponding working program is chosen whose realization enables a robot to automatically take, transport and install an object on a previously chosen place (another conveyer, container, etc.). The time of infor-

mation processing by the proposed optimal sensor is estimated as $IOms^3$. The specified resolution of the photoreceiver
with a radial electric field is 10 lines/mm.

The proposed device is simple, small-size and reliable,
it is acceptable for industry parameters and may be successfully applied in automation of loading/unloading and
sorting operations on the suspended conveyers with the order
of objects supply and exact location unknown.

REFERENCES

1. V.D. Zotov, Semiconductor devices for sensing optical information, Energya Publisher, Moscow (1976).
2. V.D. Zotov, O.K. Arobelidze, A visual system for an
   industrial robot, in proceedings of Ist International Conference on Robot Vision and Sensory Control,
   Stratford - upon-Avon, UK, (1981).
3. V.D. Zotov, O.K. Arobelidze, Position-sensitive
   photoreceiver with the radial electric field, in
   preprints of IX world congress IMEKO, Berlin (West)
   (1982).

IMAGE PROCESSING FOR INTELLIGENT LENGTH MEASUREMENTS

- THE DISCRIMINATION OF MINIMUM RADIAL ZONE CYLINDRICITY ERROR

Xiong Chuan-mei

Measurement Center
Sian Institute of Highway
Sian Shensi, China

INTRODUCTION

This paper deals with the determination of the minimum radial zone cylindricity error.

LASER FRAUNHOFER DIFFRACTION OF A SINGLE SLIT AND THE MEASUREMENT OF THE WIDTH OF A NARROW SLIT

Diffraction is sometimes defined as "the bending of light around an obstacle". In most diffraction problems some light is found within the region of geometrical shadow.

Laser Fraunhofer diffraction of a single slit is an infinitely distant diffraction, i.e. both light source and observing screen are in two infinite distances from the diffracting screen. Parallel laser rays illuminate the narrow slit AB of width X, a diffraction pattern can be observed on the observing screen located at a distance L from the narrow slit /see Fig.1/, when $L > X^2/W$. The wavelength of helium neon laser rays W is 0.6328 /um.

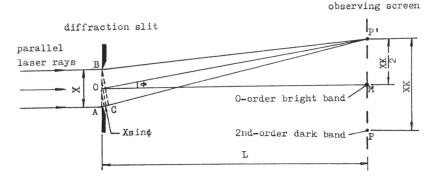

Fig.1. Laser Fraunhofer diffraction of a single slit

351

Let X represent the width of the narrow slit AB, O represent the center point of the narrow slit AB, and consider a point P on the observing screen in a direction making an angle Ø with the axis of the system. A straight line BC is perpendicular to the straight line OP'. If the distance L from slit to screen is large in comparison to the width X of the slit, BC can be considered a straight line at right angles to P'B, P'O and P'A. Then the triangle ABC is a right triangle, similar to P'OM, and the distance AC equals Xsinø. This latter distance is the difference in path length between the waves reaching P'from the two points A and B. Diffraction effects at point P'are determined by the overlap of all the wavelengths on the propagating wavefront AB. According to the half waveband method, the distance AC is divided into K equal strips by W/2, and the wavefront AB is also divided into K equal half-wavebands, correspondingly. Hence

$$AC = Xsinø = K(W/2)$$

where $K = 0,1,2,...$

The number of half wavebands K represents the order of diffraction fringes.

When having no difference in the path length of light, the number of half wavebands $K = 0$, the point M /i.e. point P'/ is a 0-order bright band.

When there is a difference in the path length of light, because the phase difference of two adjacent half wavebands $\theta = \pi$ radiant, there is a dark band at point P'. Therefore when the number of half wavebands K = an even number, there is a dark band at point P', i.e.:

$$AC = Xsinø = \pm 2K(W/2) , \qquad \text{/dark band/ /1/}$$

where $K = 1,2,...$

When K = 1, and the difference in path length of light corresponds to formula /1/, $AC = \pm W$, there is the 1st-order dark band at point P'.

When K = 2, and the difference in path length of light corresponds to formula /1/, $AC = \pm 2W$, there is the 2nd-order dark band at point P'.

Measuring the center distance XK of two same order dark bands to compute the width X of a narrow slit

From formula /1/:

$$Xsinø = KW,$$

$$X = KW/sinø,$$

because angle Ø is small,

$$sinø = tgø = XK/2L.$$

therefore,

$$X = 2KWL/XK \qquad\qquad /2/$$

After measuring the center distance XK of two Kth order dark bands, a microcomputer can compute the width X of a narrow slit by using formula /2/.

Example- Let's suppose that the wavelength of helium neon
laser rays W=0.6328 um, the distance of the observing screen
from the diffraction slit L=1000 mm, measure two 2nd-order dark
bands, K=2, the practically obtained center distance of the
two 2nd-order dark bands XK=X2=23.01 mm.
  Then, the width of a narrow slit
  X = 2KWL/XK
    = 2x2x0.0006328x1000 /23.01
    = 0.11 mm.

Enlarging the center distance XK of two same order dark bands

  The center distances of two-sided diffraction bands, either
bright bands or dark bands, of Zero-order bright band are many
times the width X of a narrow slit. Enlarging XK/X of the center
distance XK of two same order dark bands:

$$XK/X = 2KWL/X^2 \hspace{3cm} /3/$$

LONG DIFFRACTION BANDS ARE THE EXPANDED IMAGES OF THE VARIATION
OF THE RADII OF ALL POINTS ON ONE MEASURED GENERATOR OF A
CYLINDRICAL WORKPIECE

  One measured generator of a cylindrical workpiece and the
edge of a tool /a standard/ form a narrow slit. Parallel laser
rays illuminate it, so that long diffraction bands can be
observed on the screen /Fig.2/.

The length of diffraction bands exactly equals the length of
one measured generator of a cylindrical workpiece. The form
of long diffraction bands is corresponding to the form of the
measured generator, but the error of straightness of the
center lines of long diffraction dark bands is larger. The
center lines of long diffraction dark bands are the expanded
images of a cylindrical workpiece. We can measure minimum radial zone cylindricity error of a workpiece by using the center
lines of long diffraction dark bands.

parallel
laser rays

edge of a tool   cylindrical   observing screen
                   workpiece

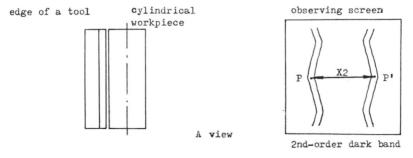

                        A view   2nd-order dark band

_Fig.2 Long diffraction bands on the observing screen

DETERMINATION OF THE MINIMUM RADIAL ZONE CYLINDRICITY ERROR

Minimum radial zone cylindricity error f_{cy}

The minimum radial zone cylindricity error zone is limiter to two coaxial cylinders, which contain the considered surface and the radial difference of which is minimum . The radial difference of the two coaxial cylinders is the minimum radial zone cylindricity error f_{cy} /Fig.3./

Determination of the minimum radial zone cylindricity error

With an idel straight line as an axis, measure all the generators If the maximum one among maximum radial differences of all the generators is minimum, the ideal straight line is the exact direction of the ideal axis of the minimum radial zone cylindricity error range. When keeping the exact direction of the ideal axis unvarying, with an ideal straight line as an axis, measure the radii of all points on the considered surface. If the maximum radial difference of them is minimum, the ideal straight line is the exact position of the ideal axis of the minimum radial zone cylindricity error range. With this ideal axis with exact direction and position as a datum line, measure the radii of all points on the considered surface, then the maximum radial difference is the minimum radial zone cylindricity error.

Determination of the exact direction of the ideal axis of the minimum radial zone cylindricity error range. It may be determined according to one of the following cases:

When each generator direction of every section including axis is parallel, the exact direction of the ideal axis, which makes ah maximum one among maximum radial differences of all the generators minimum, parallel to the same generator direction of every section including axis /Fig.4./. With the ideal axis of such direction as a datum line, measure all points on every generator; then each of all maximum radial differences of the generators is minimum.

When each generator direction of every section including axis is not parallel each other, the exact direction of the ideal axis, which makes the maximum one among maximum radial differences of all the generators is minimum, it is a direction of the ideal axis, which makes two maximum radial differences of two generators of one section including axis equal

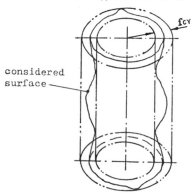

considered
surface

Fig.3 Minimum radial zone cylindricity error

354

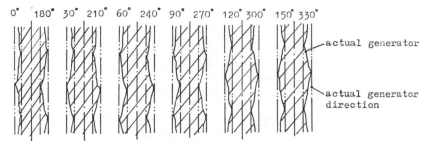

0°   180° 30° 210° 60° 240° 90° 270° 120° 300° 150° 330°

actual generator

actual generator direction

Fig.4. Each generator direction of every section
including axis

/i.e. minimum/, the sum of two maximum radial differences
of two generators is maximum /Fig.5./.

Let's suppose that the sum of two maximum radial differences of two generators of the $60° \square 240°$-section including axis
is maximum, i.e.

$$\left(60°R_{max} - 60°R_{min}\right) + \left(240°R_{max} - 240°R_{min}\right) = max$$

adjust the direction of the ideal axis. When two maximum radial differences of the $60°$-generator and the $240°$-generator
are equal, i.e. when

$$60°R_{max}' - 60°R_{min}' = 240°R_{max}' - 240°R_{min}' \quad ,$$

the direction of the ideal axis is the exact direction of the
ideal axis, which makes the maximum one among maximum radial
differences of all the generators minimum.

Determination of the exact position of the ideal axis of
the minimum radial zone cylindricity error range. When keeping
the exact direction of the ideal axis unvarying, i.e. keeping
two equal maximum radial differences of two generators of one
section including axis minimum and unvarying, the sum of two
maximum radial differences of two generators is maximum /Fig.6./
Displace the position of the ideal axis, hence vary the absolute radial size of every generator. When two equivalent
maximum radii and two equivalent minimum radii of the measured
cylindrical surface alternately take place around the circumference, this position of the ideal axis is the exact position
of the ideal axis of the minimum radial zone cylindricity
error range /Fig.7./.

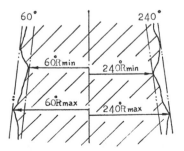

60°   240°

$60°R_{min}$ $240°R_{min}$

$60°R_{max}$ $240°R_{max}$

Fig.5 Exact direction of ideal axis

exact direction
of ideal axis

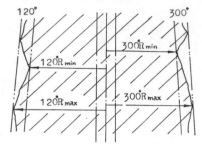

Fig.6. Keeping the exact direction of the ideal
axis unvarying

Assuming in Fig.6. that the $120°$-generator and the $300°$-generator are two generators of one section including axis, the sum of two maximum radial differences of which is maximum, and two maximum radial differences of the two generators are equal /i.e. minimum/, namely

$$120°R_{max} - 120°R_{min} = 300°R_{max} - 300°R_{min} = \text{minimum.}$$

Displace the position of the ideal axis, hence vary the absolute radial size of every generator, if the exact direction of the ideal axis in unvarying, the maximum radial differences of the generators /or the result from the calculation above/ are unvarying.

In Fig.7. displace the position of the ideal axis to the exact direction, so that on the $96°$-generator and the $274°$-generator of the measured surface two equivalent maximum radii appear, $96°R_{max} = 274°R_{max}$, and the $0°$-generator and the $178°$-generator two equivalent minimum radii appear, $0°R_{min} = 178°R_{min}$. This position of the ideal axis is the exact position of the ideal axis of the minimum radial zone cylindricity error range zone of the measured surface. The difference of maximum radius and minimum radius of the error range is the minimum radial zone cylindricity error, i.e.

$f_{CY} = 96°R_{max} - 0°R_{min} = 274°R_{max} - 178°R_{min}$.

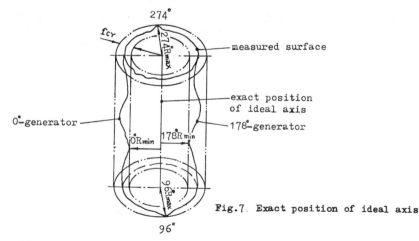

Fig.7. Exact position of ideal axis

HIGH ACCURACY INSTRUMENT FOR ROUNDNESS MEASUREMENT AND ANALYSIS WITH THREE DISPLACEMENT SENSORS

Lu Deju, Wang Baoyu
Zhang Quigping, Fan Tianquan

Institute of Optics and Electronics
Academia Sinica
Chengdu, China

INTRODUCTION

At present the roundness instrument by the axis standard is widely used to measure roundnesses of precision shafts all over the world. Its overall uncertainty is largely dependent on the rotating accuracy of the main spindle. It is rather difficult to further increase the rotating accuracy of the main spindle. If the roundness is measured with three component isn't needed. So the error of the standard component doesn't exist and the accuracy can be further increased. Based on the work done by other scientific workers[1] a high precision roundness instrument with microcomputer has been developed.

BASIC PRINCIPLE

The roundness error $r(\theta)$ is the function of the rotating angle θ and can be expressed as Fourier series

$$r(\theta)=\sum_{n=2}^{\infty} C_n \cdot \cos(n\theta+\varphi_n) \qquad (1)$$

where C_n the amplitude of the nth harmonic,

φ_n the original phase angle of the nth harmonic.

Three displacement sensors are set along the circumference of the cross section of the shaft to be measured /shown in Fig.l./, the included angles between the sensors are α and β, respectively. The three displacement sensors detect the roundness information simultaneously while the shaft is rotating. The information obtained from the displacement sensors contains the setting extremes/. By combining the information obtained from three sensors in a certain way, the function $Y(\theta)$ can get rid of the influence of shaking.

then
$$Y(\theta)= \sum_{n=2}^{\infty} C_n^1 \cdot \cos(n\theta+ \varphi_n') \qquad (2)$$

where
$$C_n^1 =C_n \cdot K_n$$
$$\varphi_n' = \varphi_n+\phi_n \qquad (3)$$

357

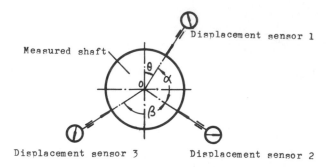

Fig.1. Position of three displacement sensors

After determining the values of angles α, β, the amplitude distortion coefficient K_n of the nth prism circle and the phase angle distortion value ϕ_n of the nth prism circle may be calculated. And then C_n' and ϕ_n' can be found by formula (2). Putting them into formula (3), C_n, ϕ_n can be calculated. After putting them into formula (1) the roundness error is obtained.

COMPOSITION OF THE INSTRUMENT AND CONSIDERATIONS ABOUT DESIGN

The instrument consists of a positioning system, a micrometer system, a driving system, a microcomputer system and the applied software. The computer controls the driving system to make the shaft to be measured rotate while gathering the information coming from three sensors. The roundness error and the frequency spectrum of the measured shaft are obtained by computation. All data are printed out by a printer controlled by the computer.

In the design of the instrument, emphasis is put on how to acquire the small error of the instrument. The micrometer system being the most important part adopts the high precision inductance displacement sensors whose resolving power is 0.005 um and its accuracy is 0.02 um within 2 um. And the micromeasurement system, the positioning system, etc. have high stiffness and stableness.

Meanwhile the angles α and β should be properly selected so as to obtain the small measurement error. The used angles α and β should not make the K_n become zero, especially for the low-order harmonics. Otherwise the roundness can't reflect the harmonics, the measurement error is introduced into the result. In addition, the used α and β should make it convenient for the computer to process the data fast.

UNCERTAINTY OF THE INSTRUMENT AND COMPARISONS

Generally the overall uncertainty of roundness instruments by axis standard is given only in individual technical specifications. We belive that the general uncertainty should be provided for users to compare, select and calculate the error.

This instrument is an automatic one, the main source of the error is the error in the displacement sensor's readings. Through analysis and calculation, it follows that the uncertainty /3σ/ of the instrument is 0.03 μm if the average of the three times measured values is taken as the result of the measurement.

We have also verified the uncertainty of our instrument by comparison and have measured the roundness of 3 short shafts using this instrument and Talyrond Model Roundness Instrument respectively. The measured roundnesses /P + V/ are listed in Table 1. The difference between the two sets of results is within the range of the uncertainty. So, the uncertainty given by our instrument is reliable.

Tab.1.   Comparison of roundness   unit: μm

	2# shaft	3# shaft	4# shaft
Talyrond - 73	0.04 ~ 0.07	0.20	0.06
Our instrument	0.05	0.19	0.04

DISCUSSION

Based on the three displacement sensors method we know that the measured Y (Θ) shouldn't contain the fundamental harmonic /let its amplitude be C_1 /. In fact Y (Θ) still includes quite a small residual C_1 /in comparison with the roundness error/. How was C_1 produced? What does its value show? So far we have not read any papers discussing this issue. We believe C_1 is produced by the measurement errors. When using the three displacement sensors method, the value C_1 should be calculated although it is not used in calculating the roundness errors. If the first value of C_1 is too high, it shows that the measurement error is comparatively large. So, we must repeat the measurement after finding out the cause for that.

CONCLUSION

The instrument is mainly used to measure a long shaft and a shaft whose position to be measured is comparatively far from its ends. By changing the measuring systems of the instrument we can achieve on-line measurement, measure the roundness of sjafts on the machine tool and analyse the roundness error and the frequency spectrum of the rotating shaft to improve the technology and to increase the manufacturing precision.

ABSTRACT

In this paper the principle of a high accuracy instrument for measuring and analyzing roundnesses with three displacement sensors introduced. Composition of the instrument, considerations about design and measurement error are described.

The roundness error and its frequency spectrum can be automatically measured, calculated, displayed on CRT and printed.
When the instrument measures the precise shaft, the overall
uncertainty of the roundness errors /P + V/ is 0.03 um.

REFERENCE

1. Yasuo Aoki, Shigeo Ozono,: On a New Method of Roundness
   Measurement Based on the Three-Point Method. Journal of
   the **Japan** Society of Precision Engineering, Vol. 32,
   No.12, 1966.

EVALUATION OF DEFLECTIONS OF
COMPONENT FORMS IN COORDINATE
MEASUREMENTS

Liudvika Nagineviciene,
Aldona Vitkute

Machine Production Engineering
Department and Department of
General Mathematics
Kaunas Polytechnic Institute,USSR

INTRODUCTION

Coordinate measurement methods are coming into use in the all-round check of the dimensions, the shape and the relative surface position of machine parts and devices.

A coordinate instrument measures the points of a geometrical element of a part. With the respect to these points a real element model is determined by means of mathematical methods. Using a suitable criterion of optimization the geometrical shape of a part is described and its deflection is determined by the measured points of an evaluated reference element. The computing algorithms are set up and no restrictions in measuring and mounting the parts on a coordinate instrument are needed.

DETERMINATION OF ELLIPSE AND CONE PARAMETERS

When a part is being mounted on the coordinate instrument its position is not aligned. The computing algorithms and programs should correct the errors relevant to the adjustment of a part automatically. For instance, to check the circle of cylindrical surfaces the points are measured in one plane, and due to the slope of a cylinder axis relative to the coordinate axes of an instrument the measured points lie on an ellipse. The central circle may be computed small slope angles being in the first approximation. In this case, the slope of an axis is determined by the parameters of the central circles obtained in two cross-sections (1-3). We want to suggest a method for a more accurate determination of the deflection from both the circle and the position of an axis of cylindrical surfaces.

To check the circular form of a part in its any transverse plane $Z = c$, we measure coordinates N of points $\vec{r_i} = (x_i; y_i; Z)$, $i = 1,2,\dots,N$. Substituting them into a general equation of an ellipse, the set of N equations is obtained

$$BA = C_N \qquad (1)$$

in which the ith row of matrix B is

$$b_i = (2x_i y_i \quad y_i^2 \quad 2x_i \quad 2y_i \quad 1),$$

unknown solution.

$$A' = (a_1 \quad a_2 \quad a_3 \quad a_4 \quad a_5),$$

' - transposition sign,
absolute terms are

$$C_N' = (-x_1^2 \quad -x_2^2 \quad -\cdots - x_N^2).$$

When $N = 5$, the set (1) is solved by Gaussian, Cramer method or by an inverse matrix.
When $N > 5$ we suggest to seek for the solution which satisfies the condition of the least square method making use of this recurrent formula

$$A_{n+1} = A_n + P_n b_{n+1}' (b_{n+1} P_n b_{n+1}' + 1)^{-1} (x_{n+1}^2 - b_{n+1} A_n),$$

in which

$$P_n = P_{n-1} b_{n-1}' (b_{n-1} P_{n-1} b_{n-1}' + 1)^{-1} b_{n-1} P_{n-1}$$

for all $N > 5$.
P_5 and A_5 are defined from the first five equations of the set (1)

$$P_5 = B_5^{-1},$$
$$A_5 = B_5^{-1} C_5.$$

Here matrix $N \times N$ of the inversion problem is reduced to the division by a scalar and by some multiplications and additions. When is large this method is efficiant. The new assessment of the ellipse coefficients is defined as the sum of the old evaluation and the linear correction term based on both the new information b_{n+1}, x_{n+1} and the old one P_n.
    The coordinates of the ellipse center are

$$\vec{r_c} = ((a_1 a_4 - a_2 a_3)/d \; ; \; (a_1 a_3 - a_4)/d \; ; \; c), \quad d = a_1^2 - a_2.$$

If the origin of the coordinate system of a part is in the ellipse center, the Z-axis coincides with that of rotation and the X-axis is perpendicular to the Z-axis of a coordinate instrument. Then the coordinates of measured points in the coordinate system of a part are

$$\vec{r_{id}} = \Pi_x^{\gamma} \Pi_z^{\alpha} (\vec{r_i} - \vec{r_c}) = (x_{id} \; ; \; y_{id} \; ; \; z_{id}), \qquad (2)$$

where Π_x^{γ}, Π_z^{α} - transformation matrices of a revolution around the Ox and Oz coordinate axes by angles γ and α, respectively,

$$\cos \gamma = \left(\frac{a_2 - 1 - \sqrt{D}}{a_2 - 1 + \sqrt{D}}\right)^{0.5},$$
$$D = (a_2 - 1)^2 + 4 a_3^2,$$
$$\cos \alpha = \sqrt{2} a_1 (D + (a_2 - 1)\sqrt{D})$$

In the coordinate plane Oxy the part points $\vec{r_{i0}} = (x_{i0} \; ; \; y_{i0})$ are in the circle the center of which is in the point $\vec{r_0} =$

362

= (0;0). For the reference of deflections from the circular forms, the mean circular form determined by Gaussian square error minimum, contiguous surfaces or the circular form corresponding the criterion of a minimum distance between two concentric circles may be taken.

To measure bevel gears by a coordinate instrument the coordinates of the vertex of a cone and the angle at it are to be known. To find these values we measure n_j points in k generating lines $n_{ij} = (x_{ij} ; y_{ij} ; z_{ij})$. If merely the intersection points of the cone surface with curvilinear teeth can be measured, then the measured points are turned round the cone axis by angles

$$\psi_{ij} = \frac{\pi}{2}(1 - sgn\, x_{ij}) + arctg\, \frac{y_{ij}}{x_{ij}} - \psi_{ij}.$$

$\vec{s} = (cos\,\alpha_x ; cos\,\alpha_y ; cos\,\alpha_z)$ is used to denote the unit directrix vector of the cone axis (the third row in the transformation matrix (2) when the ellipse points are in the cone hole),

$$\vec{r_j} = \frac{1}{n_j}\left(\sum_{i=1}^{n_j} x_{ij} ; \sum_{i=1}^{n_j} y_{ij} ; \sum_{i=1}^{n_j} z_{ij}\right),$$
$$\vec{q_j} = \vec{r_j} - \vec{r_c} ,$$
$$\vec{s_{ij}} = \vec{r_j} - \vec{r_{ij}} = (s_{ijx} ; s_{ijy} ; s_{ijz})$$

vector product of vectors

$$\vec{q_j} \times \vec{s} = (b_{1j} ; b_{2j} ; b_{3j}).$$

By the method of least squares the unit guide vector of the jth generating line $\vec{l_j}$ is determined

$$\vec{l_j} = \frac{1}{\Delta_j}(M_{1j} ; -M_{2j} ; M_{3j}),$$
$$\Delta_j = (M_{1j}^2 + M_{2j}^2 + M_{3j}^2)^{0.5},$$

where M_{1j}, M_{2j}, M_{3j} — minors of the third order of the matrix T_j

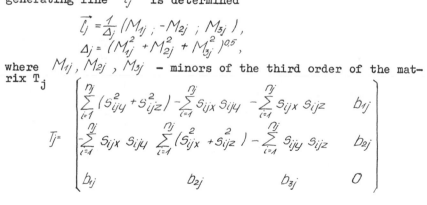

The coordinates of the vertex of the cone are

$$\vec{r_v} = \vec{r_c} + t\vec{s} , \qquad t = \frac{1}{k}\sum_j \frac{|\vec{l_j} \times \vec{q_j}|}{|\vec{l_j} \times \vec{s}|}.$$

Cosinus of the angle δ when the vertex of the cone is

$$cos\,\delta = \frac{1}{k}\sum_j (\vec{l_j}\cdot\vec{s}),$$

where $(\vec{l_j}\cdot\vec{s})$ — scalar product of vectors.

To determine the deflections of the above-mentioned and the other geometrical elements, the programs and algorithms are developed. For instance, the deflections of points $\vec{r} = (x;y;z)$ from the surface of a cone are determined by formula

$$d = ((\vec{r_v} - \vec{r})\,\vec{s})\sin\delta - ((\vec{r_v} - \vec{r})^2 - ((\vec{r_v} - \vec{r})\vec{s})^2)^{0.5}\cos\delta$$

CONCLUSIONS

By means of the least square method the position of a part in the coordinate system of a coordinate instrument is determined by measured points on cylindrical and conical surfaces. The formula of the transition to the coordinate system of a part is derived. The recurrent formula for solving the system of incompatible linear equations is suggested.

REFERENCES

1. Lotze, W. Ausgleichkreis in der Koordinatenmesstechnik. Feigerätetechnik 30 (1981) 12 S. 538-542
2. Peters, R.D., Wollersheim, H.-R. Algorithmen zur Ermittlung der Formabweichung mit Koordinatenmessgeräten. VDI-Z 126 (1984) 1/2 S. 37-42
3. Bressel E. Messung der Formabweichung vom Kreis auf Koordinatenmessgeräten. Feingerätetechnik 33 (1984) 1 S. 14-17

A NEW SYSTEM FOR TESTING THE PROFILE

OF LARGE INVOLUTE GEAR TEETH

Wang Yan-Fang, Zhu Zuo-Hang

Department of Precision Instrumentation
Shanghai Institute of Mechanical Engineering
Shanghai, China

INTRODUCTION

In order to improve the manufacturing technique of large gears, people
have been trying to find out ways to measure them effectively. Most of the
traditional gear measuring machines now available for testing tooth profiles
are stationary, although they may have a high measuring accuracy, but they
can't meet the needs of inspections for especially big and medium-sized large
gears because of their narrow measuring ranges, complicated constructions and
high cost of manufacture. The transportable instruments for testing profiles
of large gear tooth have still some problems in measuring accuracy and practical application. Therefore it is necessary to pay more attention to finding an accurate, simple and reliable measuring method. In this paper, we
would introduce a new system for testing profiles of large involute gear
tooth and a correlative apparatus.

MEASURING PRINCIPLE

It is well known that the more the number of gear tooth, the closer the
profile to a line and the smaller the difference between the two. According
to this fact, we have designed a measuring principle which includes means
for positioning a measuring unit on a gear via ball rests contacting with
tooth spaces and for setting up a reference line tangent to the profile at
a point near the pitch circle instead of the theoretical profile. In the
course of inspection, let probe move along the direction of the reference
line and scan on the actual profile to generate a measuring signal. The
difference between the line and the theoretical profile in measuring direction and other systematic errors would be compensated and corrected by means
of the software system. Thus, the profile deviation can be separated from
the measuring signal and be recorded automatically.

COMPUTATION OF THE THEORETICAL INVOLUTE PROFILE

In the XOY measuring coordinates system (Fig. 1), the involute profile
can be given in the form of parametric equation:

$$X = r_b (\cos (\phi - \phi_0) + \phi \sin (\phi - \phi_0) - 1),$$

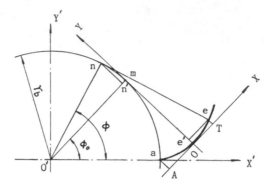

Fig. 1 Computation of the coordinates X, Y of
the theoretical involute

$$Y = r_b (\sin (\emptyset - \emptyset_o) - \emptyset \cos (\emptyset - \emptyset_o) + \emptyset_o).$$

For getting the precise solution of Y when X is equal to any value, the parametric equation may be converted into a first order differential equation as follows:

$$dY / dX = tg (\sqrt{(Y / r_b - \emptyset_o)^2 + (X / r_b + 1)^2} - 1 - \emptyset_o),$$

where the initial condition: $Y_{(X=0)} = 0$.

The computing result shows that the differential equation can be solved quickly and precisely with numerical solution by the computer. At the same time, we can also prove that all the involutes with different base circles are similar, they can be expressed as follows:

$$Y_i / r_b = Y_{r_b = 1},$$

$$X_i / r_b = X_{r_b = 1}.$$

Put the compensating function $Y_{r_b=1}$ into the computer, we may easily get the difference values Y_i between the reference line and the theoretical involute profile for any base circle. (i.e. compensating values.)

ORIENTING OF THE PROBE

Positioning the measuring unit relative to the gear tooth is performed via ball rests. The key quantity determining the actual position of the probe in X direction is the chord length S (Fig. 2). The mathematical relationships for orienting the probe are

$$\Psi = \arcsin (H / S),$$

$$\delta = K\pi / Z,$$

$$D = S \cos (\delta - \Psi) / (2 \sin \delta) - r_b,$$

where Z -- number of teeth of gear, H -- constant, K -- number of teeth between the ball rests.

366

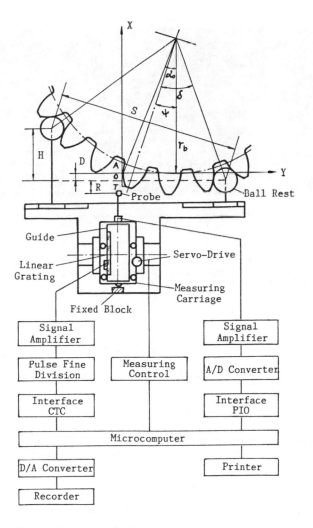

Fig. 2 Diagram of the measuring system

Referring to the input data, the computer solves the above equations and determines the distance D. Then according to the structural constant R of the apparatus and the measuring circle, the probe will move automatically to the starting point A under control of the computer.

CONSTRUCTION OF THE APPARATUS

The apparatus is composed of two parts (Fig. 2, 3). The first part is a measuring unit which is mounted on the gear. It contains a measuring carriage, a precise guide on which the carriage moves forwards or backwards, an incremental measuring system, a servo-drive for travel in X direction and ball

Fig. 3 Profile measurement with the new apparatus

rests. A transducer fixed on the head of the carriage for measuring in Y direction. The second part is a control–operation unit with a software system. The block diagram is shown in Fig. 2.

MEASURING

In the period of inspection, the probe moves along the X direction on the guide which forms a reference line and then scans on the actual profile starting at point A and goes as far as point T. On passing through the stored values X_i, the relevant measuring signals P_i generated by the transducer are detected and immediately stored. The relevant deviation values E_i are given by the formula:

$$E_i = P_i - Y_i - V_i,$$

where Y_i — difference values when $X = X_i$, V_i — systematic errors. After testing, the deviation values E_i are calculated, plotted and printed out.

MEASURING RESULTS AND COMPARISON

We have got profiles of a large involute spur gear tested. A series of deviation curves (Fig. 4) are compared with those (Fig. 5) taken down by PH – 100 Gear Measuring Machine whose involute master gauge has been checked by the Chinese Academy of Metrology. The accordance is satisfactory, showing that the measuring system is feasible and fairly available.

CONCLUSIONS

1. The accuracy and reproducibility of the apparatus have been improved owing to the use of new measuring principle. The measuring process can be finished in a simple and reliable way thanks to the optimum combination of the idea of the error compensation and correction with the software technique.

2. A reference base can be set up more easily by a precise guide than by a theoretical involute. It may have a high straightness. In addition, we need only a small measuring range transducer to obtain the measuring signals. As a result these features will make the mechanical structure of the apparatus

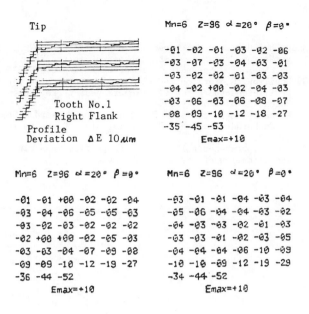

Tip

Mn=6 Z=96 α=20° β=0°

```
-01 -02 -01 -03 -02 -06
-03 -07 -03 -04 -03 -01
-03 -02 -02 -01 -03 -03
-04 -02 +00 -02 -04 -03
-03 -06 -03 -06 -08 -07
-08 -09 -10 -12 -18 -27
-35 -45 -53
        Emax=+10
```

Tooth No.1
Right Flank
Profile
Deviation ΔE 10 μm

Mn=6 Z=96 α=20° β=0°

```
-01 -01 +00 -02 -02 -04
-03 -04 -06 -05 -05 -03
-03 -02 -03 -02 -02 -02
-02 +00 +00 -02 -05 -03
-03 -03 -04 -07 -09 -08
-09 -09 -10 -12 -19 -27
-36 -44 -52
      Emax=+10
```

Mn=6 Z=96 α=20° β=0°

```
-03 -01 -01 -04 -03 -04
-05 -06 -04 -04 -03 -02
-04 -03 -03 -02 -01 -03
-03 -03 -01 -02 -03 -05
-04 -04 -04 -06 -10 -09
-10 -10 -09 -12 -19 -29
-34 -44 -52
      Emax=+10
```

Fig. 4. Measuring results

very simple and the cost of manufacture quite low.

3. The positioning method via ball rests contacting with tooth spaces is helpful to overcoming the structural limitations of the gear measuring machines and enlarging the measuring range of the gears in size and weight to the greatest extent.

4. The transportable measuring system is suitable for inspection of tooth profiles on the gear production machine. Based on the measured profile characteristics, corrective information can be fed back directly during the machining process.

5. A high economic efficiency can be achieved because of the short operating time, avoidance of the machine standstill and the simple construction of the apparatus. This system is particularly fit to inspect especially big and medium-sized large gears.

Tip

Tooth No.1
Right Flank

Profile
Deviation Δf_f 10 μm Fig. 5 Profile deviation curves compared

REFERENCES

1. Maag-Zahnräder und Maschinen Aktiengesellschaft, Verfahren und Prüf-gerät zum Prüfen des Zahnflankenprofils von Zahnrädern grossen Durchmessers, Europäische Patentanmeldung 0 019 075 A1 (1980).
2. G. Bouillon, G. Tordion and G. Tremblay, Apparatus for testing the profile and the pitch of involute gear teeth, United States Patent 3 757 425 (1973).
3. R. Wiechern, Meßtechnische Bestimmung der Werkstück-Geometrie von Großzahnrädern auf der Verzahnmaschine, Dissertation, TH Aachen (1980).
4. Mobile computer-controlled involute profile measuring system ES-430, specification of the factory Maag.

MICROCOMPUTER CONTROLLED POSITION

SENSITIVE MEASURING SYSTEM

Holker Schott ; Heiner Herberg

Academy of sciences of the GDR
Central Institute of Nuclear Research Rossendorf
DDR-8051 Dresden , German Democratic Republic

1. INTRODUCTION

In the last years the development of microelectronics has resulted in decreasing costs of integrated circuits. Therefore, the possibility is given to construct electronic systems with more intelligent character. This development is hindered, if on the hand the high intelligent information processing systems are available and on the other hand no sensors to detect nonelectrical input signals exist with acceptable performance price ratio. The presented position-sensitive measuring-system (PSS) permits to move samples from one position to another. The actual position coordinates are determined and stored. Every registrated position (x,y) of the sample can be adjusted again. Several manipulations of the sample can be carried out, which are implied in the software. The computer controlled measuring system is a so called "learnable system", i.e. at the beginning the system stores a given moving process, than it can be corrected and repeated every time. An optical control of the sample movement is possible by a stereo microscope. The cross table is driven by step motors. It moves the sample in the x-or y-direction with an accuracy better then 10 um. The whole data collection is carried out by a microcomputer system. The operation process of the system is subdividend into two cycles. The first cycle comprehends the registration of the coordinates, the analog to digital covertion and the following storing. In the second cycle, the reproducting cycle, the step motors are controlled by the computer in order to approach a sample position which was selected by the operator. The movement of the sample can be observed on a graphical display and the actual position coordinates are represented simultaneously. The languages used for the software are ASSEMBLER and BASIC.

As the mainpart of the PSS a hybridintegrated optical sensor circuit (HIPPD) for position measurement is used. It determines the centre position of a light spot /1/. The position measurement is realized using a full-area two-axis position sensitive photodiode (PPD), which is characterized by a high linearity in the correlation between displacement of a light beam and the change of the electrical output signal. Figure 1 shows the complete principle block diagram of the unit PSS.

2. COMPONENTS OF THE PSS

2.1 Description of the hybrid integrated silicon sensor for position measurement (HIPPD)

2.1.1. Sensitive element

In the hybrid circuit a full-area two-axis position sensitive photodiode is used. The electrical output signal of the PPD gives an information about the position (x,y) and the intensity Po of a light spot on the sensitive area of the PPD. The coordinates (x,y) represent the intensity centre of the light beams, there are no special demands on homogeneity or geometric form of the light spot. The position measurement is realized by the electrical subdivision of the photo current generated by the light beam to the lateral contacts. The four partial photo currents are the output signals of the PPD and they are further processed by the analogous electronics. Standard silicon was used to produce the PPD. The chip dimensions of the PPD are 14mm x 14mm. The active area of 10mm x 10mm is located inside of the four lateral contacts. The PPD is light sensitive in the visible range of the spectrum. The maximum of the spectral sensitivity is situated in the wavelenght region between 750 and 900 nm ; the essential quality criterion of the PPD is the linearity between the optical input signal and the electrical output signal. The linearity error of the duolateral PPD is better than 1% over the full area.

2.1.2. Analogous electrical circuit

Figure 2 shows the principle circuit diagramm of the hybrid integrated sensor (HIPPD). The lateral partial photo currents are converted to voltages. Two difference amplifieres form the output signals $U(x) = U(x)1 - U(x)2$ and $U(y) = U(y)1 - U(y)2$. In general this signals depend on the light intensity Po. By an external calculation of the quotients $U(x)/$ U and $U(y)/$ U we can get the output signals independent on the light beam intensity. The possibility of the zero point variation for both channels is given by two direct voltage sources. The complete hybrid integrated position circuit is situated in a 32-pin hermetically sealed metal glas housing. The voltage supplies are +-15V /2/.

2.2. Description of the control equipment

2.2.1. Microcomputer system

The modular 8-bit microcomputer system MPS 4944 (produced by Central Institute of Nuclear Research Rossendorf) represents the main part of equipment. It based on the processor Z 80. This system consists of: basis modules with the CPU, dataway display, PROM and RAM, I/O-boards, alpha-numeric keyboard with display and board for external memory. For process handling it is necessary to sample the coordinate datas with analogous multiplexer and an analogous - digital converter and to give the central information to the actuators I/O-unit, in form of number of steps.

2.2.2. Positioning system

The mechanical equipment includes a stereo microscope for visual control of the object movement and a cross table with the fixed object. The cross table is driven by two step motors in x-and y-direction with an accuracy of 5...10 um. A light source is fixed on the cross table and directs a light beam on the sensitive area of the HIPPD. The step motors are driven by power units which are also hybrid integrated circuits, followed by power transistors. The information about the number of steps and directions is provided by the microcomputer.

2.3. Software

2.3.1. Basis software

The microcomputer system MPS 4944 implies two basis software systems for this application : the standard monitor for starting, handling and control functions and the BASIC interpreter for treating arithmetical functions and datas.

2.3.2. Software for the positioning system

In the software are combined the advantages of ASSEMBLER and BASIC programmes. Therefore, the leader programme is written in BASIC and the handling programme in ASSEMBLER. The connection is realized very simple by the commands"CALL" or "SCALL". In the "normal position" mode the system is started by man-computer dialog to define parameters (e.g. accuracy, number of points etc.). The second step is the calibration of the system and after that the system is ready to carry out the treatment of the specimen on the cross table (e.g. bonding an integrated circuit). The position system is also able to work in the "teach in" mode. The movement of the specimen is started by the help of the keyboard functions. It is possible to select the coordinates x,y and the direction, than the step motors drive the cross table. The registration is also started by the keyboard. The sampling data are registred in a memory field. This cycle is repeated for every given point. The point registration or repeat mode operates as follows. After definition of the position error limit the microcomputer controls the step motors and compares the actual data with the coordinates of the stored points. At the aimed position manipulation functions are possible (e.g. bonding of the circuit). Than the next point will be adjusted. These manipulations may be carried out at all points of an area 10mm x 10mm .

3. SUMMARY

The presented PSS has an absolute reproducibility of about 5...10um. The maximum number of stored points in the memory is about 1000. The described PSS represents only one example of application for a computer controlled system of two dimensional measurement of movement in a plane. Essential applications of such a system are e.g. robotics and position sensing in production equipments, machine tool alignment, angle sensing, three dimensional position measurement etc .

4. REFERENCES

1. Schmidt ,B. ; Schott ,H.:
   Proc. 11th Intern. Symp. Photon Detectors , Weimar 1984
2. Schmidt ,B. ; Schott ,H. ,Just ,H.-J.:
   Hybridintegrated Silicon Sensor for Position Measurement.
   Proc.10th IMEKO Congress , Praha 1985

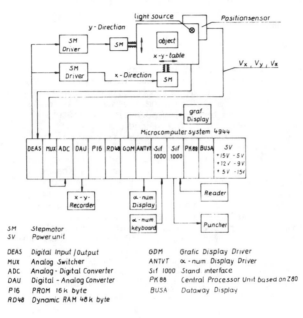

| SM | Stepmotor |
| SV | Power unit |

DEAS	Digital Input /Output		GDM	Grafic Display Driver
MUX	Analog Switcher		ANTVT	α-num Display Driver
ADC	Analog-Digital Converter		Sif 1000	Stand interface
DAU	Digital - Analog Converter		PK 88	Central Processor Unit based on Z80
P16	PROM 16 k byte		BUSA	Dataway Display
RD48	Dynamic RAM 48 k byte			

Figure 1   Complete principle block diagram of the unit PSS

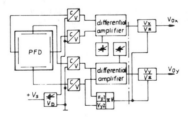

Figure 2   Principle circuit diagram of hybridintegrated sensor HIPPD

AUTOMATED PROCESSING AND DISPLAY

OF NAVIGATION MEASUREMENT SIGNALS

Herbert Strickert and Peter Hoffmann

Maritime Academy Warnemünde/Wustrow
GDR

INTRODUCTION

For ship's tracking, measured values out of different navigation systems (e. g. dead-reckoning navigation system consisting compass and log, radio navigation system with hyperbolic lines of position such as DECCA, LORAN-C, OMEGA, Transit
satellite navigation system) are to be processed in such a way
that in certain time intervals the marine navigator obtains
actual measurement information on important state values, for
instance on position and speed of vessel. Time interval between
actualizing the information as well as permitted range of measurement uncertainty depend upon respective navigational situation.

In many cases the marine navigator is not any longer able
to guarantee a safe ship's guidance by means of traditional sea-
chart operating. Here information oversupply and time factor are
of importance.

By methods of optimum statistic processing of measured
values algorithms can be defined which result in reducing the
measurement uncertainty by software-means. At this, by use of
processor-near microcomputers one achieves computation dura-
tions, which make possible intervals of actualizing down to
one minute. At the same time the subjective weighting of measured values, got out of the different navigation systems,
will be replaced by a statistic weighting under use of a-priori-
information.

PROCEDURE OF SIGNAL PROCESSING

An efficient procedure for getting the required measurement information consists in the following: In each case at
the forthcoming measurement instant k a state value $\underline{x}^*$ (k)
(position, speed) is forecasted on the basis of preceded measurements as well as by the equations of motion of the measurement object. In general this forecasted value does not coincide
with the correct value because of system disturbances (influence upon state parameters due to sea, wind, current) and

since the preceded measurement values were not error-free.
After coming in of the actual measured values $\underline{y}(k)$ out of the
cooperating navigational measurement systems, which due to unavoidable measurement disturbances also differ from the correct
value, forecasted value and actual measured values are to be
combined to a plausible estimate $\hat{\underline{x}}(k)$ of the instant state
(v. fig. 1). At this it is to be decided on the weights which
are to be given to forecasted and actually measured values.
Under linearizable measurement equation the estimate can quite
generally be determined after eq. (1), in which the weights are
established by the choice of the gain matrix $\underline{K}$

$$\hat{\underline{x}}(k) = \underline{x}^*(k) + \underline{K}(k) \; \left[\underline{y}(k) - \underline{C}(k) \; . \; \underline{x}^*(k)\right] \tag{1}$$

with $\underline{C}$ as measurement matrix. Usefully $\underline{K}$ is to be chosen to be
relatively large if the intensity of system disturbances is
large in comparison with measurement noise. That is to say in
such case the extrapolated estimate is relatively uncertain,
and the difference between real and fictitious measured values
is with large weight to be applied to correction. In borderline
case $\underline{K}$ is equal to $(\underline{C}^T \cdot \underline{R}^{-1} \; . \underline{C})^{-1} \cdot \underline{C}^T \cdot \underline{R}^{-1}$ with $\underline{R}$ as covariance matrix
of measurement noises concerned (MARKOV filtering[1]). Reversely
$\underline{K}$ has to be relatively small if the measurement noise intensity
is large compared to system disturbances. Now the extrapolated
value is relatively accurate, and the difference between real
and fictitious measured values is only allowed to enter into
correction with small weight. In borderline case $\underline{K}$ is equal
to $\underline{0}$ (pure dead reckoning). Within the two borderline cases a
useful decision on weights for extrapolated and actually measured values can be obtained by KALMAN filtering if the motion
equations of measurement object are linearizable and if the
statistical properties of the random fluctuations in the system
state as well as in the measurement signals can be characterized
by the covariance matrixes $\underline{E}\{\underline{w}(k) \cdot \underline{w}^T(l)\}$ and $\underline{E}\{\underline{v}(k) \cdot \underline{v}^T(l)\}$,
resp.

The used recurrent estimation algorithm proves to be
favourable to processing the navigational measurements in real
time because not any measured values belonging to the past
have to be stored. All information on system past is kept in
the last estimate, in each case. This preference is significant
because on long-time voyages dates in large quantities would
result which could not be stored in ship-borne computers.

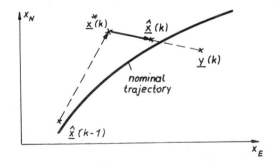

Fig. 1   Procedure of signal processing

The strategy taken as basis for eq. (1) can likewise be applied to running determination of the error of particular measurement systems out of the navigation set. To this the redundant measured values are processed according to the scheme after fig. 2: An autonomous base system (dead-reckoning system) with low-frequency error spectrum is combined with an external support system (radio navigation system) of wide-band error spectrum. The difference of the navigational parameters from base and support systems is available as measurement vector $\underline{y}(k) = \underline{e}_S(k) - \underline{e}_B(k)$ at the filter input. For the measurement error of the base system (dead-reckoning error) an estimate $\hat{\underline{x}}(k) = \hat{\underline{e}}_B(k)$ is determined at the filter output. Such an integration of single measurement systems is useful in cases where KALMAN filtering for the system state itself is not possible (non-linear system dynamic, lacking a-priori information on random fluctuations in system state). The estimation error of KALMAN filter in the integrated measurement system is simultaneously the error of this system. Therefore the covariance matrix of the estimation error is equal to the error-covariance matrix of the integrated measurement system after

$$\widetilde{\underline{P}}(k) = \underline{P}^*(k) - \underline{K}(k) \, . \, \underline{C}(k) \, . \, \underline{P}^*(k), \tag{2}$$

where $\underline{P}^*$ is the covariance matrix of the prediction error $\underline{x}^* - \underline{x}$.

The matrix $\widetilde{\underline{P}}$ delivers the n-dimensional probability-density distribution of state-estimation error as (n-number of state components)

$$p(\underline{x}) = (2\pi)^{-\frac{n}{2}} \, . (\det \widetilde{\underline{P}})^{-\frac{1}{2}} \, . \, \exp\left[-\frac{1}{2}(\hat{\underline{x}} - \underline{x})^T . \widetilde{\underline{P}}^{-1} \, . \, (\hat{\underline{x}} - \underline{x})\right] \tag{3}$$

The loci of constant probability density are always hyper-ellipsoids because the covariance matrix $\widetilde{\underline{P}}$ is positively semi-definite. From eq. (3), among others, the error figure of position estimation as an important navigational decision criterion can be derived: The twodimensional density distribution of the position is a marginal distribution regarding the n-dimensional random variable $\underline{x}$. For instance, if the state vector consists of two position components and two speed components, $\underline{x} = (x_N x_E v_N v_E)^T$, then eq. (3) delivers the joint-density distribution of the two position components according to

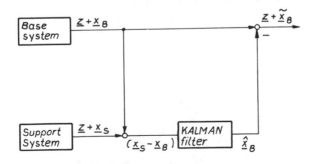

Fig. 2 Principle of system integration

$$p(x_N, \ x_E) = \int_{-\infty}^{+\infty} \int_{-\infty}^{+\infty} p(x_N, \ x_E, \ v_N, \ v_E) \ dv_N \ dv_E \qquad (4)$$

Then the design elements a(k), b(k) and Θ (k) of the well-known error ellipse can be determined[2], v. fig. 3.

EXAMPLE

Fig. 4 shows a typical navigational problem as well as the approach to defining the state model if a dead-reckoning system and a radio navigation system are used.

The desired ship's track may be given. According to the nominal course steering the dead-reckoning track r_d is equal to the nominal track. Because of the current δv the ship drifts. This current cannot be measured with log for determining the ship's speed through the water.

On an average the ship travels the drift track (dotted line). Because of environmental influences like wind and sea the true ship's track r_t fluctuates with w_d regarding the mean drift track. The radio navigation measurements with noise w_r are available as external measurement information r_r.

The state model will be designed in such a way that successively estimates are determined for the ship's drift in north- and east-directions, for components of current as well as for the fluctuations of true ship's track regarding the drift track.

If after run of a number of filter cycles a sufficient accuracy is obtained, then the drift angle will be computed from the state vector. This drift angle is used for correcting the steering of direction, so that after it the drift track coincides with the nominal track, approximately.

By suitable choice of sampling interval one obtains with $\hat{\underline{e}}_B(k)$ a quasi-continuous estimation of the dead-reckoning error. Therewith in addition to the current position indication a predicate is possible on the vector of the drift current, which cannot be detected sensorially.

Realizing an integrated navigational measurement system is done by use of a microcomputer of suitable performance characteristics

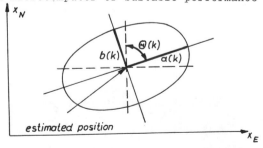

Fig. 3 Design of error ellipse

in on-line operation. State models as well as a-priori-information on random fluctuations structured accordingly are stored in the computer. Proceeding from start values fed by keyboard a complete cycle of KALMAN's filter algorithm will be performed each time when the actual measured values come in. In this connection transformations of measured and estimated values into hyperbolic and MERCATOR coordinates, respectively, are to be performed befor and after the filtering. Some problems of practical realizing an integrated measurement system are commented upon more detailed in other place[3].

At use of such measuring system in navigational practice there is often a compromise to be made between computation expenditure and estimation accuracy. For instance, the state model of six components, shown in fig. 4, is a simlification compared to the reality: The wide-band measurement noise of the radio navigation system was thought to be white noise, approximately. A more realistic model should consider the autocorrelation of the process, described by an correlation function of second order. Then further shaping filter were to be introduced and the state vector would increase. Because of inversion of matrices the expenditure in the case of complicated models could increase in such a way, that real-time requirements are not any longer to be accomplished. By simulation investigations it is previously to be cleared up, if sub-optimum models will yield sufficiently accurate estimates.

DISPLAY OF INFORMATION

The drawbacks of traditional navigation practice - connected with extensive manual chart operating - can be overcome by up-to-date possibilities of electronic information display. Especially for control of navigation process instead of conventional paper sea-chart an electronic sea-chart can be used which is displayed on a oolour CRT-screen in MERCATOR's projection with appropriate mapping scale, fig. 5.

Besides the geographic grid and coast lines further important navigational information can be faded in, for instance traffic separation areas, sea marks, shallows, planned and travelled ship's track. Decision relevant information, like

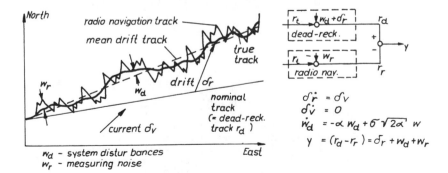

Fig. 4 Model of integrated dead-reckoning/
radio navigation system

date, time longitude and latitude, speed over ground, current,
next course-change position and so on can digitally be indicated
on a special screen field. The error ellipse will successively
be computed in the ship-borne computer and can be superimposed
to the actual ship's position as well as displayed on the screen
by call. Orientation and sizes of error figure change on the
ships way permanently. Thus, the marine navigator can clearly
recognice the direction of largest measurement uncertainty.

By means of appropriate control elements nautical-chart
sections can be either turned over page by page or shifted
within a large-scale chart. The mapping scale is variable in
discrete steps so that more or less detail information can be
represented if required.

Through this automated decision aid the marine navigator
will be relieved of routine work so that he can concentrate
his efficiency fully upon tracking process especially in critical navigational situations such as sailing in closely navi-
gated sea areas, in areas with shallows and obstacles, or
sailing with zero visibility, furthermore in positioning operations. Safety and effectiveness of the tracking process will
be essentially increased by it:

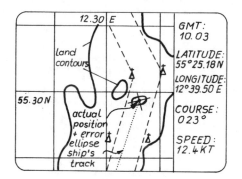

Fig. 5 Display of navigational information

REFERENCES

1. H. Strickert, Messen, Beobachten, Schätzen,
   Wiss. Beiträge der IH für Seefahrt Warnemünde/
   Wustrow, 8 (1981) 2, 81-103
2. P. Hoffmann, Untersuchungen zur optimalen Verar-
   beitung von stochastisch gestörten Navigations-
   meßwerten durch KALMAN-Filterung, Dissertation A,
   IH für Seefahrt Warnemünde/Wustrow (1979)
3. H. Strickert und P. Hoffmann, KALMAN-Filter in inte-
   grierten Meßsystemen, messen,steuern,regeln,
   Berlin, 27 (1984) 10, 453-456

POSITION FINDING IN MINES WITH ELF WAVES FOR ARBITRARILY

ORIENTED TRANSMITTERS

Norbert H. Nessler

Institute for Experimental Physics
University of Innsbruck
Innsbruck, Austria

INTRODUCTION

In a mine disaster it often occurs that miners can survive the first explosion, but are captured somewhere in the mine with no possibility to make their position known to the rescue team, since all standard communication systems are interrupted and passages blocked. It is of vital importance to locate them as quickly as possible by means of independent methods of position finding.

Several location techniques have been published, most of them use a vertically oriented source dipole (loop antenna horizontal) and require a fan-shaped search on a flat surface /1,2,3,4,5/. The disadvantage of these methods is the necessity of a free, flat surface and the source being powerful enough to penetrate the overburden. A laterally inhomogeneous overburden may cause a significant shift of the "zero point" and the field becomes elliptically polarized /8/. An inclined surface is taken into account in /2/. The relative phase of the received signal in two points of reception is used in /9,10/ to calculate distance data from the propagation delay. Only in /11/ field direction measurements are used to find lost boreholes of vertical direction inside the mine (this method is a special case of the one described in this paper). A sophisticated method published in /7/ is based on one three-axis underground sensor and several powerful three-axis sources on the surface to detect the position of the former.

However, none of the cited methods are applicable to locate an arbitrarily oriented source dipole with both source and sensor inside the mine. The method described here has no restriction or prerequisite regarding the orientation of the source dipole. Using the field structure that is calculable in any point it is possible to "recalculate" the source position from a limited number of field direction measurements rather than performing a fan-shaped search for a certain field direction or a field strength maximum. The points of reception may be located at fixed points inside the mine so that there is no restriction regarding the overburden.

THEORY AND LIMITS OF APPLICATION

For signals in such low frequency ranges as are necessary for underground location magnetic antennas only provide the required efficiency and are still small enough to be portable /15/. Thus only the magnetic components of the electromagnetic field are required which are calculated from basic principles /12,13,14/. In spherical coordinates the field components are

$$H_r = \frac{m}{2\pi} \cdot \frac{\cos\theta}{r^3} \cdot (1 + ikr) \cdot e^{-ikr} \tag{1}$$

$$H_\theta = \frac{m}{4\pi} \cdot \frac{\sin\theta}{r^3} \cdot (1 + ikr + k^2 r^2) \cdot e^{-ikr} \tag{2}$$

$$H_\phi = 0 \tag{3}$$

with $m = n.I.A$ ... magnetic moment

and $k^2 = i.\omega\mu\sigma + \omega^2\varepsilon\mu$ .... complex wave number $\tag{4}$

with the complex wave number containing all the medium parameters. The first term describes the conduction current, the second one the displacement current. The latter is negligible for the conditions found in underground location problems.

It is evident that any medium influence on the field structure should be avoided, which is equivalent to the condition

$$kr \ll 1. \tag{5}$$

Since the distance range, r, is fixed by the location problem and the rock conductivity, σ, cannot be varied, the only independent variable is the signal frequency, $\omega = 2\pi f$. To find the upper frequency limit the medium influence has to be calculated. (The lower limit for f is conditioned by technical problems like noise and interference with mains harmonics). For a more general description $x = |k.r|$ is used as a "distance" measure (standardized distance).

The attenuation of the field strength in equs. 1 and 2 can be divided into a "purely geometric" attenuation (r^{-3}, θ) summarized as

$$G_1 = \frac{m}{2\pi} \cdot \frac{\cos\theta}{r^3} \qquad \text{(for } H_r) \tag{6}$$

$$G_2 = \frac{m}{4\pi} \cdot \frac{\sin\theta}{r^3} \qquad \text{(for } H_\theta) \tag{7}$$

and the influence of the conducting medium. The field components equs. 1 and 2 then read:

$$H_r = G_1 \cdot \underbrace{\sqrt{1 + 2x + 2x^2}}_{\text{magnitude}} \cdot e^{-x} \cdot \underbrace{e^{i(\omega t + p_1 - x)}}_{\text{phase}} \tag{8}$$

$$H_\theta = G_2 \cdot \underbrace{\sqrt{1 + 2x + 2x^2 + 4x^3 + 4x^4}}_{\text{magnitude}} \cdot e^{-x} \cdot \underbrace{e^{i(\omega t + p_2 - x)}}_{\text{phase}} \tag{9}$$

with the phase shift values $p_1 = \arctan \frac{x}{1+x}$ and $p_2 = \arctan \frac{x + 2x^2}{1 + x}$.

The phase shift is plotted in fig. 1 (showing the minor influence of the displacement current). Since the field components have different phases, the field is elliptically polarized causing an angular deviation of the field direction (direction of maximum field strength) compared to the undisturbed case. The angular error is plotted in fig. 2. Depending on the required measuring accuracy the maximum value for x can be found from fig. 2. For practical use an angular accuracy of less than 1° is required, which corresponds to the accuracy of the bearing head or of the orthogonal antenna system. Even for the worst case of $\theta = 30°$ this can be achieved for $x_{max} = 0.5$. In fig. 3 the relation between distance, r, and frequency, f,

is plotted for the standardized distance, $x = 0.5$, for various conductivity
values. The maximum frequency for a given distance range can be found using
the area below the respective conductivity curve as "allowed range" for negligible medium influence. The electric rock conductivity measured in
Austrian mines is appr. 3.10^{-4} S/m /13,14/, in some US coal mines appr.
3.10^{-3} S/m /9/. Test measurements in mines showed that even iron rails and
tubes have a negligible influence (if source and sensor, resp., have a sufficient distance from them (> 2 m)), since their volume is small compared to
the rock masses.

Under the above mentioned preconditions the field strength components
read as follows:

$$H_r = m/2\pi \cdot \cos\theta / r^3 \tag{10}$$

$$H_\theta = m/4\pi \cdot \sin\theta / r^3 \tag{11}$$

$$H_\phi = 0. \tag{12}$$

The field strength measured with a sensor antenna (e.g. a loop or ferrite
rod antenna) at any point of reception is the vectorial sum of the components equs. 10, 11, 12.

$$H = m/4\pi r^3 \cdot \sqrt{(1 + 3.\cos^2\theta)} \tag{13}$$

The undisturbed dipole field has two main features which can be used for a
location technique: Because of equ. 12 and the symmetry of the problem
there is no tangential field component, H_ϕ, so every field line lies in a
meridional plane of the source dipole. Therefore the direction of the field
strength, H, points into the dipole axis ("condition of intersection"),
but not to the source itself (fig. 4). The field direction relative to the
radius vector source/sensor is given by

$$\tan\psi = H_\theta/H_r = 0.5 \cdot \tan\theta \tag{14}$$

which is independent of the distance ("angular relation").

POSITION FINDING METHOD

Using feature 1 ("condition of intersection") a precision location technique for mining applications can be developed. Every miner has a
small transmitter consisting of an oscillator, operating on his "personal
frequency" (within the frequency range discussed earlier) with his head lamp
battery as a power source. In an emergency a piece of wire is laid out on
the floor to form a loop of as large an area as possible and is connected
to the oscillator. For economic use of the battery power the oscillator may
be on-off-keyed. This forms the source dipole; the orientation is arbitrary
so that no attention must be paid to forming a horizontal loop.

For the search of a lost borehole a solenoid with an outer diameter
small enough to fit into the borehole is used as a source antenna. To compensate for the small area the number of turns is increased. To avoid irregular fields emitted from the feeding lines it is recommended to position the
oscillator circuit inside the solenoid or close to it, so that the feeding
wire carries only the dc-current from the power source.

For the measurement of the field direction in various points of reception ferrite rod antennas are used with a narrow band selective levelmeter.
The direction measurement can be performed either with a single ferrite

antenna pivot-mounted on a precision bearing head /16,17/, whereby the spatial adjustment to the field direction is performed manually, or with an orthogonal antenna system, consisting of three ferrite antennas perpendicular to each other /18/. The latter method has the advantage that the antenna system need not be moved at all; the field direction is calculated from the voltage measurements in the respective antenna. These voltages are proportional to the direction cosines. The calculation with the absolute values contains an eight-fold ambiguity, since the signs of the direction cosines are supressed. This problem is solved by measuring the relative phase between the field components, H_x, H_y, and H_z, resp. Since in a linearly polarized field the phase difference can only be 0° or 180° (in phase or antiphase), a complex phase measurement requiring three exactly matched narrow-band amplifiers and a phasemeter can be replaced by a simple analog adder /18/: for in-phase components the measured sum is equal to the algebraic sum of the respective components, antiphase components "sum" up to the difference of the absolute values.

Four such field direction measurements in arbitrarily located points of reception inside (or outside) the mine selected according to the available galleries or caves are sufficient to solve the location problem for the arbitrarily oriented source antenna.

The mathematical calculation is divided into two steps:
1st step: Calculation of the source axis as the common line of intersection for each of the four field directions using the condition of intersection. The calculation is performed by means of Plücker's straight line coordinates. The ambiguity (system of second order equations) can be eliminated by choosing all points of reception on one line or by measuring a fifth direction, calculating another pair of solutions for the source axis and selecting the coinciding ones. This result is sufficient to find a lost borehole; for the location of a trapped miner, however, the exact position of the source is calculated in a
2nd step: The angle between the measured field direction and the source axis (from step 1) is the sum $\delta = \psi + \vartheta$ (cf. fig. 4). From the angular relation, equ. 14, the angle between the radius vector and the source axis is calculated.

$$\tan\theta = -\frac{3}{2.\tan\delta} \pm \sqrt{\frac{9}{4.\tan^2\delta} + 2} \tag{16}$$

The mathematical formalism is remarkably simplified if the direction of the source antenna is known: If the direction of a (lost) borehole is known to be vertical, only two field direction measurements are required to find the location of that borehole. (This is the above mentioned special case.)

SUMMARY

The field distribution of a magnetic dipole imbedded in a weakly conducting medium (rock) depends on the electric rock parameters and on the frequency of the transmitted signal. The frequency can be chosen low enough for the field distribution within a limited distance range being independent of the rock parameters so that the field strength and the field direction at a given point of reception are only a function of geometric variables. Any field direction (direction of maximum field strength) points into the axis of the source dipole, but does not point to the source itself. This feature is used to determine the location of a source antenna by measuring field directions in a limited number of points of reception. In the first step of calculation the source axis is determined, in the second one the position of the source dipole is found /19/.

The method can be used to find trapped mine workers after a mine disaster, to find lost boreholes, or quite generally, to solve location problems in underground engineering.

REFERENCES

1. Geyer R., Keller G., Constraints affecting through-the-earth electro-
    magnetic signalling and location techniques, Radio Science, Vol 11,
    4, pp 323-342 (1976).
2. Olsen R. G., Farstad A. J., Electromagnetic direction finding experi-
    ments for location of trapped miners, IEEE Trans. Geosci. Electron.,
    Vol GE-11, pp 178-185 (1973).
3. Kavels D., Farstad A. J., Trapped miner through the earth communication
    in coal mines, Proc. of the 4th WVU Conf. on Coal Mine Electrotechno-
    logy, 28/1-12 (1978).
4. Durkin J., Electromagnetic detection of trapped miners, IEEE Comm. Mag.
    (USA), Vol 22, 2, pp 37-46 (1984).
5. Guster A., Patterson J., Tunnel searcher location and communication
    equipment, US Patent 3,906.504 (1975).
6. Wait J. R., Locating an oscillating dipole in the earth, Electr. Lett.
    (GB), Vol 8, 16, pp 404-406 (1972).
7. Raab F. H., Quasi-static magnetic-field technique for determining posi-
    tion and orientation, IEEE Trans. Geosci. Remote Sensing, Vol GE-19,
    4, pp 235-243 (1981).
8. Stoyer C. H., Wait J. R., Analysis of source location errors for a mag-
    netic buried dipole in a laterally inhomogeneous conducting earth,
    Pure Appl.Geophys., Vol 114, 1, pp 39-51 (1979).
9. Hopkins W. G., A phase difference of arrival technique for locating
    trapped miners, IEEE southeastcon, Proc., pp 542-6 (1983)
10. Tsao C. K. H., Radio location system for mine-rescue, Asilomar Conf. on
    Circuits, Systems and Computers, Western Periodicals, North Hollywood,
    XIII+738 (1974).
11. Sacks H. K., Electromagnetic technique for locating boreholes, US Dep.
    of Int., Bureau of Mines, TN23.U7 No 8302 (1978).
12. Sinah A. K., Bhattacharya P. K., Vertical magnetic dipole buried inside
    a homogeneous earth, Radio Science, Vol 1 (New Series), 3, pp 379-95
    (1966).
13. Bitterlich W., Theoretical and experimental studies of VLF and LF waves,
    Scientific report, US-Governm. Contr. F44620-71-C-0052 (1971).
14. Nessler N. H., The propagation of LF-waves through homogeneous and in-
    homogeneous solid media, Thesis, University of Innsbruck (1967).
15. Bitterlich W., Magnetische Dipolantennen für Feldstärkemessungen im LF-
    und VLF-Bereich, Int. El. Rundsch., pp 225-8 (1967).
16. Nessler N. H., Position finding with ELF-waves,Proc. of 9th IMEKO
    Congress Berlin, pp 337-344 (1982).
17. Nessler N. H., Seeberger R., Position finding in mines with ELF waves
    for arbitrarily oriented transmitters, 5th IMEKO Symp., Intelligent
    Measurements, Vol 1, pp 232-5, Jena (1986).
18. Nessler N. H., Seeberger R., Automatic position measurement in mines,
    Proc. of the 8th IASTED Symp. on Robotics and Aritificial Intelli-
    gence, Vol 1, pp 293-304, Toulouse (1986).
19. Nessler N. H., Ortungsverfahren, Austrian Patent AT 374 595 (1983).

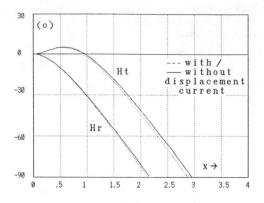

Fig. 1 Phase difference between H_r and
       H_θ caused by the rock influence

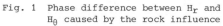

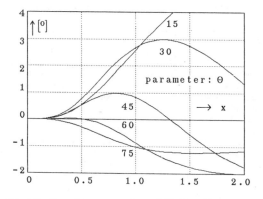

Fig. 2 Angular error for elliptically
       polarized field

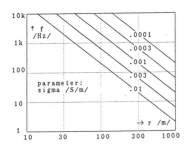

Fig. 3 "Allowable" range for f and
       r below the respective σ-line

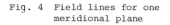

Fig. 4 Field lines for one
       meridional plane

386

ADAPTIVE SELFCALIBRATING RASTER SYSTEMS AND GRATINGS

Ramutis Bansevitchius ,   Vytautas Giniotis

Kaunas Politechnical   Institute of in-service
Institute   training of national
                              economy specialists of
                              Lithuanian SSR,
                              Vilnius, USSR

INTRODUCTION

The present report discusses some aspects of developing
selfcalibrating raster systems and tratins together with the
particularities of their empolyment in the adaptive displacement measurement systems. These gratings may be employed in
the transducers of angular and linear displacement and are
widely used in computer-controlled machines and devices.

ADAPTIVE SELFCALIBRATION

The advantages of gratings in the intelligent measurement
and displacement control devices are disclosed by using piezoelectrical materials as the base of grating where the physical
phenomenon of reverse piezoclectrical effect is empolyed. The
behaviour of the piezoelectrical plate with the grating coated
on its surface is characterized by the following:

$$E_x = \frac{1}{eb_1(x)} \quad b_2(1 - k^2) \, C_1{}^D \quad \frac{du}{dx} \, , \tag{1}$$

where e — piezoelectrical constant, $b_1(x)$— width of activated
piezoelectrical plate, u — applied voltage to the plate, b_2—
general width of the plate, k^2 = eh — coefficient of the
electromechanical link, h — piezoelectrical constant of deformation, $C_1{}^D$— stiffness modulus under constant electrical
displacement.

The pitch error control in the usage of piezoelectrical
materials is expressed by the equation:

$$u(x) = - k_u \delta(x) \, , \tag{2}$$

where u(x)_supplied voltage to the piezoelectrical grating

in the length of coordinate displacement, $\delta(x)$ – the value of systematic pitch error at the same displacement, k_u – transition coefficient.

A case of theusage of raster scale on piezoelectrical material is illustrated in Fig.1.a. Fig.1.b. shows the pitch error of the scale without voltage where δ_{ox} expresses the value of systematic pitch error. The correction is performed employing voltage supply of various magnitude at the defined positions of the coordinate /Fig.1.c./

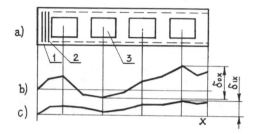

Fig.1. Pitch error correction: 1 – piezoelectrical scale, 2 – raster, 3 – electrodes for voltage supply

The correction of the whole length of raster scale creates great difficulties for improvement of transducer accuracy. It could be achieved by providing the indicating part of raster transducer with the intelligent selfcalibrating system. A case of such technical realization is shown in Fig.2.

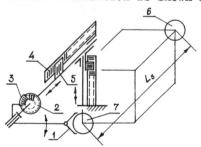

Fig.2. Multidimensional intelligent selfcalibrating system: 1 – sensing grip of the robot arm, 2 – angular transducer, 3 – indicating part of the transducer, 4 – linear transducer, 5 – indicating part, 6,7 – ball bodies

The sensor grip of the robot arm /1/ is located towards the center of the ball body /7/. The similar ball body /6/ is positioned in space at the standard distance L_s. The linear displacement of the movable parts of the robot is measured by means of photoelectrical raster transducers /4/ and by angular transducer /2/ of the same type. The grip of the robot is directed to the point /6/ in space according to the command from the computer. The gauges in the sensible grip indicate the disproportion in the real location in comparison with the standard location of the point /6/. The computer calculates the coefficients of the correction k_x, k_z and $k\varphi$ by the following:

$$k_x = \frac{Lxi}{Lxs} \quad ; \quad k_z = \frac{Lzi}{Lzs} \quad \text{and} \quad k_\varphi = \frac{\varphi_i}{\varphi_s} . \qquad (3)$$

The correctional extension or distraction of the indicating parts of the transducers is equal to $l_i = k_i \cdot \Delta a/a$, where k_i - the correctional coefficient, Δa - the linear or angular length of the indicating part of the transducer and a - the general length of the raster scale.

The computer controlled selfcalibrating system employing piezoelectrical raster scales presents new possibilities in performing accuracy correction and gives adaptive features to such automatic devices, as high accuracy assembling robots, measuring robots and other equipment. This helps to avoid the unnecessary overloading of computer with great quantities of information and enables to perform the calculations more easily.

INTERACTIVE COMPUTER AIDED TESTING CODE

FOR THE IDENTIFICATION OF VIBRATING SYSTEMS

Rinaldo C. Michelini and Giovanni B. Rossi

Istituto di Meccanica Applicata alle Macchine

Università di Genova   Italy

Summary

The opportunities opened by a computer program, purposely developed for the identification of multimass vibrating systems, are presented. The efficiency of the procedure depends on the modularity of the processing op tions, that allows for a systematical sweeping of the observation field, with subsets of homogeneous operations.

INTRODUCTION

The parametrical modelling and identification of the dynamical beha- viour of multimass vibrating systems are extensively performed, in the pre sent day engineering, through the computer aided testing (CAT) facilities offered by several instrumentation manufacturers. The approaches, based on conventional observation schemata, and executed via "canned" programs, de- serve growing interest, specially when applied to the automated diagnosis through the recognition of the plant running situations, by means of the "signature analysis". These standard CAT procedure, actually, give satisfac tory results, whenever "good"conditions of appropriateness are met by the dri ving inputs. Unfortunately the forcing terms (actuations and/or disturban- ces) cannot be mastered in several situations of practical interest. For the industrial diagnosis purposes, the ill-conditioning situations are over ridden, referring to synthetical indices rather than to individual process parameters.

The identification procedure becomes critical, if it represents an in termediate step of an informational loop, introduced to improve the accura- ry of the measurements. The IMIMS program, here presented, is a modular CAT code developed for collecting   "independent" information , useful for the data conditioning of a "principal" measuring chain of a test rig emplo yed in an experiment of mechanical metrology [1]. It is arranged to provide high flexibility, when used interactively, enabling the selection of the mo delling and of the operational environment, with different analysis methods, say: independent channel analysis (mode I); interconnected channels analy-

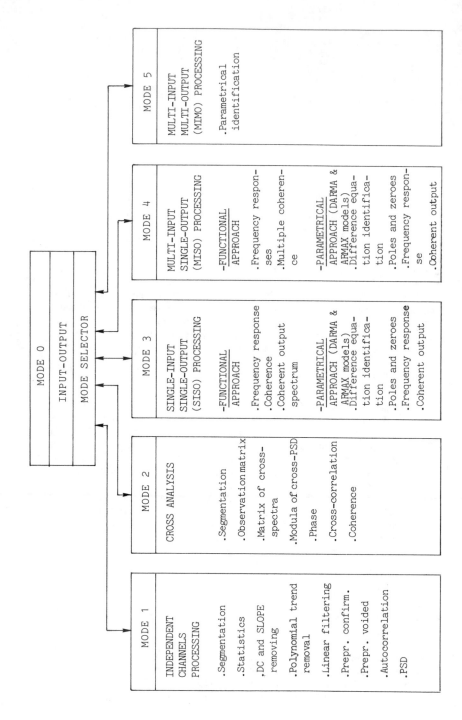

MODE 0
INPUT-OUTPUT
MODE SELECTOR

MODE 1
INDEPENDENT CHANNELS PROCESSING

.Segmentation
.Statistics
.DC and SLOPE removing
.Polynomial trend removal
.Linear filtering
.Prepr. confirm.
.Prepr. voided
.Autocorrelation
.PSD

MODE 2
CROSS ANALYSIS

.Segmentation
.Observation matrix
.Matrix of cross-spectra
.Modula of cross-PSD
.Phase
.Cross-correlation
.Coherence

MODE 3
SINGLE-INPUT SINGLE-OUTPUT (SISO) PROCESSING

-FUNCTIONAL APPROACH
.Frequency response
.Coherence
.Coherent output spectrum

-PARAMETRICAL APPROACH (DARMA & ARMAX models)
.Difference equation identification
.Poles and zeroes
.Frequency response
.Coherent output

MODE 4
MULTI-INPUT SINGLE-OUTPUT (MISO) PROCESSING

-FUNCTIONAL APPROACH
.Frequency responses
.Multiple coherence

-PARAMETRICAL APPROACH (DARMA & ARMAX models)
.Difference equation identification
.Poles and zeroes
.Frequency response
.Coherent output

MODE 5
MULTI-INPUT MULTI-OUTPUT (MIMO) PROCESSING

.Parametrical identification

392

sis with: (mode II) unassigned cross- conditioning, or (modes III, IV & V) previously established conditioning.

PROGRAM RUNNING STRUCTURE

The Fig 1 schematically shows the "horizontal" structure of the program, that allows for the operational environment selection, established on ce the homogeneity between the data processing modes is acknowledged. The classification provides an unifying approach between the "identification" and the "signal analysis" procedures, showing the effects of the assumptions introduced to solve individual "optimality" problems (e.g. the control over the driving inputs). Few comments on the previously defined operational modes introduce   the conceptual foundation references of the program.
The mode I concerns the independent channel processing and includes the traditional "preprocessing" operations (i.e.: local statistics computation, stationarity tests, frequency selective filtering, polynomial trend removal, etc.), with additional options (i.e. the estimation of the autocorrelation, of the PSD, etc.) computed by the special purpose module SACEM^2 for the fea tures extraction from individual signals.
The mode II provides second-order (time and frequency) properties directly computed with previously-unassigned cross-conditioning. In addition data statistical filtering with measurement channels redundancy, is performed for the in-field zero and drift setting .
The modes III, IV and V perform the SISO (single input single output), MISO (multiple input single output) and MIMO (multiple input multiple out put) analyses. Identification is performed via both spectral and parametri cal techniques, for a complete consistency analysis. In the time domain DARMA and ARMA^8 models are assumed as reference and, respectively, least squares and non linear optimization algorithms   are implemented.
The different operational modes interact and the individual outputs (typically the mode I preprocessing operations or the mode II spectral matrix) can be used as input for a different analysis: for that purpose local evolutive data-bases are realised connected with a general management procedure. The connection between the operational modes allows for simple iteration capabilities, avoiding redundancy.

PROGRAM OPERATIONAL CAPABILITIES

For each operational environment, different modelling options are pos sible, built on parametrical or on functional bases; the operational environments are interconnected through shared data-base doubled by local evolutive data-bases.
The modularity expands the IMIMS-possibilities, giving a "vertical" architecture to the program, over different levels, say: (see Fig 2)
. a general governing level, for selecting the operational environment, and the global i/o management;
. the local governing level , for establishing the identification structure and coordinating the options within each individual mode;
. the shared data-base interface level , for presetting the conditioning actions, interacting both with the shared and the local data-bases;
. the processing level , for the identification trimming and the obervation schema performance evaluation.

The processing subroutines are totally independent from the data struc‗
tures and present modularity to be shared by the different operational modes.

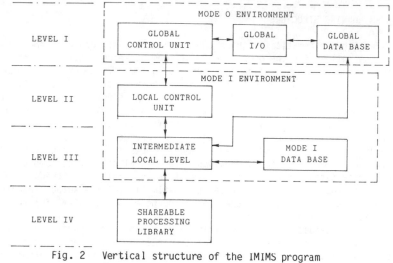

LEVEL I

MODE O ENVIRONMENT

GLOBAL
CONTROL UNIT

GLOBAL
I/O

GLOBAL
DATA BASE

LEVEL II

MODE I ENVIRONMENT

LOCAL CONTROL
UNIT

LEVEL III

INTERMEDIATE
LOCAL LEVEL

MODE I
DATA BASE

LEVEL IV

SHAREABLE
PROCESSING
LIBRARY

Fig. 2   Vertical structure of the IMIMS program
for the environment setting

REFERENCES

1. R.C. Michelini and G.B. Rossi, Computerized test schema for high accu‗
      racy measurements. IMEKO Symp. "Intelligent Measurement - INQUAMESS
      '86", Jena (1986).
2. R.C. Michelini and G.B. Rossi, The PSD measurement of ARMA processes
      correlation by exponential modelling, IASTED Intl. Symp. Applied Si‗
      gnal Processing & Digital Filtering, Paris (1985).
3. Digital Signal Processing Committee (Ed) " Programs for Digital Signal
      Processing", IEEE Press, New York (1979).
4. STI Signal Technology Inc.,"ILS Technical Manual",Galeta Cal. (1983).
5. R. Isermann, Practical Aspects of process Identification, Automatica,
      vol. 16, pp. 575-587 (1980).
6. M.S. Leaning et alii, Validity of Measurements obtained by Model Iden-
      tification and Parameter estimation, Measurement, vol.2, No.4 (1984).
7. L. Ljung and T. Soderstrom,"Theory and practice of recursive identifi‗
      cation", MIT Press, Cambridge, Massachusetts (1983).
8. G.C. Goodwin.and K.S. Sin,"Adaptive filtering, prediction and control",
      Prentice Hall, Englewood.Cliffs, New Jersey (1984).

Note: The IMIMS code has been developed supported with a Italian Ministry
       of Education grant, within the M.P.I. 40% - GNA.22/1206.03230/03240
       research program "Industrial Diagnostic & Reliability" headed by the
       project leader professor P.M. Calderale.

SOFTWARE FOR COMPUTER MEASUREMENT OF DIAGNOSTIC SIGNALS OF NUCLEAR STEAM SUPPLY SYSTEM

Vladimir Grof, Vladimir Hůzl

Power Machinery Plant
ŠKODA Concern Enterprise,
Plzeň, Czechoslovakia

INTRODUCTION

The first Czechoslovak diagnostic system for the VVER-40 nuclear steam supply system /the primary circuit/ is based on vibroacoustic diagnostics. This diagnostic system consists of the following two main parts:
- a subsystem for vibration monitoring /VM/
- a subsystem for loose parts monitoring /LPM/.

The diagnostic signals processing and evaluation is carried out using two sets of instrumentation:
- a spectral analyzer
- a digital computer
There is a data transfer between both devices.

The software for the operation of the whole diagnostic system contains:
- basic system software including the real time disc operating system of the computer
- utility software for the main measurement and storage of the diagnostic signals.

The utility software for determining the actual state of the primary circuit components /that is, software for diagnosis/ is still under development.

A BRIEF DESCRIPTION OF THE DIAGNOSTIC SYSTEM HARDWARE

As mentioned above, the hardware consists of two parts /VM and LPM/. As sensors, absolute and relative inductive displacement sensors and accelerometers are used. Special electronic blocks and cabels create the measuring channels. The computer system, the spectral analyzer and the measuring tape recorder are used for diagnostic signal processing /Fig.1./

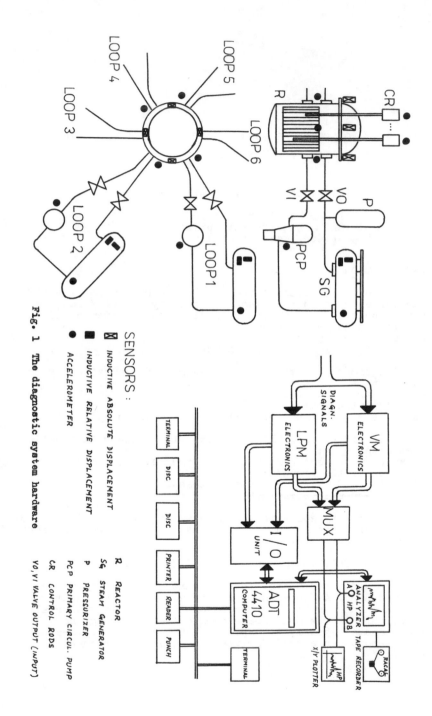

SENSORS :

⊠ INDUCTIVE ABSOLUTE DISPLACEMENT

◼ INDUCTIVE RELATIVE DISPLACEMENT

● ACCELEROMETER

R   REACTOR
SG STEAM GENERATOR
P   PRESSURIZER
PCP PRIMARY CIRCUL. PUMP
CR CONTROL RODS
VO,VI VALVE OUTPUT (INPUT)

Fig. 1 The diagnostic system hardware

396

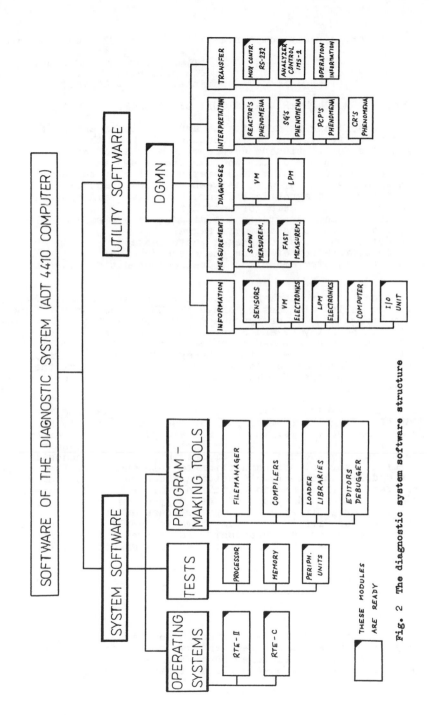

Fig. 2 The diagnostic system software structure

397

Primary circuit compnents: R = reactor, CR = control rods, SG = steam generator, PCP = primary circulating pump, P = pressurizer, VI,VO = inlet and outlet valves

SOFTWARE FOR COMPUTER MEASUREMENT OF DIAGNOSTIC SIGNALS

The software is an integral and very important part of the diagnostic system. It is necessary to keep the following principles for continuous development of the software:
- modular structure of the whole software system
- operation under real time operating system of the computer
- interactive mode of work between user and computer
- dividing the diagnostic software into five main blocks according to the kind of activities:
  - information concerning the diagnostic system
  - measurement of diagnostic signals
  - determination of diagnoses /actual state/
  - interpretation of found phenomena /events/
  - data transfer among individual components of diagnostic system

The software package containing the main control program - the so called diagnostic manager /DGMN/ which controls the other programs - was designed and verified for the first three units of the Dukovany and Bohunice nuclear power plants. This package has been included into the disc operating RTE-II system. The structure of this package can be seen in Fig.2. The needed functions of software can be picked out by the user with the help of DGMN-menus on the display of the computer terminal.

The software is first of all used for measuring the diagnostic signals from outputs of the electronic blocks of VM and LPM subsystems. The signals are processed either by the spectral analyzer or by the computer. When processing the signals by the analyzer, signals are remotely choosable by multiplexer for the connection to this analyzer. Then software modules execute remote setup and control of the analyzer and transfer the measuring data /in time or frequency domain/ from the buffers /display/ of the analyzer to the disc files and vice versa. In case of computer measurement, the signals are connected direct from subsystems to the I/O unit and the signal measurement and processing is carried out by software modules for slow /100 Hz/ or for fast measurement /10 kHz/.

Further, a special measuring and diagnostic program Monitor on the base of RTE-C operating system is used for continuous variables processing and evaluation.

CONCLUSION

The first part of the diagnostic software, which is for diagnostic signal measurement and saving and for the spectral analyzer and the multiplexer control determinded, has been in the first three VVER-440 blocks installed and experiences are collected.

REFERENCES

1. J.Kott, L.Haniger, V.Grof, J.Liska and J.Majer: Automated
   diagnostics of nuclear steam generation unit, _Automatizace_
   5, 27: 120 /1985/
2. V.Grof: Utility software conception of the diagnostic
   system of the nuclear steam supply system _Automatizace_,5
   29: 122 /1986/
3. J.Vitous: ADT 4410 computer program for control of NPP
   diagnostic system funcions, _Report_ Skoda ZES, Plzen /1985/
4. B. Ststná: ADT 4410 computer program "Monitor" /RTE-C/
   for diagnostic system of NPP, _Report_ Skoda ZES, Plzen
   /1984/.

MOBILE MEASUREMENT AND MEMORY UNIT FOR VEHICLE VALUES -

ANALYSIS BY A MICROCOMPUTER SYSTEM

Günther Engler
Christian Bujack

Zwickau University of Engineering
Department of Automotive Engineering Zwickau GDR

INTRODUCTION

The objective for the mobile measurement and memory unit is the design and improvement of electronically controlled combustion engine control systems. Such kinds of engines have a lower fuel consumption and less toxic exhaust gases. Hence they help to urgent problems of the mankind.

For improving electronic control devices in a moving vehicle it is necessary to have some measured values to check its operation. This way it is possible to locate break-downs and to increase the reliability of the whole system. The analysis of these values during the test is impossible, because the driver has to attend the road. That's why it has been decided to use a memory system and to analyse the values after the test.

Analoque memories (e. g. measurement tape recorders) are not capable of handling the high information rate of the digital data from the control unit. Therefore digital memory units are used. The use of solid state memories (RAM) has some disadvantages, and floppy disc memories cannot operate in a moving car because of the mechanical accelerations. That's why it has been decided to apply magnetic tape cassette recorders. They have the best resistance to mechanical loadings and the measurement time is practically unlimited. Digital recorders are common with any kind of computer and hence the analysis is simple when using them. Analogue values must be a/d-converted before being recorded.

PRINCIPLE OF THE MEASUREMENT AND THE MEMORY UNIT

The unit has the following parts (figure 1):
primary electronic circuits for each measurement value (up to 20)
a/d-converter and digital memory for each value
two magnetic tape cassette recorders (ROBOTRON K 5200), which operate sequentially

main control unit
power supply unit (12 V airborne supply system).

Table 1 shows all measured values and the kind of sensors employed. The kind of primary electronic circuit depends on them. For some values the circuit has only to eliminate electromagnetic disturbances (e. g. 2,4). For most of the analogue values, however, an amplifier is required, too. It may be a d. c. amplifier (e. g. 5,19) or a chopper amplifier (e. g. 7,9, 11) or a charge amplifier (e. g. 13, 15, 17). The thermoresistive sensor, a special measurement thermistor, must be supplied with a constant current (400 μA) by the primary electronic circuit, too. The circuits for most of the digital values (e. g. 1, 6, 8, 10, 12, 14) are well known counters, registers and so on. All analogue circuits have an output level from 0 to 5 V which corresponds to the measurement range. All digital circuits have an TTL-output level.

The non-linear characteristic of some sensors is not corrected by the primary electronic circuits, but by the analysis computer. For the thermocouples there is a thermostat (50° C $\pm$ 1 K) which also may be used for calibration of temperature sensors.

The range of the a/d-converter (successive approximation) accounts for 8 bit like that of all digital memories. This corresponds to the common 8-bit-processors which may be used for the analysis (e. g. Z 80, U 880).

Because of the high information rate (20 values in 70 ms) the recorders must run at a high speed (38 cm/s). So already after about 3 minutes one cassette is fully recorded. To reach a longer recording it is possible to have stops after each measurement cycle (0,1 s up to 1 s). Its duration is recorded to have a real time scale for the analysis.

One cassette being recorded the main controller switches the other recorder on. While the second recorder runs the operator (test driver) must change the first cassette.

The main controller (microprocessor control under design) is necessary for all information processing. It stores the recording formate (ECMA 34) which is compatible to the common ROBOTRON-computer which may be used for the analysis. It also checks the recorded data (read after write) and performs the cyclic redundancy check (CRC) necessary to obtain less data errors.

PRINCIPLE OF THE ANALYSIS

For the analysis specific software is required. In general all values are read from cassette and checked by CRC. If there is a data error the computer eliminates it automatically by interpolation. The computer calculates the real values. Therefore the exact values of the characteristics of the calibrated sensors must be integrated into the software. If one sensor must be changed (breakdown) this part of the analysis program must be changed, too.

All values are shown on the monitor and the operator can get a hardcopy (plot or print). It is also possible to have the time-dependence of one value only. By kind of cumulation the computer calculates the total fuel consumption, the total way run during the test and some other values (e. g. maximum speed, maximum fuel consumption). The output by computer-graphic is possible, too.

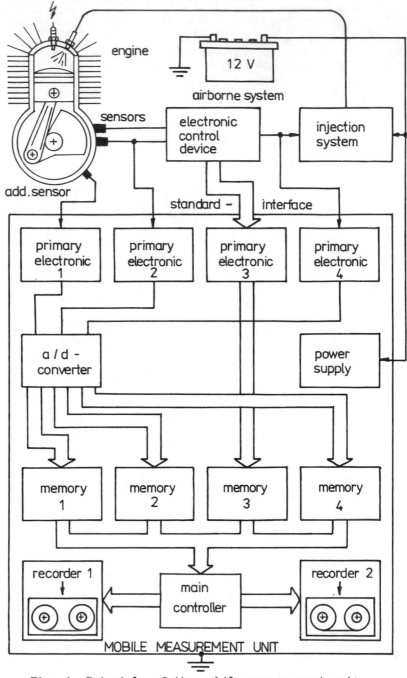

Fig. 1 Principle of the mobile measurement unit

Table 1 Measured values and the kind of sensors
employed

No.	a/d	type of value	kind of sensor
1	d	crankshaft speed	inductive sensor at the starter toothed wheel
2	d	address "crank-shaft speed"	from the electronic engine control device
3	a	sucked-in air temperature	sensor in the intake mani-fold, thermistor
4	d	address "throttle valve position"	from the electronic engine control device
5	a	voltage	attenuator of the air-borne system
7,9	a	cylinder head tem-perature cylinder 1 and 2	thermocouples in the spark plug gasket ring
6,8	d	period of injec-tion cylinder 1 and 2	voltage at the injection pump device
10,12	d	begin of injec-tion cylinder	voltage at the injection pump device
11	a	exhaust gas temperature	thermocouple in the exhaust port
13	a	injection system operating pressure	piezoelectric sensor with charge amplifier
14	d	fuel consumption	flowing through turbine with optoelectronic sensor
15,17	a	injection pressure cylinder 1 and 2	piezoelectric sensor with charge amplifier
16	d	gear speed	inductive sensor in the gear
18	d	stop period	from the unit
19	a	electrical energy consumption of the injection system	measurement resistor in the airborne system

INTELLIGENT REVOLUTION MEASURING DEVICE FOR SPARK IGNITION ENGINES

Vladislav Pajdla, Petr Sládeček

Motor Car Research Institute

Prague, Czechoslovakia

INTRODUCTION

At present the spark ignition engine revolution measurement represents a problem in case when high accuracy, high resolution and relative high response are required and if the input signal is the course of ignition voltage.

The solution of this problem is the application of intelligent measuring device, based on single-chip microcomputer.

MEASURING METHOD

Intelligent measuring device gives the possibility to use advantageous measuring method, consisting in counting pulses derived from the ignition course during the measuring time interval T_m /Fig. 1/ given by a defined number of reference frequency pulses t_m plus time between the last pulse of reference frequency series and the first following pulse derived from the ignition t_w. The length of the measuring interval is put in the microcomputers memory as a number N of reference pulses during that interval.

From both values the microcomputer calculates the average value revolutions per minute during the measuring interval T_m

$$n = C \frac{M}{N} \quad \left[\text{min}^{-1} ; \text{min}^{-1}, 1, 1\right] \qquad / 1 /$$

n   revolutions per minute $\left[\text{min}^{-1}\right]$

M   number of pulses derived from the ignition course during the measuring time interval $[1]$

N   number of pulses of the reference frequency during the measuring time interval $[1]$

C   constant $\left[\text{min}^{-1}\right]$

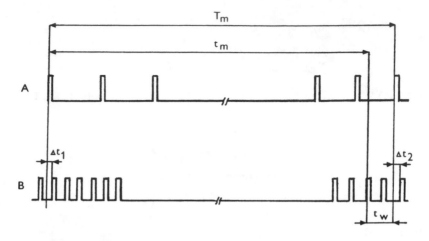

$$T_m = t_m + t_w$$

T_m Measuring   interval

A   Pulse   course   derived   from   ignition

B   Pulse   course   of   reference   frequency

Fig. 1

The static measuring error is given first of all by the magnitude of the time intervals Δt_1 and Δt_2 and the calculation accuracy of the equation /1/.

SYSTEM CONFIGURATION

The principle device scheme is shown in Fig. 2. The device is based on the single-chip microcomputer 8035 with 256 bytes of RWM internal and external memory and 2k bytes of external EPROM memory, which forms a memory circuit 2716. The signal from the ignition is brough on the microcomputer interrupt input via the shaping and monostable circuit SAK 215. The value of the revolution per minute and the auxiliary informations /e.g. error indication/ are displayed on the four digit LED readout /two seven segments two-digit displays VQE 24/. The analog rpm value is on the device voltage output, which form the I/O circuit 8155 and D/A converter with filter and amplifier.

SOFTWARE

The device function is done first of all by software. The principle software diagram is shown in Fig. 3. Just after the

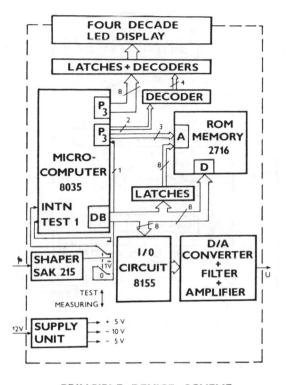

PRINCIPLE DEVICE SCHEME
FIG. 2

system set-up the device performs the initial self-test and
then, if it is choosen, jumps the programm on the full test
execution and in dependence on the result goes the programm
on the error display or it returns to the measuring mode. In
this mode the device first measures and then calculates the
rpm value as it is shown in the measuring method paragraph with
the repetition rate about 0.6 s. After the finishing each measu-
suring and calculating cycle new rpm value is displayed in the
form of the four decade number /0000-9999/ with resolution
1 rpm and converted in output voltage. If the time interval
between two ignition pulses crosses over a certain value adequate to level deep under the minimal working rpm /about 100
rpm/, zero is displayed and the calculation is not carried out.

CONCLUSION

Tests and calibrations proved that the device had necessary
characteristics and the measured error corresponded with theoretical value ± 0.02% rdg + 1 d.

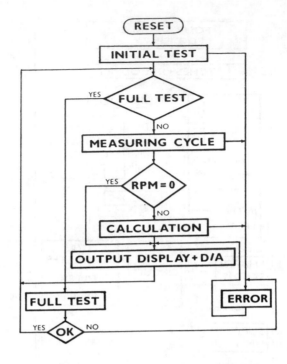

PRINCIPLE SOFTWARE DIAGRAM
FIG.3

PRESSURE MEASUREMENT WITHIN A FIBREBOARD

DURING PULTRUSION

Jan Cieplucha, Slawomir Wieczorkowski

Institute of Fluid Flow Machinery
Technical University of Lodz, Poland

INTRODUCTION

Many processes of the paper industry involve some mechanical separation of water from the cellulose fibres. In
production of the fibreboards the pultrusion process is used
during which the paper pulp is pulled through a convergent
slot where the water is separated. A measurement of the
pressure distribution within the fibreboard during that
pultrusion process has been made to determine optimum separation of the water. Such a measurement appears, however,
difficult because of the small thickness of the fibreboard
/$g = 5$ mm/, its continuous movement and due to the pulp environment.

Following criteria of the pressure measurement have been
established:
- a pressure probe shall be inserted into the required
  position within the fibreboard before the pultrusion
  starts;
- minimum thickness of the pressure probe and the monitoring equipment shall be provided to allow an uniter-
  rupted measurement during the whole pultrusion cycle.
- long conductors between the pressure probe and the
  monitoring equipment shall be provided to allow an
  uninterrupted measurement during the whole pultrusion
  cycle.

To minimize the distortion of the measured pressure due to
the presence of the pressure probe, the criterium of the
minimum thickeness of the probe was taken as a guiding rubber, for the probe design. Thus, an electrically conducting
rubber, for which the specific resistance varied with the
applied strain, was used in first designs. Two versions
of the pressure transducer were constructed:

a/ a pressure transducer with the thickness of $g = 0,4$ mm,
in which the rubber was vulcanized between two flat metal
plates. An applied pressure caused a variation in the
electrical resistance of the rubber layer;

b/ a pressure transducer with the thickness of $g = 0,6$ mm,
in which, by special shaping of the rubber pad, a pressure
applied to the transducer caused a variation of the rubber-

metal contact area.
No satisfactory results were achieved with those transducers.
In spite of a large voltage, the characteristics of the transducers had no repeatibility. The transducer output voltage
varied with the applied pressure to the same order of magnitude
as it did with time.
As results, the final version of the pressure transducer was
built as a two strain gauge system in which foil strain gauges
were used.

DESIGN OF THE PRESSURE PROBE

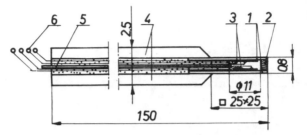

Fig.1.

A cross-section of the pressure probe is shown in Fig.1.
The pressure transducer consists of two membranes /1/ with
foil radial-spiral gauges /3/ attached with a glue. The membranes are distanced with a spacer /2/, and the resulting air
chamber is connected to the surroundings through a small diameter tube /5/. The membranes and the spacer are connected
with epoxy glue and the transducer is mounted to a holder
/4/ enclosing also the electrical connectors /6/. The strain
gauges /3/ form the full Wheatstone bridge with a constant
voltage supply of U = 5 V. The Wheatstone bridge with a constant voltage supply of U = 5 V. The Wheatstone bridge unbalance
signal is measured with the V 540 digital voltmeter or the
Riken-Denshi recorder.

STATIC CALIBRATION

Static calibration of the transducer was performed in a
specially prepared pressure chamber.

The calibration rig, shown schematically in Fig.2., consisted of the actual transducer /1/, voltage supply device,
output signal meter /digital voltmeter/, measurement chamber
/2/ provided with an electric heater /3/, mixer /4/ and a
temperature measurement circuit /a thermocouple with the digital voltmeter VC 2/.
The chamber pressure varied from 0 to 1 MPa, whereby the
outlet of the tube /5/, comp. Fig.1., was kept outside the
pressure chamber /2/.
The temperature of the liquid inside the pressure chamber
varied in the range 20°C to 50°C.
The transducer calibration curve was determined to have the
form:
$$U = 2,611 \ p + 0,006$$

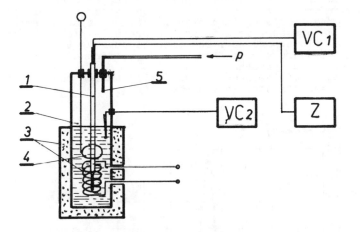

Fig.2.

where   U [mV] – Wheatstone bridge unbalanced voltage
        p[MPa] – calibration pressure

with the temperature zero shift being 0,3 %/10 K and the
temperature sensitivity 0,5 %/10 K.

APPLICATION OF THE PRESSURE PROBE

Measurement of the pressure within the fibreboard during
pultrusion were made on a rig shown schematically in Fig.3..

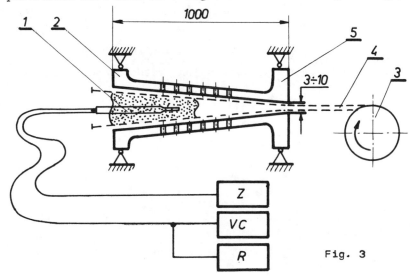

Fig. 3

A primarily shaped fibreboard /2/ is fitted between two pul-
ling wire gauges /4/ which are winded up ton the reeler /3/.
The fibreboard /2/ is thereby pulled along between the two
rigid, perforated plates /5/, where the separation of water
and the press moulding take place. Continuous measurement of
the process pressure is made by the pressure probe /1/ located
within the fibreboard /2/. The time to pull up the full length
of the fibreboard varies from 2 to 20 s, depending on the input material and process regime.
An example of the pressure variation during the fibreboard
by the pressure probe, is shown in Fig.4.

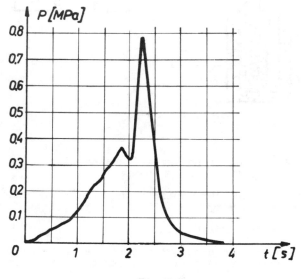

Fig.4.

Analysis of that pressure variation enables an optimum
selection of the convergence of the plates /5/, of the plate
shape, as well as the determination of the number and size
of the water separation holes.

414

AN 8-BIT MICROPROCESSOR AS A BUILT-IN

MEASURING SYSTEM OF A DISPATCH BOARD

Jan Piecha

Electronics Division, Department of Technology
Silesian University in Katowice
Poland

INTRODUCTION AND TECHNOLOGY

A measuring device for installation in the dispatch-board of an extraction level transportation system in a coal mine is described. The fundamental tasks confronting the designer of an intelligent measuring system are presented. The dispatch room of the transportation system is located close to the pit bottom. There are several loading stations where the gotten is loaded to the mine cars. Full cars are sent to the pit bottom where they wait for emptying. Standage room is provided at the pit bottom for empty and full cars. Cars run on single or double tracks with semaphore signaling for control of track occupancy.

This dispatcher is responsible for ensuring: (i) delivery of empty trains to the loading stations at the face, (ii) discharge of full cars at the pit bottom, (iii) manriding to the faces and back, iv reporting output.

The designer of the system must be initially: (i) study the technology of the transportation process, (ii) select suitable indicators and determine their installation points, (iii) simulate the process on a full-size computer system.

From the dispatch board it is possible to monitor the transport situation over the whole extraction level. The display indicates the states of discharge systems, track occupancy, states of track switches, and light signals.

Mine cars have three duties: manriding, haulage of the gotten, materials transport. In conflict situations manriding always takes priority. Transport duty can be considered in three time periods: (i) beginning and end of a day shift, when manriding has priority, (ii) main part of a day shift, (iii) night shift, when materials transport has priority.

INDICATORS

Practical experience showed the need to design very simple and reliable indicators for train location. A block diagram of this kind is given in Fig.1. Train position is indicated by a signal coming from the pantograph pulser, generated by the trolley jumper. The indicator is stimulated by the power line voltage via the pantograph (channel a). Channel b of the indicator is used when it is installed in a shunting area. The puls coming from the pantograph opens the time window for photoelectric pulses. When the number of pulses generated by the photoelectric pulser is smaller then the preset number this means that a single locomotive or/and persons

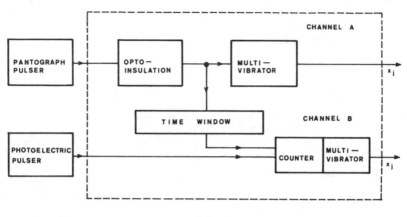

FIG. 1

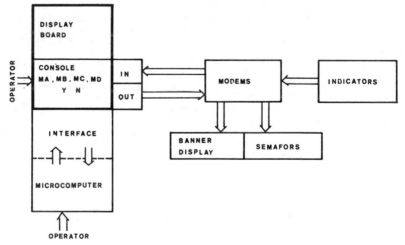

FIG. 2

has passed.

ARCHITECTURE OF THE SYSTEM

The measuring system is part of the dispatch board on which the state
of transportation network is displayed. A block scheme of the system is presented in Fig.2. Signals of track occupancy, states of loading stations and
loading devices are set on a system interface via a long-distance transmission block modems . Signals are collected and demodulated in an analog
interface. There is also a digital interface block enabling organization
of a method of entering a data feed from the board to the microcomputer and
data display on the board.

The measuring system comprizes the following blocks: (i) data transmission block, (ii) digital interface. allowing data formats to be converted to a form more suitable for the microcomputer, (iii) operator keyboard,
used for selecting the data generated by the computer and for choosing system operation mode, (iv) microcomputer, consisting of: 8085 microprocessor,
serial I/O interface - 2x 8251, parallel I/O Interface - 2x 8255, timer-
-8253, master-slave interrupt block - nine 8259 controllers, hexadecimal
and operational keyboard and the 7-segment LED display.

ORGANIZATION OF THE MEASURING PROCESS

The transportation process considered in this paper is described by
about sixty binary input variables. They describe track occupancy, gotten
load of loading stations and states of switches at main points on the tracks.
All this data is displayed on the dispatch board.

The computer has three internal real-time clocks. The main one (T_1)
is used to mark the values of input variables. The second (T_2) is used to
recall the main program of calculations the so-called proposals for the
dispatcher. The third (T_3) selects the time for reporting. The organization
of system operation is shown in Fig.3. A rising edge of any input variable
i registered on one of the sixty-four I_i of the interrupt block. INTR of
the interrupt block recalls the program, allowing the number of the input
signal to be identified and to set its time marker T_1. This data is fed to
temporary states matrix after a defined period of time - T_2 the all contents
of the temporary states matrix are modified as required for calculations.
The T_3 marker selects the time period for reporting the production per shift
or per 24 hours. Program priorities are as in the block diagram in Fig.3.

OPTIMISATION OF PROPOSALS

One of the most difficult problems in such a system is reducing time
required for calculations, since 8-bit microprocessor is unable to cope
with so many tasks.

The transportation process described here was simulated on a full-
-size computer as a multivariable dynamic model. This model allows us to
select from a set of ready-made solutions to the transport problem the one
which best suits the current state as signalled on the dispatch board.
A sequence of proposals is selected from an EPROM memory and shifted to
a computer RAM. The subprograms are linked as shown in Fig.4. The computer
gives the dispatcher feasible proposals for deploying the trains and equipment. The dispatcher has several buttons on the board for mode selection.
Pressing the key MA the mode A is selected and the computer reads only
coming data. No calculations are made, as the computer has lost its contents
mainly after a power-off state. The system then begins to monitor the states of the transportation process. Mode B is selected by pressing key MB.
The computer waits for data to be entered by the operator, using the computer key-board. This mode is used when the system has some faulty information and the operator needs to make corrections. Mode C is selected by
pressing the key MC and can be called the regular work mode of the system.

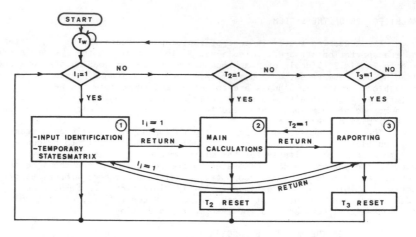

FIG . 3

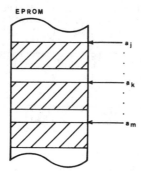

EPROM

$$a_k = f_k (x_{1k}, x_{2k} \ldots, x_{Nk}, a_{k-1})$$
$$f_k \in F_k$$

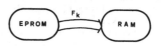

FIG . 4

418

This program offers proposals for the dispatcher. Pushing the button MD
mode D is selected. This mode is used during manriding to and from the face.

RECAPITULATION

    The paper presents only the conceptual scheme of the system, but cer-
tain difficult problems had to be solved. One major problem is organization
of proposals selection from the simulated model of the transportation process
and another is to find an effective way of linking the "ready-made" answers.

REFERENCES

    1. J.Piecha, An 8-bit microprocessor in a programmable control
        logic system. Proc. of 8th International Conference - ICAMC'86.
        Dubrovnik, October 1986. Yugoslav Association of Societies for
        Measurement, Regulation and Automation.
    2. J.Piecha, H.Holon, The algorithm for an empty trains disposition
        on an extracting level of a coal mine. Projekty, Problemy,
        Budownictwo Węglowe Vol.XXIII 1978 Nr 2 257 , GBSiPG Katowice
        (in Polish).

4
ON-LINE MEASUREMENTS AND
QUALITY CONTROL

INTELLIGENT MEASUREMENTS FOR CARS

Diplom-Physiker Alfons Happe

VOLKSWAGEN AG

Research and Development, Division of Metrology
D-3180 Wolfsburg 1, Federal Republic of Germany

To meet high demands for the quality of measurements and for the efficiency
of test procedures computer aided measuring equipment is necessary. Unfortunately, the instrumentation industry does not provide the adequate equipment for the necessary road run tests under severe enviromental conditions.
The paper takes a survey of Volkswagen's own development on the field of
instrumentation for intelligent measurements.

INTRODUCTION

The requirements for the development of cars with regard to environmental
protection, fuel economy, reliability, safety, handling etc. are steadily
growing.

On the other hand, the introduction of CAD and CAM methods has decreased
the time to develop a new car by a considerable amount. However, the number
of test data is increasing and the time necessary to test the cars at different development stages is still too high. It has to be decreased by at
least the same amount by means of computer aided testing (CAT).

ENVIRONMENTAL CONDITIONS FOR CAR MEASUREMENTS

Testing of cars on the road is often done under extreme weather conditions
for instance in the polar region in winter time or in the Death Valley in
Arizona in summer time. Therefore, all components of the test equipment
have to work under severe environmental conditions for instance under temperatures reaching from $-40^{o}C$ to $+85^{o}C$.

Unfortunately, the instrumentation industry does not provide the adequate
equipment for the necessary road run tests under these conditions. Therefore, the automotive companies have to develop their own test equipment.

THE FARES/MEDACS-SYSTEM OF VOLKSWAGEN FOR COMPUTER-AIDED CAR-TESTS

On the one hand we have developed a special multiprocessor system in CMOS-technology which is called MEDACS. Figure 1 shows the MEDACS-system structure. The name MEDACS stands for measuring data acquisition and storage system. It has a small volume and a low power consumption and is adapted to preprocessing and storing the measured data under extreme enviromental conditions. Of course a special programm is necessary for each measuring task, which has to be developed by means of a special μp-development system.

MEDACS System Structure

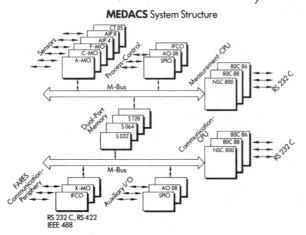

Fig. 1   MEDACS System Structure

On the other hand, we have developed a more powerful computer-system based on commercially available LSI 11-components of Digital Equipment Corporation. This additional computer-system is not based on CMOS-technique and can only be used within temperature ranges from 0° to $+55^{\circ}$C. This system is called FARES.

Figure 2 shows the FARES system structure.

FARES System Structure

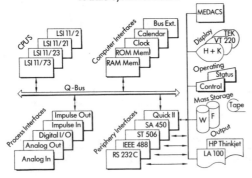

Fig. 2   FARES System Structure

424

The name FARES stands for the German word "Fahrzeugrechnersystem" = car computer system. This system can be programmed on board using high level languages as Fortran or Basic. A combination of these different systems shows Figure 3. A survey of the three typical configurations makes Figure 4. A brief technical description follows.

Intelligent Measurement System **FARES/MEDACS**

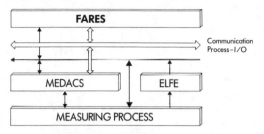

Fig. 3   Intelligent Measurement System FARES/MEDACS

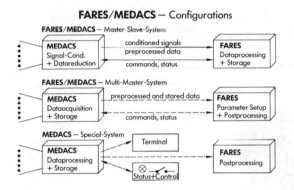

Fig. 4   FARES/MEDACS - Configurations

FARES WITH MEDACS - FRONTEND

Multipurpose measuring system with large storage capacity (Winchester drive). Standard Software FLEX 11 with menue technique affords great convenience with respect to data acquisition, data processing, graphics and documentation for test reports. Applicable examples: vehicle handling tests, brake tests, dynamic top dead center measurements, time performance tests etc.

MEDACS PREPARED VIA FARES FOR A SPECIAL TEST

MEDACS works disconnected from FARES as a self-reliant system for signal conditioning, data acquisition, preprocessing and storing. The stored data were post-processed by FARES immediately after the test. Applicable at ambient temperatures from -40°C to $+85^{\circ}$C. Application example: test of air conditioning systems.

425

MEDACS-BASED SPECIAL MEASURING EQUIPMENT FOR LONG-TERM-TESTS

In this case MEDACS is PROM-programmed for a special measuring task. A very compact housing with integrated solid state memory is typical for this version. Postprocessing of the data is done by FARES for instance.

Applicable at ambient temperatures from -40°C to $+85^{\circ}$C. Application examples: controlling of catalysts in long-term tests, fatigue load measurements in long-term test vehicles.

EXAMPLES OF FARES/MEDACS-APPLICATIONS

MULTICOMPONENT WHEEL DYNAMOMETER WITH FARES/MEDACS INSTRUMENTATION

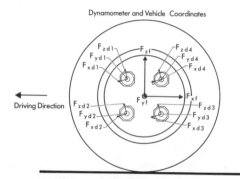

Fig. 5 Output signals of the four 3-component transducers of the wheel dynamometer and the vehicle coordinate system

Fig. 6 Photograph of the multicomponent wheel dynamometer. It can easily be mounted on each standard Volkswagen drive shaft

The multicomponent rotating wheel dynamometer measures directly the three components of all forces and moments acting on a wheel. Four 3-component quartz transducers are used for this task. They deliver together twelve outputs (Fig. 5): The force components $F_{xd1..4}$, $F_{yd1..4}$ and $F_{zd1..4}$.

426

These can be reduced to eight components by summing up all components working in the same direction.

From these eight outputs, six resultant values can be derived:

o the 3 components of the resultant force
o the 3 components of the resultant moment vector,

both relative to the ISO-coordinate-system of the vehicle.

For this the eight force components measured in the rotating wheel coordinate system have to be transformed into the ISO-coordinate system for cars with regard to the momentary angle of wheel rotation. This transformation should be possible with a high resolution of 1 degree of the angle of rotation at a speed of 200 km/h. For the 16 multiplying and the 6 summing operations in this worst case we have 90 μs time. To fulfill this high demand with respect to the computing power we have developed a special coordinate transformation computer with multiplying digital/analog converters. Fig. 7 shows the block diagram of the complete measuring equipment for wheel dynamometers. In this form the system has been tested on a test stand with a built-in not rotating wheel dynamometer. The data of both dynamometers agree very well.

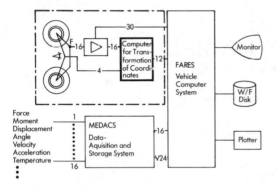

Fig. 7 Block diagram of the complete measuring equipment for wheel dynamometers

The figures 8 - 11 show the results of these tests.

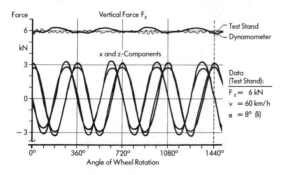

Fig. 8 The normal force f_z measured with a static load of 6 KN

427

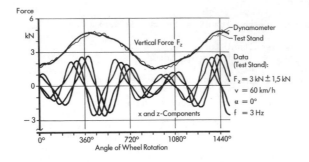

Fig. 9   The normal force f_z measured with a static load of 3 KN and a
superimposed modulation of ± 1.5 KN

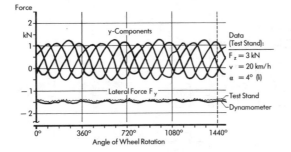

Fig. 10   Lateral force F_y measured at a slip angle $\alpha = 4^o$ and a
speed of $v = 20$ km/h

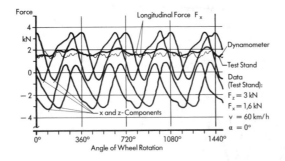

Fig. 11   Longitudinal force F_x measured with a static normal load of
$F_z = 3$ KN at a speed of $v = 60$ km/h

MEDACS Temperature Recorder Type 101 A, Diagram of Structure

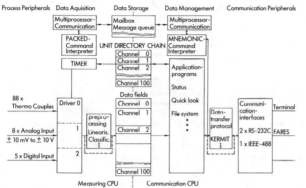

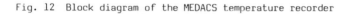

Fig. 12 Block diagram of the MEDACS temperature recorder

Fig. 12 shows a diagram of the software structure of the MEDACS-tempera-
ture-recorder. This system has 88 analog inputs for thermocouples, 8 analog
inputs for signals from $\pm$ 10 V and 5 digital inputs. The parameter setting
of this system is done by FARES via the data transfer protocol KERMIT.
After that the system works self sufficient within the termperature range
from -40°C to $+85^{\circ}$C.

MEDACS-MONITORING RECORDER FOR CATALYSTS IN EXHAUST EMISSION CONTROL
DEVICES

The monitoring recorder for long-term tests of catalysts is a special data
acquisition and storage system based on MEDACS hardware with 16 input channels. Fig. 13 gives a survey of the points of measurement. Fig. 14 shows
how the system works.

Monitoring of Catalysts in Long-Term-Tests.

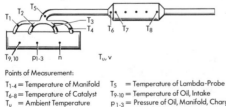

Points of Measurement:

T_{1-4} = Temperature of Manifold T_5 = Temperature of Lambda-Probe
T_{6-8} = Temperature of Catalyst T_{9-10} = Temperature of Oil, Intake
T_u = Ambient Temperature p_{1-3} = Pressure of Oil, Manifold, Charger
n = Engine Speed   v = Driving Speed

Fig. 13 Points of measurement for the longterm test of catalists

429

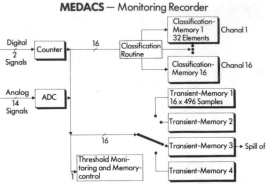

MEDACS — Monitoring Recorder

Fig. 14 Block diagram of the MEDACS monitoring system

The digitized signals of the 14 analog channels and the signals of the 2 digital channels were stored according to two parallel working procedures:

- transient recording
- time at level classification with 32 classes.

At the transient recording mode all data were at first stored in 4 pipeline memories with different trigger levels. The 3 temperatures inside the monolith of the catalyst (T_{6-8}) are used as trigger signals. If the temperature exceeds the trigger level of the currently operating pipeline the content is frozen so that 250 data of each channel before and after the triggerpoint are stored.

The long-term load history during the operation of the vehicle can be monitored by classification of data in respect to level at time into 32 classes.

REFERENCES

1. Boßmann, G. E., Tscheuschner, Chr.: MEDAKS - ein Meßdatenakquisitions- und Speichersystem für extreme Umweltbedingungen.
    VDI-Bericht 553, VDI-Verlag, 4000 Düsseldorf 1,

2. Mäge, B., Reif, P.-D., Walloscheck, P. U.: Koordinatentransformation und Meßdatenverarbeitung beim rotierenden Raddynamometer.
    tm - Technisches Messen, 1985, Heft 12

NEW SYSTEMS FOR OPERATIVE CHECK OF RESONANCE FREQUENCIES OF MACHINE ROTOR BLADES DURING ROTATION

Alexander Bozhko and Alexander Fyodorov

Institute for Problems in Machinery
Ukr. SSR Academy of Sciences
2/10 Pozharsky St.
310046 Kharkov-46, USSR

INTRODUCTION

When designing turbine rotors, compressors, ventilators and other high-speed machines it is important to isolate blades from vibration effects caused by unbalance and other factors. Generally when designing rotor blades of these machines it is necessary that their natural frequencies should be out of the zone of vibratory load frequencies action. Different authors have given priority to theoretical investigations of vibration reliability of rotor blades for energy converting machinery. However, conclusions of papers require experimental validation on the basis of blade natural frequencies and excitation load frequencies control.

The most complete estimation of blade vibratory reliability is that based on test results at operation and equivalent conditions. Up to now there were no sufficiently effective systems for blade vibration excitation at rotation and systems for operative check of their resonance frequencies.

CHECK METHODS AND SYSTEMS

In the Institute for Problems in Machinery, Ukrainian SSR Academy of Sciences original methods and systems for vibration excitation of machine rotor blades during their rotation and operative check of their resonance frequencies have been developed.

The principle of methods is to provide impulse excitation on rotating blades, to record and analyse vibrations of the latter. The shaping of impulse excitation at blades is carried out by specially developed vibrational exciters.

The first group of methods is based on introduction into resonance and resonance frequencies sustaining of machine rotor rotating blade.

Methods of the second group are based on recording and analysis of rotating blade damped vibrations caused by an impact momentum on a rotating blade.
Rotating blade resonance frequency is determined in the first case by the maximum amplitude of its resonance vibrations from vibrational spectrum.

As to the first group of methods aerodynamic exciters of rotor blades resonance vibrations have been created. Particularly, for rotor blades of centrifugal ventilators, an exciter is a construction of two concentric cylinders inserted one into another with an identical number of slots in each one. A number of slots in cylinders is chosen allowing for resonance vibrations of the rotating blade. A unit is installed into ventilator inlet so that slots of cylinders should be located along rotor blades. A similar unit for testing rotor blades of axial ventilators consists of two contacting discs with an equal number of slots in them. The unit is installed at ventilator inlet in front of a rotor. Disc slots are also located along rotor blades.

The blade resonance control is carried out by a microphone in acoustic resonance or by the use of vibration gauges and metering equipment.

Aerodynamic vibration exciters can be used to excite vibrations of rotating blades made, practically, of any material. An amplitude of blade resonance vibrations is regulated by partial overlapping of unit slots as well as by changing pressure of air or gas fed to blades when an external compressed air (gas) source is used. In fact, in assumption of cosine relationship of disturbing force change along exciter slot width

$$p(t) = P_0 \cos \omega t , \quad \omega = \frac{2\pi}{T} , \qquad (1)$$

$$0 \le t \le T_0/2 ,$$

where P_0 – disturbing force amplitude;
T – time for a blade to pass a span between exciter slots;
T_0 – time for a blade to pass an exciter slot;
the expression for the maximum amplitude at resonance according to the first mode of blade vibrations is

$$P = P_0 \frac{T_0}{T} \left\{ \frac{\sin \frac{\pi}{2}\left(\frac{2T_0}{T} - 1\right)}{\frac{\pi}{2}\left(\frac{2T_0}{T} - 1\right)} + \frac{\sin \frac{\pi}{2}\left(\frac{2T_0}{T} + 1\right)}{\frac{\pi}{2}\left(\frac{2T_0}{T} + 1\right)} \right\} , \qquad (2)$$

which is a result of disturbing force expansion into a Fourier series.

As seen from (2) the maximum amplitude of the disturbing force depends on air-pressure P_0 , fed to blades and relative overlapping of unit slots T_0/T . Using the above units automated systems have been developed to search and maintain blade resonance frequencies.

During investigations a decrease of natural vibration frequencies of rotating blades of centrifugal machine rotors was found and proved with respect to natural vibration frequencies of non-rotating blades

$$f_d = \sqrt{f_s^2 - n_z^2} \; , \qquad\qquad (3)$$

where f_d , f_s - natural vibration frequencies of rotating and non-rotating rotors, respectively;
n_z - rotation frequency of machine rotor.

As to methods of the second group, for the blades made of conducting materials a vibration impulse exciter has been created which operates using an impulse magnetic field impact against the rotating blade.

A unit consists of a magnetic work-coil connected to a power source via an impulse discharger. During rotor rotation when a blade passes by a work-coil, the former is subjected to impulse magnetic field impact resulting in forced damped vibrations of the blade.

An impact against a blade is synchronized with rotor rotation using a blade-impulse discharger feedback. Blade vibrations are recorded either by contactless pick-ups or by strain gauges.

The well-known methods are employed to analyze recorded blade vibrations.

NEW SENSOR SYSTEMS FOR ROTATING OBJECTS

Günther Engler, Werner Kindl

Zwickau University of Engineering
Department of Automotive Engineering
Zwickau, GDR

INTRODUCTION

In development and production processes as well as in
product information on angular velocity, angular acceleration
and retardation at rotating objects must often be obtained
and processed. In this case digital measuring procedures are
especially suitable because the information is directly
available as data word for being processed in microprocessors.

The period of measuring and processing in microprocessors
is very important for the quality of automatic control. In the
entire speed range a period of no more than 10 ms is necessary.

USUAL MEASURING PROCEDURES

Each digital measuring procedure is absed on counting
pulses. Concerning the measurement of the rotating motion it
is useful to procced from the circumference of the measuring
object and to provide it with regularly divided marks. The
measurement of these marks is possible with optoelectronic,
inductive, capacitive or other sensors and leads to a frequency
f_1 proportional to the rotating frequency of the measuring
object.

Amplifiers and consecutive pulse former modules provide
a sequence of rectangular pulses with correct amplitude and
shape which is suitable for the subsequent processing.

Furthermore it is possible to generate a constant frequency
f_2, for instance on the basis of the constant clock frequency
of a microprocessor.

With these two frequencies f_1 and f_2 a digital measuring
procedure can be realized.

To get the necessary measuring information some well-
known measuring procedures are used, for instance the constant
timing measurement and the constant position measurement.

435

The following paragraph contains some important characteristics of these prodedures:
Due to the real dimensions of the measuring objects the
number of possible pulses per revolution is limited.
This leads to a high digitizing error.
Due to the principle of all these measuring procedures
two cycles are necessary.
This leads to a long measuring time, which in general
cannot be pressed under 10 ms.

Constant timing measurement

Concerning the constant timing measurement a time basis
is necessary /Fig.1./. This time basis can be realized by a
backwards timer which begins to count down backwards with the
frequency f_2 during the time T_s. During the time T_s the pulses
of the frequency f_1 proportional to the rotating frequency
are counted. This results in the number of pulses Z_1.

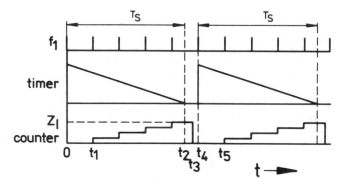

Fig.1. Constant timing measurement

The following paragraph contains some important characteristics:
The given possible digitizing error results from the
necessary minimum time T_s of a measuring interval.
T_s is constant and does not depend on the rotating
frequency.
Resulting from the small number of marks per revolution
- nearly 100 - a great time T_s is necessary.
The measured number of pulses Z_1 is proportional to the
rotating frequency and guarantees a good evalution.

Constant position measurement

Concerning the constant position measurement a distance
basis is necessary /Fig.2./. This distance basis can be realized by a counter 1 which counts the pulses of the frequency
f_1 beginning with the number of pulses Z_1. The therefore necessary time is T_s. During the given distance with the time T_s
the counter 2 counts the pulses of a high constant frequency f_2.
The counted number of pulses at the end of the measurement Z_F

is inversely to the rotating frequency and makes it possible
to calculate the rotating frequency.

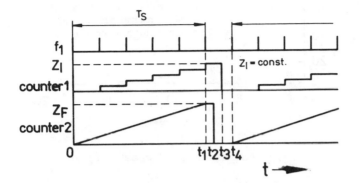

Fig.2. Constant position measurement

The following paragraph contains some important characteristics:

The given distance results from the maximum rotating frequency to be measured regarding the digitizing error and the
given constant frequency f_2 as well.

At the same time the corresponding time T_s is smallest,
but increases with decreasing the rotating frequency.
T_s is only sufficiently small at high rotating frequencies.

The measured number of pulses is inversely proportional
to the rotating frequency and that requires difficult calculating operations for a very long time.

Due to its hyperbolic increase in connection with reduction
of rotating frequency the measuring procedure is unfavourable
at low rotating frequencies.

To calculate the time depending change of the rotating
frequency it is necessary to measure the number of two subsequent intervals. This requires the time $2T_s$.

Based on the following input conditions $-f_2$ = 1.22 Mcps,
100 pulses per revolution and possible digitizing error
20 rad/s^2 concerning the time depending change of the rotating
speed- Fig.3. shows that both measuring procedure are not
capable to meet the requirement concerning a low measuring
time of lower than 10 ms. Therefore the development of a
new measuring procedure was necessary.

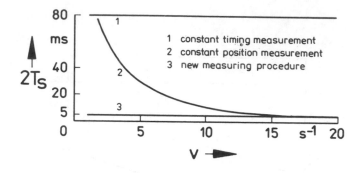

Fig.3. Comparisons of measuring procedures

NEW MEASURING PROCEDURE

The function of this new procedure is shown in the pulse diagram [1,2] /Fig.4./. A CTC is used as timer which counts from a programmed counter-level down to zero. At point zero the timer triggers an interrupt and turns back to the programmed counter-level. Then the timer counts down again. The timer begins to count down at the time O. Starting at the time t_1 the counter 1 receives every pulse of the pulse-sequence f_1. At the time t_1 the timer has the counter-level zero and the given time T_s for one measuring interval is finished. An interrupt-demand arises which is delayed up to the time t_4 by means of hardware. At the time t_2 the left counter 2 begins to count pulses of a high frequency $f_2 = 1.22$ Mcps up to the time t_3. The interrupt-programme gets active by a pulse of the pulse-sequence f_1 at the time t_3. This programme transmits the pulse-quantity Z_1 from the counter 1 and Z_F from the counter 2 to the memory cells of the microprocessor. The counters are cleared and a new measuring interval begins. To avoid the time-consuming division a correction function is elaborated. For every correction value the measured pulse-quantity Z_F is assigned for address. The angular velocity and its change as a function of time can be calculated by means of relatively simple arithmetic operations on the basis of the measured pulse-quantities Z_1 and Z_F if the correction function is used.

The hardware conception is shown in Fig.5.

The following paragraph contains some important characteristics:
The new measuring procedure is characterized by a very short and nearly constant measuring time in the entire speed range. With respect to the input conditions mentioned above this time is nearly 5 ms.
The word length is several times smaller thus decisively improving the processing in real time.
The procedure is especially suitable for a high signal resolution with a subsequent processing of information, for example for determining the changes of rotational speed as a function of time in microprocessors.

438

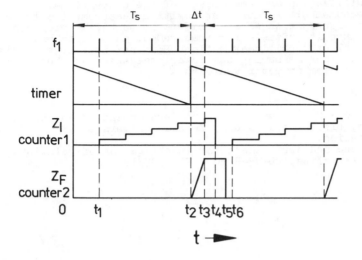

Fig. 4 New measuring procedure

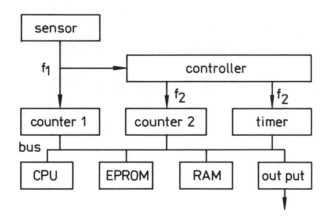

Fig. 5 Hardware conception

SUMMARY

The utilization and comparison of the conventional constant timing measurement and constant position measurement with the developed procedure shows a considerable technical progress. The procedure is suitable for measurement at objects having a rotational or rectilinear motion. For the first time it has been applied for measuring the rotational speed of wheels for antiblock system in motor vehicles.

REFERENCES

1. Kindl, W.: Contributions to the development of antiblock system, Dissertation paper, Zwickau University of Engineering, 1984
2. Technique and arrangement for investigation of rotational characteristics, DD-WP B 60 T 275 207/1.

SOME INTELLIGENT MEASUREMENT METHODS IN QUALITY CONTROL AND TESTING

Prof. Dr.-Ing. Dieter Filbert

Institut für Allgemeine Elektrotechnik

Technische Universität Berlin, Berlin(West)

1. Introduction

Powerful test methods are gaining interest in production and maintenance. The application of intelligent measurement methods in quality control and testing leads to
- shorter test times
- reliable and reproducible test results.

Intelligent measurement is the combination of a-priori information with measured quantities (a-posteriori information) /1/. Faults in a technical system usually cannot be measured directly. Their impact on external measurable state variables can be used only. Sophisticated measurement methods are necessary to solve the problem. They are possible because of the powerful signal processing capacity of modern microcomputers and microelectronic circuits.
Two different applications of intelligent measurement methods to the test of electric drive systems will be shown and compared. The two methods are:
- vibration analysis and
- parameter estimation.

2. Testing electric drive systems

An electric drive system consists of an electric motor, a gear and a mechanical load. The aim of the test is to find all kinds and the locations of faults. Faults are possible in the electric and the mechanical parts of the system. In order to keep the cost of the test low and the test time short only easily measurable state variables should be measured and a simple test stand should be used. Easily measurable quantities of an electric drive system are voltage, current, shaft-speed and vibration.

2.1 Vibration analysis

Application of vibration analysis to the test of rotating machines is a widely approved method /2/,/3/. Fig.1 shows the structure of a vibration analysis system. The pick-up measures the vibration acceleration. Signal parameters like spectrum or cepstrum will be processed by the feature extraction unit and a classifier compares the signal parameters with its normal values in order to decide if the system under test (S.U.T.) has a fault or is in good condition. The a-priori information is the number of rollers or balls in the bearings and the number of teeth of the gear wheels. The impact rates f give the vibration frequency as a function of the dimensions of the bearing and the roller frequency f_r.

For Outer Race Defect: $$f = \frac{n}{2} f_r (1 - \frac{BD}{PD} \cos\beta) \qquad (1)$$

For Inner Race Defect: $$f = \frac{n}{2} f_r (1 + \frac{BD}{PD} \cos\beta) \qquad (2)$$

For Ball Defect: $$f = \frac{PD}{BD} f_r |1 - (\frac{BD}{PD} \cos\beta)^2| \qquad (3)$$

With n=number of balls or rollers and f_r=relative revolution between the inner and outer races, PD=pitch diameter, BD=ball diameter, β=contact angle. In the case of a fault of the bearings those impact rates appear in the spectrum. Fig.2

shows the spectrum of a bearing in good condition (a) and the spectrum of a fault in the outer races (b). It can be easily understood that the fault classification using such a disturbed spectrum is a difficult task. It will be easier if the shaft speed is measured as well and if the harmonics within the spectrum are used too. Combining the measured quantities like vibration and speed with the a-priori information is an intelligent measurement procedure.

2.2 Parameter estimation

Applications of parameter estimation methods to the test of electric drive systems have been reported in recent years /4/,/5/. Fig.3 shows the structure of a model based test system. The S.U.T. is excited by the input signal. The sensor system measures the state variables and the feature extraction unit delivers the system parameters using the a-priori information given by the model. The classifier compares the system parameters with its normal values and decides whether the system is in good condition. The a-priori information is the structure of the system given by two non-linear differential equations:

$$R\ i + L_A\ \frac{di}{dt} + \frac{d\phi_F}{dt} + c_1\ \phi_F\ \omega_M - u = 0 \qquad (4)$$

$$\phi_F = a_1 i + a_3 i^3 \qquad (5)$$

$$c_2 \phi_F i - \phi\ \frac{d\omega_M}{dt} - c_4\ \omega_M^2 - T_b = 0 \qquad (6)$$

with R=copper resistance, L_A=motor inductance, ϕ_F=field flux, c_1, c_2=motor constants, ϕ=inertia factor, c_4=torque constant of fan, T_b=friction of brushes and/or a constant load. To estimate the parameters in (4) to (6) only voltage u, current i and shaft speed ω_M have to be measurable. It follows directly that no special test stand is necessary. But, because of the derivative of the speed in (6) the test needs an external excitation carried out by voltage variations.

Fig.4 shows a test protocol. It can easily be seen that in the case of a faulty fan the parameter c_4 has changed about 60% whereas the other parameters and quantities remain within a ± 5% margin. Therefore the classification of the fault of the fan is easy.

Combining the measured quantities with the model of the system and comparing the estimated parameters with their normal values enable an intelligent measurement method applied to the test of the drive system.

3. Comparison of the methods

The test methods described here have some common peculiarities and some differences. Both methods are used in testing electric drive systems.

Measured quantities: Using the vibration analysis method, the measured quantity is the vibration acceleration. This vibration is produced by the fault itself. The test method does not need an external excitation. The test is possible even if the S.U.T. is in steady state. The vibration generation is not related to a substantial amount of the energy flow within the S.U.T.

Using the parameter estimation method, the measured quantities are voltage, current and speed. The method needs external excitation. A fault is detectable if it has a measurable impact on the voltage balance, eq.(4) or the torque balance, eq.(6). That is related to a substantial change of the energy flow in the S.U.T.

Models used: The relationship of the dimensions of bearings and gears to the generated vibration frequencies is used in the vibration analysis method.

The parameter estimation method applies a physical model describing the structure of the S.U.T.

Analysis methods: Pattern recognition will gain an information about faults from measured vibrations. Signal parameters in the frequency domain are usually classified.

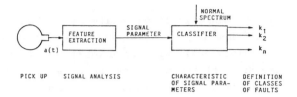

PICK UP   SIGNAL ANALYSIS   CHARACTERISTIC   DEFINITION
OF SIGNAL PARA- OF CLASSES
METERS   OF FAULTS

Fig. 1 Schematic of the Vibration Analysis

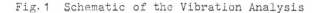

Fig. 2 Spectra of a Bearing

a) bearing in good condition   b) bearing deteriorated
(from Brüel u. Kjaer lectures)

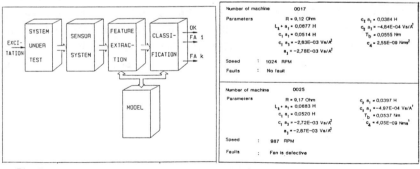

Fig. 3 Test System   Fig. 4 Test Protocol

The parameter estimation method delivers process parameters that can be classified by nearest neighbourhood procedures etc.

Detectable faults: The vibration analysis method is able to detect faults in bearings, gears and other rotating units. These units are mainly auxiliary parts of the drive system.

The parameter estimation method is able to detect faults related to the drive and load units. The detectable faults are those in the windings of the stator and rotor, in the magnetic path, in the fan, the brushes and other mechanical loads.

Sensitivity to faults: A very important question is the sensitivity to faults.

Using the vibration analysis method, a fault has to produce a vibration signal that is separable in frequency. The signal level has to be high enough to be measurable by the pick-up.

Using the parameter estimation method, the sensitivity is the deviation of the measured state variables due to a fault. Results of sensitivity studies of the parameter estimation method have been reported in /5/. The results show that the parameter estimation method is not very sensitive to faults in bearings but is an excellent method to detect faults in the drive and load systems.

4. Conclusion

Two intelligent measurement methods and their application in quality control and testing have been presented. Their comparison gives evidence of the similarities and the differences of the methods. The estimation method is able to detect faults related to the function of the system, whereas the vibration analysis method is able to detect faults in auxiliary units. A combination of both methods should obtain better test results. Further research in this area has to be done.

REFERENCES

1.   P. Eykhoff: Identification Theory; Practical Implication
     and Limitations. IMEKO Symposium on Measurement and
     Estimation. Bressanone (1984)

2.   D. Barschdorff: Diagnostic System with Distributed
     Processing for Monitoring of Rotating Machines.
     IMEKO Symposium on Technical Diagnostics. London (1981)

3.   N.N.: The Application of Vibration Measurement and
     Analysis in Machine Maintenance. Technical Lecture.
     Brüel + Kjaer (1982)

4.   D. Filbert: Fault Diagnosis in Nonlinear Electromechanical
     Systems by Continuous Time Parameter Estimation.
     ISA Transactions. Vol. 24, No. 3 (1985)

5.   D. Filbert, E. Dreetz: Empfindlichkeitsuntersuchungen bei
     der modellgestützten Fehlerdiagnose.
     GMR-Bericht 1. Langen (1984)

INTELLIGENT MEASUREMENT FOR QUALITY TESTING OF AC MOTORS

N. Polese, G. Betta, C. Landi

Department of Electrical Engineering
University of Naples, Italy

1. INTRODUCTION

Induction motors have to be tested in different operating conditions to verify that the performances stated in national and international standards are satisfied.
The tests need many technicians to manage and control the test procedures. The introduction of intelligent techniques allows:
  i) to reduce the number of the operators engaged by these operations;
 ii) to avoid all errors connected with the human element;
iii) to optimise the time taken by the whole test procedure.
At the Department of Electrical Engineering of the University of Naples, an intelligent apparatus to test a.c. motors has been set up, based on the implementation of the control of an automatic test procedure, obtained through a microprogrammed computer system. This paper synthetically relates:
  i) the architecture of the station suited to such induction motor intelligent testing, with particular reference to the measurement and control units;
 ii) the various feasibilities implemented to manage and control the test flowing;
iii) some experimental results showing the extreme flexibility of the facility and its good performances.

2. TEST STATION

The block diagram of the realized test station is shown in fig. 1; the main functional links between the different blocks are put in evidence. In no-load tests, the motor is fed by a booster and the inverter is directly connected to the main. The components of the station can be grouped in two different sections: i) power section; ii) signal section.

2.1 Power section

This section is basically composed by the motor under test, an electrical dynamometer and an inverter (SIEMENS AQG 1092-4).
The dynamometer is a DC shunt-wound machine which can be used in 4-quadrant operation by means of a converter unit: in this application the dynamometer is connected to the inverter by means of which the electric

449

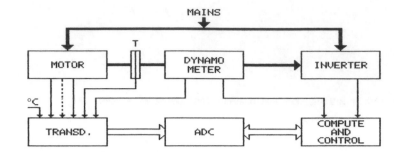

MAINS

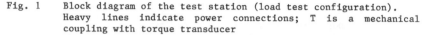

Fig. 1   Block diagram of the test station (load test configuration).
Heavy lines indicate power connections; T is a mechanical
coupling with torque transducer

energy produced by the dynamometer is recovered and sent back to the
mains.

The FFT analysis of the inverter waveform has shown a deviation
factor negligible compared with that of the mains. Therefore, in load
tests, the recovered energy is directly injected in the motor terminals,
while the mains furnishes the total amount of the losses.

The inverter is driven by the control section through a D/A channel
which varies the operating angle of the thyristors. So the control of the
dynamometer armature current is realized and consequently the torque
control is performed.

2.2 Signal section

All the electric signals necessary for the test have been transduced
in voltage signals using Hall effect probes, which present: (i) good
dynamic performaces with a large passband; (ii) galvanic insulation; (iii)
immediate availability of only voltage signals for the A/D conversion;
(iv) satisfactory precision.

The torque values are obtainable either from the dynamometer or,
in dynamic working conditions, from the PR99 torque transducer by
Philips (input range $0.1 \div 10^5$ kp m, rise time < 0.25 ms, linearity error
< 0.1 %), whose output signal (frequency) is converted into a voltage
signal to make it compatible with the measurement unit.

The angular speed of the motor is obtained from the voltage output of
a tachometer.

The acquisition, elaboration and control operations are performed by
an IBM PC-XT computer with a DT 2801-A data acquisition board plugged into
one of its expansion slots. It includes (i) one 12-bit A/D converter with
a maximum sampling rate of 27.5 kHz; (ii) 16 single-ended or 8
differential input channels with software selectable gains; (iii) 2 analog
output channels with two 12-bit D/A converters which can operate
independently or simultaneously.

This acquisition and control tool has been provided with dedicated
software suited to this application, allowing to manage the whole test
procedure as shown in the next paragraph.

3. INTELLIGENT MANAGEMENT OF THE TEST

The only external data, that the station needs as input, are the
rated values. At present, also the values of stator and rotor resistances
at a reference temperature have to be input, but as soon as possible these
measurements will be implemented in the system as automatic routines. The
system evaluates from the rated values the first measurement point

(depending on the test that has to be performed and on its standards). The wanted working conditions are reached through either a torque control (i.e. for load tests) or a motor voltage control (i.e. for no-load tests). Such controls are realized utilizing one of the D/A channels for the torque and a digital port to drive the booster for the voltage.

The measurements have to be carried out in steady state conditions hence a steady state condition criterion is required: the motor is considered in a steady state after a torque step when the dumped oscillation of the input current becomes internal to the $\pm$ 0.5 % of the rated current. Once the new wanted steady state condition has been reached, the system selects the gains of the input amplifier, that permits the maximum sensitivity, and performs the first measurement.

A prefixed number of samples are acquired for every electric quantity to evaluate the average value and so remove the random errors.

After each measurement the control section performs the following checks: (i) comparison between the computed dispersions and those fixed as limits for a valid measurement; (ii) control of the following factors, according to the IEEE 112/84 standard: voltage waveform deviation factor (<10%), input frequency variations (<0.5.%), voltage unbalance (<0.5%). If one of the previously mentioned checks gives a negative outcome the measurement is repeated. A second negative attempt causes the system to access the implemented test rules in order to decide if the test can continue or has to stop. In both cases, the system will keep in memory what happened and the output will report both the problems met and the reasons for the continuing or stopping. If no problem affects the measurement the system fixes the subsequent working conditions and manages all the operations as already mentioned.

Each test is followed by further general controls, which determine the beginning of the following quality test.

After each measurement the computation unit provides for the systematic error correction, by means of the correction tables, previously evaluated and stored in the system memory, obtained from the characteristics of the transducers and of the converters. The ambient temperature, periodically acquired during all the test flowing, is also taken into account for the correction of the measures.

Finally the parameters of interest are computed and presented in output, along with a comment of the results. Losses and efficiency are compared with those recommended by standards or committant and consequently possible penalties or rated values corrections are suggested.

Together with the protections already employed in the various apparatus, some software protections were created to preserve the test station and the motor from any wrong working condition. With this aim, for instance, torque and speed control routines were implemented.

4. EXPERIMENTAL RESULTS

The realized test station is able to perform the complete test of L.V. three-phase induction motors, with power up to 30 kW and one or more pairs of poles. Fig. 2 shows an example of the output characteristics that the test facilities can offer. They concern the no-load and load tests of a TIBB MS64 a.c. motor, 5.3 kW, 260 V, 17 A, 50 Hz, 4 poles.

As shown in fig. 2a, together with the no-load current versus the voltage, the system performs the separation of the core losses from the friction and windage losses. Fig. 2b reports efficiency and input current versus output mechanical power, along with the torque-slip curve.

The input current subsequent to a torque step equal to the 100 % of the motor rated torque reached the steady state condition, as previously defined, in less than 0.5 s.

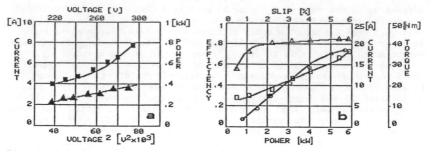

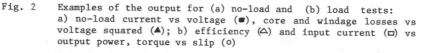

Fig. 2   Examples of the output for (a) no-load and (b) load tests:
a) no-load current vs voltage (■), core and windage losses vs
voltage squared (▲); b) efficiency (△) and input current (□) vs
output power, torque vs slip (○)

5. CONCLUSIONS

The intelligent apparatus for quality testing of a.c. motors, set up
at the Department of Electrical Engineering of the University of Naples
and described in this paper, allows:
i) to establish the correct working point in which the measurements have
to be carried out, once the rated characteristics have been input;
ii) to handle both the measurement and control systems toward an optimi-
sation of test performances as regards sensitivity, errors and energy
consumption (an energy recovery system is also implemented);
iii) to control through automatic diagnostic routines the correct flowing
of the test according to the standard recommendations;
iv) to organize the acquired data, to elaborate them according to suited
algorithms and to present the output in the most appropriate way
(graphic and/or tabular).

The realized test facility is going to be developed through the definition of knowledge-based programs, that will allow to draw consistent
conclusions from the performed tests to improve the existing memory
data-base.

COMPUTERIZED PARAMETER ESTIMATION FOR QUALITY TESTING
OF INDUCTION MOTORS

Dipl.-Ing. Dagmar Elten

Institut für Allgemeine Elektrotechnik

Technische Universität Berlin, D-1000 Berlin (West)

1. Abstract

This paper deals with the quality testing of three-phase squirrel cage induction motors. The fault detection is achieved by diagnosis procedures using parameter identification methods. The check implies the measurement of terminal currents, voltages and speed. Using a simple mathematical model of induction machines these parameters are estimated by least-squares methods. An analysis of these parameters and the values of currents and voltages delivers a criterion for fault detection.

2. Usual test methods of induction machines

There may be many test steps during production and operation of machines. Visual and electrical checks are done during production. A final test is also necessary.

Careful checks include the measurement of phase winding resistants, speed and RMS values of currents and voltages during idle and nominal-load conditions. The torque may be investigated, too. This causes a test system requiring a load machine, expensive sensor equipment and long test time. For reasons of effectivity and economy larger production runs are often dealt by checking current and speed under idle conditions. Incoming inspections are done the same way. Early fault detection may be required for reasons of continues work and production. Therefore torque, speed, vibration, voltage and current can be monitored.

3. A new test method

Increasing capability and decreasing costs of microprocessors deliver the opportunity to use computerized test methods. Taking a-priori information into consideration the attained information about the machine can be increased and the check time can be reduced /1/,/2/,/4/.

The following test system shows an example of an automatic method for squirrel cage induction machines.

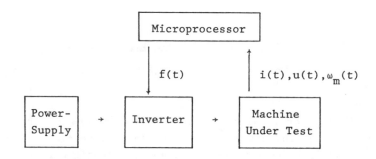

Fig. 1   Test system

The system consists of an inverter, a microcomputer and the machine under test without any additional load. The microcomputer controls the inverter, picks up the momentary values of the three-phase currents and voltages and of the speed. After computing the necessary parameters these values are shown and compared with that of a machine in good condition.

3.1 Diagnosis procedure

At first the motor works in idle-state and steady-state conditions by nominal supply frequency. The actual currents $i(t)$, voltages $u(t)$ and speed $\omega_m(t)$ are measured. If currents and voltages are not symmetric or the amplitudes are not equal within a tolerance band there might be an inter-turn or earth short-circuit or an open circuit of the windings.

In this case the test algorithm will stop and the fault will be indicated. Otherwise the parameters of the electrical part of the machine are estimated, i.e. resistance R_s and inductivity L_s of the stator and dispersion $_\theta$. If there is no extraordinary deviation from the nominal values the driving torque will be evaluated and examined whether it oscillates. Continuing the check the mechanical part of the machine has to be investigated. The supply frequency $f(t)$ is periodically caused to small changes. Due to idle-conditions internal losses, friction and the internal fan are the only effective loads.

After attaining new test data the mechanical part of the machine is examined by a new estimation cycle. Now, the inertia factor and parameters concerning the friction, the load of the fan and other losses are available. An analysis will give information about the iron losses, the bearing friction or the fan.

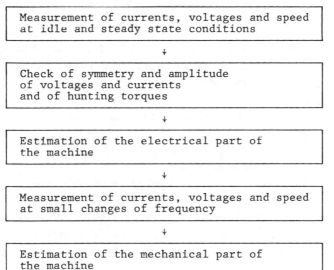

Fig. 2   Diagnosis routine

3.2 Parameter estimation

The parameter estimation routine bases on a simple model of the squirrel cage induction motor, delivered by its differential equations. Current and voltage are represented as complex stator fixed values. Because of idle operation during the test skin effect and saturation can be neglected.

The electrical equations lead to a linear estimation model during steady state operation (ω_m = constant) /3/.

$$\dot{\underline{u}}(t) = a_0 \, \underline{u}(t) + a_1 \, \underline{i}(t) + a_2 \, \dot{\underline{i}}(t) + a_3 \, \ddot{\underline{i}}(t)$$

with

$\underline{u}(t)$ - stator voltage; $\dot{\underline{u}}(t) = \underline{du}/dt$

$\underline{i}(t)$ - stator current

$\omega_m(t)$ - mechanical angular frequency

$a_0 = - R_s/L_s$

$a_1 = - R_s/L_s \, (R_s - j \, \omega_m \, L_s)$

$a_2 = R_s + R_r/L_r \cdot L_s - j \, \omega_m \, L_s \, \sigma$

$a_3 = L_s \, \sigma$

For small changes of the supply frequency the non-linear balance of torques can be approximated by an equation that is linear in its parameters.

$$T_d = \theta \, \dot{\omega}_m(t) + c_0 + c_1 \, \omega_m(t) + c_2 \, \omega_m(t)^2$$

with

T_d - driving torque

θ - inertia factor

$c_0 .. c_2$ - parameters of the internal load

For that the driving torque is evaluated using the values of current and voltage. Obtaining the derivatives of the state variables by special filter algorithms the parameters are estimated by least-square methods.

REFERENCES

1. Metzger, K.: Ein Beitrag zur Prüfung von Kleinmotoren unter Einsatz eines Parameterschätzverfahrens. Dissertation TU Berlin, 1983

2. Dreetz, E.: Diagnosis of Technical Systems Using Estimated Parameters. Proc. IMEKO, Symposium on Measurement and Estimation, Bressanone /I/, 1984

3. Kovacs, K.P., Racz, I.: Transiente Vorgänge in Wechselstrommaschinen. Verlag der Ungarischen Akademie der Wissenschaft, Budapest, 1959

4. Filbert,D.: Fault Diagnosis in Nonlinear Electromagnetical Systems by Continuous Time Parameter Estimation. ISA Transactions, Vol.24, No.3, 1985

INTELLIGENT MULTI-INPUT INSTRUMENT FOR ELECTRICAL MEASUREMENTS

F. Ferraris, I. Gorini, M. Parvis

Dipartimento di Automatica e Informatica
Politecnico di Torino
Torino,Italy

INTRODUCTION

The paper deals with the design of an intelligent instrument for the simultaneous measurement of two electrical quantities (voltages and currents). The instrument has been designed in order to provide our research laboratory with a medium speed medium resolution two channel measurement system well suited for the evaluation of input signal functionals, capable to work standing alone or connected to a host computer. The main interest is devoted to the intelligent features supplied by microprocessors in modern instrumentation in order to improve automatism, accuracy and friendliness with the user.

In the paper the functional description of the instrument is given, pointing up the measuring elements in which the intelligence plays a primary role.

The main features of the instrument can be summarized as follows:
- two independent measuring channels, with a sampling rate up to 100 kHz per channel;
- a resolution of 14 bit and an overall accuracy of the order of 10^{-4};
- on-line evaluation of simple arithmetic expressions of the two measured quantities, using a high speed microprocessor (TMS 32010);
- on/off line evaluation of more complex functionals by means of a powerful microprocessor (Motorola 68010 with 1 Mbyte RAM).

The presence of the intelligence (provided by the microprocessors) and the highly integrated components allow to perform automatic tasks which make the instrument easy and robust to use. We mention here:
- The autoranging process both in amplitude and frequency, to optimize the input devices gain and the A/D converter sampling rate.
- The autodiagnostic procedures of both digital and analogue components.
- The autocalibration of the measuring channels.
- The possibility of degraded working in case of failure of some components.
- The automatic correction of the effects of influence quantities; some auxiliary measuring channels (with a 8-10 bit resolution) are provided in the instrument.
- The automatic estimation of the uncertainties of the whole instrument; the estimation algorithms are based both on the a priori information and on the results of the autocalibration and the correction procedures.
- The friendly procedures which interface instrument to operator; for

example, the command sequence, totally menu-driven, can be memorised and recalled.

THE FUNCTIONAL VIEW

The design makes reference to a general model for the functional description of multi-input intelligent measuring devices, which was previously developed by Ferraris et al.[1,2].
The instrument is made up of three main sections (Fig. 1):
a) The <u>measuring section</u>, which is devoted to condition, to sample, to A/D convert and to digitally process the measured quantities. In the figure the functions related to each measuring quantity are syntesized in the block called "main measuring system"; the functions related to the influence quantities are in the block called "auxiliary measuring system"; quantities processing functions, correction function and uncertainties estimation functions are in the block named "processing system".
b) The <u>management section</u>, which performs the management, the diagnostic procedures and the control of the whole device.
c) The <u>interfacing section</u>, which interfaces the instrument to the user, a human operator and/or a host computer.
For the sake of brevity, in the following paragraphs the most interesting features of the first section only (the most interesting under the metrological point of view) are pointed out.

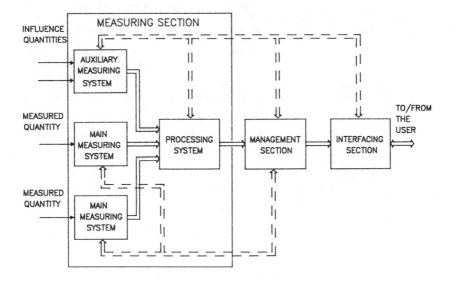

Fig. 1   The functional view of the whole measuring instrument
          Dashed lines   : commands and working information
          Single lines   : measured quantities, analogue signals
          Double lines   : measured quantities, digital signals

THE MEASURING SECTION

The main measuring system

The functional block diagram of the main measuring system is sketched in fig. 2. The intelligence is supplied primarily by the local controller which manages the whole system, schedules the tasks to be performed, carries on the closed loops and interfaces the system with the management section.

Whith reference to the measurement information flow, the measured quantity is firstly processed by the programmable conditioning devices, such as current/voltage converters, variable gain amplifiers and tunable filters. Then the signal is sampled and converted by a 14 bit A/D converter, whith a sampling rate selectable from few hertz to more than 100 kHz. Finally the measured quantity is processed by the local processor, which performs two main tasks:
- a "first step" correction of the acquired values, using the calibration parameters stored in the calibration memory;
- signal filtering and, on request, some related functionals (e.g., mean or rms values) evaluation.

The calibration parameters are settled at power up and continuously updated during the normal working of the instrument: the input stages of the conditioning devices are switched from the measured signal to the calibration circuits and measuring tasks are performed; the results, processed by the local processor, are stored in the calibration memory.

Finally, the signal analyser, built up mainly with analogue circuits (e.g., a tunable high-pass filter), manages to get information about the characteristics of the measured signal; for example, it gives information to the local controller to estimate aliasing effects.

The diagnostic functions, applicable both to analogue and to digital circuits, and the related philosophy are widely set out in another paper[3] presented at this symposium.

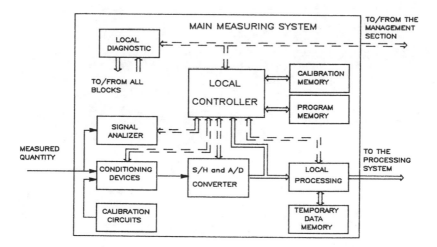

Fig. 2   The functional diagram of the main measuring system

The processing system

The main task of the processing system is to execute the algorithms related to the measured quantities and to compute more complex functionals selected by the user, e.g., transforms, correlations, etc.

As for the measurement accuracy, the processing system performs a "second step" correction of the acquired values, by means of an algorithm based on a model of the effects of environmental influence quantities; the input of the algoritm are the measured values of the influence quantities (obtained by the auxiliary measuring system), the information about the "status" of the whole instrument (stored in a permanent memory and updated by means of the diagnostics procedures[3] performed by the management section) and the time (given by a perpetual clock).

Finally, the processing system estimates the overall accuracy of the measurements performed during a given interval of time: the uncertainties are evaluated (using essentially information about the calibration and correction algorithms) and a norm synthetising the global accuracy is furnished.

REFERENCES

1. F. Ferraris, I. Gorini and M. Parvis, A general model for the metrological characterization of measuring systems, IMEKO workshop on fundam. logical concepts of meas., Torino (1983).
2. F. Ferraris, I. Gorini and M. Parvis, Un modello per una rappresentazione funzionale dello strumento di misura intelligente. 85a Riunione annuale A.E.I., Riva del Garda (1984).
3. F. Ferraris and M. Parvis, The autodiagnosis system of an intelligent measuring instrument. IMEKO symp. "Intelligent Measurement - Inquamess 86", Jena (1986).

INTELLIGENT MEASUREMENT IN COMMUNICATION NETWORKS

Ahmad I. Abu-El-Haija

Electrical Engineering Department
Jordan University of Science and Technology
Irbid, Jordan

ABSTRACT

With the advent of microprocessors and the advancement of new techniques in the area of digital signal processing, conventional measurement
methods and equipments are being replaced by more intelligent, reliable,
accurate, and faster digital-based ones. In this paper, a digital method
for measuring the frequency shift and phase jitter of telephone channels
in the voice frequency band (300 – 3400 Hz) is presented. The measurement
method can be readily implemented by microprocessors and incorporated in
telephone line measuring equipment.

INTRODUCTION

A computer network consists of two types of components: equipment
and transmission links. Today's networks vary in complexity, and a typical
large network includes several thousands of nodes, communicating with each
other as well as with a central location which contains information needed
by a large number of users. These nodes are connected with each other in
one or more of several possible topologies via transmission links which
carry data among the network nodes. One of the most common type of link is
the telephone channel, whose design, structure, and operation usually affect
data communications.

The performance of a telephone channel for transmitting data depends
upon several inherent distortions of the channel. Since these channels
carry analog signals, the performance of a channel is often specified in
terms of limits on the analog distortions present. If the characteristics
exceed these limits, they are then classified as impairments. Important
impairments of telephone channels are attenuation and envelope delay
distortions, frequency shift, phase jitter, impulse noise, and others [1].
These impairments can be measured according to standard techniques [2].

Simple, but yet fast algorithms based on microprocessors for measuring
the attenuation and envelope delay distortions on a telephone channel have
been proposed [3,4] where the measurement time is only few seconds. Similar
digital-based algorithms for measuring other channel impairments are
certainly of interest.

463

DIGITAL MEASUREMENT OF FREQUENCY SHIFT AND PHASE JITTER

Fig. 1 shows a telephone channel with a signal s(t) being sent on the channel, and due to analog distortions which are unavoidable, the received signal r(t) is different from s(t). For the purpose of measuring the frequency shift and phase jitter, it is appropriate to choose s(t) as a sinusoidal signal so that its frequency and phase can be tracked easily, thus

$$s(t) = \sin\omega_o t \qquad (1)$$

where $\omega_o = 2\pi f_o$ and f_o is the frequency of the test signal in Hz. In order to minimize the effect of channel attenuation on the test signal, f_o is chosen in the band where the channel has minimum attenuation (e.g. 1000 – 2400 Hz). The received signal r(t) can be written as:

$$r(t) = A \sin(\omega_o t + \omega_s t + \phi_j(t)) + N(t), \qquad (2)$$

where

A = signal amplitude, caused by channel attenuation,

$\omega_s = 2\pi f_s$, f_s = frequency shift in Hz,

$\phi_j(t)$ = phase jitter component,

N(t) = additive noise.

Since $\phi_j(t)$ is usually periodic, then r(t) is also periodic and it is not necessary to synchronize the receiver with the transmitter. Hence r(t) can be written as:

$$r(t) = A \sin(\omega_o t + \omega_s t + \phi_j(t) + \phi_o) + N(t) \qquad (3)$$

where ϕ_o = phase intercept, or initial phase in measurement. The problem investigated here is that of measuring f_s and ϕ_j given r(t) in (3).

Digital measurement of frequency shift and phase jitter can be accomplished using a digital phase locked loop (DPLL) to track the frequency and phase of the received signal. By comparing these with the frequency and phase of the transmitted signal (which are known exactly for the transmitted test tone), the values of the frequency shift and peak-to-peak phase jitter can be computed.

This approach is in principle illustrated in Fig. 2, and its operation can be described as follows. The received signal r(t) is first passed through an analog filter to limit its bandwidth to that of the telephone channel and to prevent aliasing errors. A sampler then samples the analog signal and passes it through an analog-to-digital converter. Note that the amplitude of the transmitted signal is unity, while that of the received signal is normally different from unity because of the channel attenuation and other disturbances. Therefore, the sampled digital signal must be

Fig. 1   A telephone channel with a transmitted signal s(t) and a received signal r(t)

464

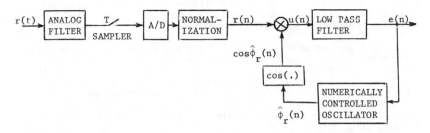

Fig. 2   Digital measurement of frequency shift and phase jitter

normalized so that its amplitude is the same as that of the transmitted
signal. The signal after normalization is

$$r(n) = \sin\phi_r(n) + \text{noise}$$

where $\phi_r(n)$ denotes the phase of the received signal in the absence of
additive noise. If T is the sampling interval then:

$$\phi_r(n) = 2\pi f_o nT + 2\pi f_s nT + \phi_j(n) + \phi_o. \qquad (4)$$

The normalized sampled signal is then passed through a loop comprising a
lowpass filter (LPF), a numerically controlled oscillator (NCO), and a
table look-up for computing cos(.).

For a simple explanation of this approach, assume that the normalized
received signal is free from the additive noise component. This signal is
then multiplied by $\cos\hat{\phi}_r(n)$, where $\hat{\phi}_r(n)$ is the estimate on $\phi_r(n)$ and is
computed by the NCO. The product

$$u(n) = \sin\phi_r(n)\ \cos\hat{\phi}_r(n)$$

$$= \frac{1}{2}[\sin(\phi_r(n)+\hat{\phi}_r(n)) + \sin(\phi_r(n)-\hat{\phi}_r(n))] \qquad (5)$$

is then passed through the LPF to filter out the term $\sin(\phi_r(n)+\hat{\phi}_r(n))$,
whose spectrum is roughly centered at $2f_o$. If $\phi_r \neq \hat{\phi}_r$, then an error

$$e(n) = \frac{1}{2}\sin(\phi_r-\hat{\phi}_r) \simeq \frac{1}{2}(\phi_r-\hat{\phi}_r)\ , \quad \text{for small } (\phi_r-\hat{\phi}_r) \qquad (6)$$

is fed back to the NCO in order to give a better estimate $\hat{\phi}_r$.

If there is no phase jitter in the received signal, it is possible to
track the frequency and phase exactly, and bring e(n) to zero. With the
presence of phase jitter, it is possible to design the DPLL [5] and other
components of the system, such that e(n) gives enough information to
compute the amplitude of the phase jitter.

NETWORK MEASUREMENT AND MONITORING

Data being carried by transmission links in a computer network are
often subject to transmission errors resulting from various phenomena
occuring on the transmission lines. In order to insure continuous and

465

smooth operation of the network, specific parameters of the transmission lines must be regularly measured and compared with predetermined thresholds to insure the correct behavior of these links.

It has been shown that monitoring the quality of data transmission over common carrier leased lines can be used in the daily operation of a computer center to increase the circuit availability. In one case [6], use of the line quality monitoring system contributed to an increase in circuit availability from 82% to 99% over a one-month period. In principle, this is possible if such a monitoring program is linked to transmission control software. When a line is discovered to be malfunctioning, the software might choose, in addition to notifying the operator, to reroute messages using a different line.

The digital part of the measurement method presented in this paper requires only arithmetic operations (additions, multiplications, and possibly divisions depending upon the implementation) and a read-only-memory to store the cosine values. Hence the algorithm can be implemented by microprocessors, and be part of telephone line measurement equipment, or part of a communication network monitoring system. Moreover, because of the intelligence available (through programming) in microprocessors, this algorithm can be used for the purpose of remote testing, measurement, monitoring, and control of complex computer communication networks. In this case, the measurement command is sent by a central control station, received and interpreted by the intelligent hardware available at a distant location, and once the measurement is remotely performed, the test results are transmitted back to the central station. Depending upon the conditions of the transmission links, the central station makes appropriate decisions to optimize the network performance.

REFERENCES

1. Analog Parameters Affecting Voiceband Data Transmission - Description of Parameters, Bell System Technical Reference, PUB 41008, Oct. 1971.

2. Transmission Parameters Affecting Voiceband Data Transmission - Measuring Techniques, Bell System Technical Reference, PUB 41009, Jan. 1972.

3. A. I. Abu-El-Haija, "Remote measurement of some telephone channel impairments using microprocessors," IEEE Trans. Instrumentation and Measurements, Vol. IM-28, No. 4, pp. 244-249, Dec. 1979.

4. R. Boite and H. Leich, "Fast measurements of the attenuation and delay characteristics of data channels," Proc. IEEE International Conference Acoustics, Speech, and Signal Processing, pp. 1770-1772, Apr. 1982.

5. W. C. Lindsey and C. M. Chie, "A survey of digital phase-locked loops," Proc. IEEE, Vol. 69, No. 4, pp. 410-431, Apr. 1981.

6. P. Bryant, F. W. Giesin, Jr., and R. M. Hayes, "Experiments in line quality monitoring," IBM Systems Journal, Vol. 15, No. 2, pp. 124-142, 1978.

COMPUTERIZED QUALITY CONTROL IN AUTOMATIC

MACHINING SYSTEMS IN BATCH PRODUCTION

Siegfried Szyminski
Horst Würpel

Technical University Otto von Guericke Magdeburg
German Democratic Republic

INTRODUCTION

The development of systems for computer aided design
and manufacturing (CAD/CAM) places equally new demands on
the implementation of computer aided quality control systems.
These demands are characterized by the process related
design of complexing and computer aided quality control
solutions in the system of external production process control of machining systems and the development of flexible
devices includ-ing programme and control solutions for
process integrated measurement technology in machining cells.
In what follows, some examples of these computerized
solutions for geometrical quality control in automated
machining systems for turning operations including
industrial robots are presented.

APPLICATION OF COMPUTERIZED EXTERNAL MEASURING STATIONS FOR
AUTOMATED DIAMETER CHECKING

One major aspect of the design of external measuring
stations for diametering checking of work pieces is the
development of quality control related measuring heads.
According to their design they are to be created as modular
systems to correspond relatively easily to the flexible
conditions of the small and medium batch production. Based
on 1-, 2- or 3-point measurement by means of measuring heads
the testing features can be determined with the presently
dominating mechanical sensing of work pieces. The relatively
narrow design of such measuring heads allows, by using
multiple arrangements the determination of several testing
features of work pieces. Some examples of measuring heads
which have been developed are given in /1/. The implementation of automated motion cycles of the elements of measuring
stations is to be considered as a further major aspect in
the development of external measuring stations. These
motions are to be carried out according to a coordinated
system and are integrated into the automatic operational
sequence of the machining cell. For this purpose a

decentralised computer aided information and control system
can be used, guaranteeing the automatic checking cycle in
connection with state indicators (compare /2/ and Fig. 1).
A third major aspect applying automatical measuring stations
results from actively influencing the machining process with
necessary tool corrections. Unattended processes require the
definition of objective criteria for tool corrections. Therefore objective control algorithm have to be developed taking
into consideration the significant parameters which impair
the quality. According to Fig. 1 and by taking the above
mentioned effects into account the following work piece and
information flow results in a machining cell.

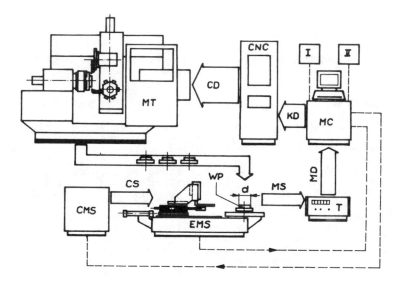

Fig. 1   Schema of the work piece and information flow in
          a machining cell

        The robot which is not shown in the Fig. takes the
machined work piece (WP) from the machine tool and places
it on the measuring station (EMS). This state is signaled
to the microcomputer (MC) which causes via a plug-in unit
programmable externally (I) the control functions of the
measuring station (CMS) for carrying out all motion cycles
required in the measuring station. The individual work piece
dimensions determined during the measuring process are
transmitted as measuring signals (MS) to the display (T) and
as measuring data (MD) to the micro computer. The micro
computer (MC) then has to accomplish two jobs by the plug-in
unit programmable externally (II).
These are:
(1) Storing, printing and evelution of measuring date according to control algorithm which must be continuously updated
and (2) passing on corrections (KD) to the machine tool control (CNC) resulting from a comparison with the quality para-

meter which are given by the work piece positioning data of
the machine tool control or the quality parameters which
have seperately been put into the micro computer. The
machine tool control then gives the new control data for
necessary tool corrections to servo drives of the machine
tool.

INTEGRATION OF EXTERNAL MEASURING STATIONS IN MACHINING
SYSTEMS FOR TURNING OPERATIONS

     For the continuous implementation of unattended pro-
cesses quality control in automated machining systems is
becoming more and more important. In this respect we have to
consider solutions for computer aided test routine as well as
scheduling measures for the quality oriented machine tool
utilization aspecially taking into account the external
production process control. An important basis is the knowledge of the geometrical manufacturing level of the equipment
used in production which can be obtained from the quotient
of machining accuracy in relation to the work piece tolerance.
Such an algorithmic method is shown in Fig. 2.
     Within the framework of a statistic cyclical test the
geometrical machining accuracy of each machine tool of a
machining system is determined. This can be realized by an
algorithm to be agreed upon either after the input of actual
work piece measurement values from the micro computer (MC)
or in the frame of preliminary data collections

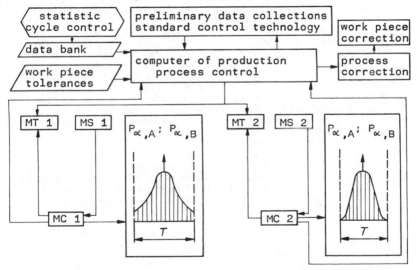

Fig. 2   Schema of quality oriented machine tool
         utilization and dynamic test scheduling

The calculated machining accuracy is given to the data bank. After this, these data including data taken from the manufacturing documentation about work piece tolerance are transmitted to the computer of the production process control. Under the condition of manufacturing level < 1 the computer selects the machine tool (MT 2). After determining the manufacturing level the desired standard control technology must be made precise by the computer. Now we have the conditions for a quality oriented manufacturing process. This process including the above mentioned checking operations then runs as shown in chapter 1. The results of the machining process as reworking- ($p_{\alpha A}$) and reject-percentages ($p_{\alpha B}$) are compared with the values allowed by the computer of the production process control. If the required value is exceeded by the actual measured value, then necessary process corrections or measures of preventative maintenance will be carned out to correspond to input data. If the necessary manufacturing level < 1 could not be attained, for example by selecting the machine tool then in principl (MT 1), the same method is to be applied as described before. But a main difference in the case of selecting the quality control technology is that 100 % checking of parts is necessary and a essentially larger amount of reject and reworking must be calculated.

SELECTION OF OPTIMIZED CONTROL ALGORITHM

For unattended processes in machining cells the development and application of computer aided and optimized control algorithm are an essential necessity. The aim is to ensure geometrical tolerances of work pieces during batch production by process optimized tool corrections. Important aspects of such control algorithm are several types of automatic control processes and coordinated control ranges. Existing typical process conditions and geometrical demands must be respected considering important determined and stochastic laws. To respect the different influence factors and their variation a model for simulating the geometrical quality behaviour of NC-lathes by using computers was developed. As an assessment criterion of the running simulation programme the attained percentage amount of pieces which must be reworked'or are rejects because they are above the tolerances was chosen. The simulation was made by using the process computer PRS 4000 and the computer EC 1040. From investigations made by process simulation without taking the influence of dead time and rounding-off the tool correction values into consideration, the effiency of control algorithm was received as follows:

- the relatively average decrease in exceeding tolerances in relation to the efficiency of individual value control amounts for a manufacturing level N_F = 0,8 with

adaptive control	80 – 90 %
average value control	50 – 60 %
individual value control with a determined account of exceeding units	30 – 40 %

- the attainable efficiencies vary in a broad spectrum in significant dependence on the proportion of process variance to the specific trend.

CONCLUSION

The erection of universal computer aided systems in design and manufacturing demands the integration of computer aided quality securing (CAQ) at the same level of automation as in all other areas of production. The present contribution shows some fields of tasks and examples of solutions.

REFERENCES

1. H. Würpel, G. Töfke and S. Szyminski,
   Gestaltungslösungen für externe Prüfstationen zur geometrischen Qualitätssicherung bei der automatisierten Drehbearbeitung
   Wiss. Zeitschrift der Techn. Hochschule Magdeburg 28 (84) 4, S. 18-23

2. H. Würpel, G. Töfke and H. Richter
   Zum Einsatz externer Prüfstationen zur Geometrieüberwachung und Meßsteuerung in der roboterbeschickten ROTA-Teilefertigung
   Wiss. Zeitschrift der Techn. Hochschule Magdeburg 29 (85) 5, S. 28-33

3. H. Würpel, B. Niehoff, S. Szyminski, H.-G. Böhme
   Statistische Regelalgorithmen für die Meßsteuerung in automatisierten Fertigungseinheiten der Teilefertigung
   Wiss. Zeitschrift der Techn. Hochschule Magdeburg 30 (86) 5, S. 33-38

4. H. Würpel and S. Szyminski
   Integrierte Qualitätssicherungslösungen - ein notwendiger Bestandteil bedienerarmer Fertigungsstrukturen
   Feingerätetechnik 33 (84) 5, S. 194-195

5. H. Würpel, E. Gottschalk and S. Szyminski
   Zur Gestaltung der Qualitätssicherung für überwachungs-arme Produktionsprozesse beim Einsatz von Beschickungs-robotern
   Fertigungstechnik und Betrieb 34 (84) 4, S. 226-229

AUTOMATIC ACCEPTANCE TESTING OF MACHINE TOOL SUBASSEMBLIES

BY USING CONDITION MONITORING

Wolfgang Weber and Achim Wolf

Department of Technology of Metal-Working Industry
Technical University of Karl-Marx-Stadt
DDR-9010 Karl-Marx-Stadt
German Democratic Republic

INTRODUCTION

The trend toward modern manufacturing concepts in engineering is being increasingly determined by automated machining equipment. It must have a high functional reliability and system availability in order to ensure process capability.

Process capability - in our sense - means to prevent the defective manufacturing of production equipment. To pursue this objective monitoring and diagnostic-devices are effective means to maintain and to increase the availability of machine tools by predetermining failures and detecting faults automatically. It can be recognized that the reliability of machine tools is greatly impeded by an almost complete lack of any reliable data on the subject. In general, the failure of one single element will result in a breakdown of the whole system due to the existing chain structure of the technological equipment. Therefore, each system component must have a high functional reliability.

Increased utilization of machine tools is after all one of the most important objectives of designers, makers, and users of machine tools. Breakdowns are only one cause of lack of utilization. The desire for common work is intended to reduce downtime caused by machine tool breakdowns. It is worth commenting here that increasing the reliability of machine tools and improving the maintenance procedures must necessarily be self-evident. In general, neither makers nor users of machine tools have authoritative objective data on either machine tool reliability or on its relation with maintenance.

Among the effects on the effort of raising machine tool availability today there is a broadened interest in an optimum performance of the production equipment, i.e., guarantee of process capability (Fig. 1).

① INCREASE OF FUNCTIONAL RELIABILITY OF SYSTEM COMPONENTS

② DECREASE IN TIME TO DETECT MODE AND CAUSE OF FAILURE

③ EARLY IDENTIFICATION OF FAILURE BY MEANS OF PROPER
MONITORING AND CONTROLLER SYSTEMS

④ DECREASE IN TIME AND REDUCTION OF RELATIVE MAINTENANCE
EFFORT (FOR RESTORATION)

PREREQUISITES

o KNOWLEDGE ABOUT INFLUENCING PARAMETERS RELEVANT TO
PROCESS CAPABILITY

o KNOWLEDGE ABOUT FAILURES, CAUSES AND EFFECTS -
REDUCTION IN WEAR STAND-BY

Fig. 1 Approach to availability

Consequently, success of untended machining in a flexible-
automated environment will depend on the continuous development
and application of automatic sensing techniques to monitor the
performance of machines and processes.[1]

According to Takata and Schwager[2] monitoring is part of
the diagnostic procedures that are executed continuously during
the operation of the machine tool or machine tool subassembly.
The extent of monitoring depends on the system, but it must
include at least failure detection procedures. In a narrow sense,
the remaining diagnostic procedures which are applied after
detecting a failure or malfunction are called diagnosis. Generally speaking, the diagnosis programme is realized by a separate
monitoring unit which executes diagnostic procedures. The identification of the actual machine state is basically carried out
by a comparison between the starting destination and the over-
running state.

From the viewpoint of the machinery user, particularly in
in practical service of flexible automated manufacturing systems,
there arises, among other things, the constraint to deal with
condition and function monitoring as well as fault diagnosis.
On the other hand, the machine-tool builder in-house automation
necessitates automated run-in and functional tests of the subassemblies if possible. In both cases the acquisition of condition and damage characteristics, e.g. by pattern recognition
as well as tolerance limit signalling and visual readout or
written record of the test results is aimed at.

Therefore, machine and process diagnostics can be structured by a variety of dependences between product, machine-tool
manufacturer, and user.

With regard to this range of problems it is the purpose
of this paper to present an automated complex testing section
within a flexible assembly station of machine-tool subassemblies
in terms of a model equipment.

APPROACH FOR PREPARING AN AUTOMATED FUNCTIONAL TESTING SYSTEM

Especially when working in shifts with reduced personnel, automatic test equipment for functional tests of mechanical components is of extreme urgency. On the other hand, such systems are problem dependent devices by now. All machine tools - whether untended, under limited supervision, or even fully attended - will have to be more reliable than the existing type. Finally, when the facilities do fail, they will have to provide the kind of diagnostic indications.

To achieve an information minimum, machine-tool companies need to learn more about the field performance of their products. In this connection requirements exist for a diagnostic system which continuously monitors machine parameters such as load, current, wear, temperature, vibration, noise levels, and a lot of other factors.

Fig. 2 shows several proper quantities for testing, diagnosis, and monitoring, which may be acceptable criteria. They can be subdivided into three categories, as Fig. 2 demonstrates. Also it should be taken into consideration that a reliable criterion to determine the process capability of a machine tool or a machining centre is not available as yet. This paper is basically focussed on techniques concerned with the condition of the machine and its functions.

Finally, a main purpose of research is to develop a predictive diagnostic system that will warn the user of impending component failure. Important factors are, in this connection, continuous data monitoring, field testing, machine logs, and data analysis.[3]

Indeed a variety of sensors already play important roles in troubleshooting, but an increasing number of sensors on a machine tool does not necessarily improve its reliability. Therefore, the focus of our research is to examine the application of picked up sensing techniques in connection with predictive diagnostic measures.

For the different research approach the main-spindle of machining centres for milling was selected for diagnostic item, i.e., the UUT (unit under test). Figs. 3 and 4 illustrate the fulfilled requirements taken into consideration for problem solution.

QUANTITATIVE-DYNAMIC	QUANTITATIVE-STATIC	QUALITATIVE
o TEMPERATURE	o FRICTIONAL TORQUE BEFORE RUN-IN	o GEAR BOX MECHANISM
o AIRBORNE NOISE	o FRICTIONAL TORQUE AFTER RUN-IN	o OIL LEVEL
o STRUCTUREBORNE NOISE		o SEALING
o MAIN-SPINDLE SPEED	o PARALLELISM	o SOUND
o MOTOR SPEED	o CONCENTRICITY	o TOOL CLAMPING
o CURRENT	o AXIAL DEVIATION	o TOOL EJECTION
o VOLTAGE	o AXIAL SLIP	

Fig. 2 Test quantities

o UNTENDED AUTOMATIC RUN-IN

o PERMANENT AUTOMATIC TEMPERATURE MONITORING

o AUTOMATIC RECORDING OF TEMPERATURE

o REPRODUCIBLE TEST CONDITIONS

o SAFETY REQUIREMENTS (INHERENT SAFETY)

Fig. 3   Dual requirements

REQUIREMENT	UNIT	MINIMUM VALUE	SET LEVEL	IDEAL VALUE
TOLERABLE MEASURING ERROR	K	± 2	± 1	± 0.1
TOLERANCE OF SPEED REGULATION	%	± 10	± 5	± 1
M. T. B. F.	h	1500	2000	5000
USE CLASS (National Standard)	-	+5/+40/ +20/80	+5/+45/ +25/80	-5/+55/ +30/90

Fig. 4   Tolerable requirements

DESIGN CONCEPT AND ASSIGNMENT OF CONTROLLER MODULES

The main-spindle of horizontal milling centres is driven by a d.c. motor which offers a wide, infinitely-variable speed range of 20 to 4000 r.p.m. for the standard model and 6000 r.p.m. for the high-speed type. The spindle runs in precision anti-friction bearings, compensated for axial play.

The life of the antifriction mounting - i.e., the smooth run of the spindle - is strongly influenced by its installation, operating conditions, and the maintenance measures. Machine reliability, efficiency, and safety depend on bearings function-ing properly. Bearings fail prematurely due to operating condi-tion, lubrication, and usage problems, shown in Fig. 5.

Metal fatigue causes every bearing to wear out and fail. Typical failures are (a) excessive torque; (b) metal damage to rolling elements or races; (c) contamination by particles; (d) improper lubrication or poor surface finish; (e) retainer hang-up; (f) improper preload; (g) axis misalignment of assembled set (shaft, bearings).

When a bearing does fail, the consequential damage to asso-ciated machine parts results in a loss of production, which is by far higher than the costs of replacing the bearings.

Replacing bearings after a set number of hours is also risky, since bearings being still serviceable are neednessly removed. Consequently, the best solution is to monitor spindle condition and schedule replacement of bearings systematically - or to carry out reconditioning. By this method production effi-ciency will be advantageously influenced. Several methods are currently applied to monitor spindle condition. Temperature, sonic and ultrasonic noise, and acoustic emission measurement techniques are preferably used.

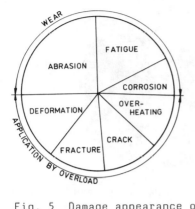

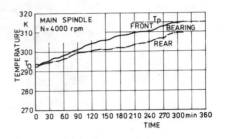

Fig. 5   Damage appearance of   Fig. 8   Temperature vs. time
antifriction bearings   during run-in test

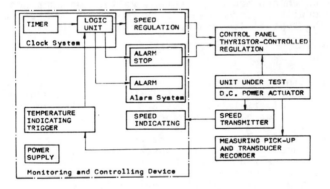

Fig. 6 Functional solution (block diagram)

     The automatic run-in testing is carried out by means of the
monitoring and controlling device including temperature and continuous monitoring and display units as well as a process control
unit which combines all parameters mentioned above in dependence
of time and speed as shown in Fig. 6.

     Vibration detecting, measuring, and analysis are addition-
ally carried out, but a discussion of them is beyond the scope
of this paper. For the required purpose an integrated control
system should be developed, which can be built from a set of
components having been designed for easy assembly being characterized by a uniform modular appearance.

     There are several levels of possible hardware variants.

     The lower-level testing device is developed for application
with commercial measuring instruments such as resistance temperature detectors combined with potentiometric systems (e.g. self-
balancing potentiometer and recording device), etc.

Fig. 7. Modular monitoring and control unit

The advanced level device was designed and developed by research at our department. It consists of a modular electronic package which can accept different voltage-based signals and may be linked with a micro-computer-system, Fig. 7. Fig. 8 illustrates the typical temperature behaviour versus time.

CONCLUSIONS

The main-spindle may be considered as a relevant UUT. The diagnosis and monitoring device to be drafted and developed should have universal character and should also be applicable, in principle, for other items and applications.

Given targets are (a) low-manpower performing both in run-in and acceptance test (validity check) of machine-tool subassemblies; (b) increasing availability of subassemblies and machines; (c) decreasing down-time by means of accelerated fault diagnosis; (d) increasing quality of statement, evidence, and reproducibility of test results com pared with manual acceptance testing; (e) provision of a standardized test record (comprising the real values of the machine); (f) releasing of assembly personnel from routine work; (g) the recommended system is suitable to indicate machine malfunctioning. In connection with the use of newer technologies the diagnostic device ascertains the actual machine state from continuous sensor signal values.

The solution described allows an untended run-in of mounted main-spindles including recording and limit indication of the parameters monitored. This system has been tested for over one year in the 'Fritz-Heckert-Plant' at Karl-Marx-Stadt during the automatic assembly of subassemblies.

Future work will focus on developing diagnostic programmes to check out system components of flexible automatic manufacturing systems in order to develop a predictive diagnostic system that will warn the user of impending component failure.

REFERENCES

1. M. Weck, Automatisierte Fertigungsüberwachung, Ind.-Anz. 56:86 (1984).
2. S. Takata, An diagnostic program generator for sequentielly controlled machines, Angew. Inform. 8:421 (1982).
3. G. Schaffer, Sensors, Americ. Mach. 7:109 (1983).

478

INTELLIGENT MEASUREMENTS IN FINE MECHANICS USING LASERS AND MICROCOMPUTERS

N. Miron*, M. Petre **, D. Sporea*

*Central Institute of Physics, Lasers Dept.

Bucharest, Romania

** Mecanica Fina Enterprise
Bucharest, Romania

INTRODUCTION

Technological advances in today fine mechanics require sophisticated measuring equipment. Some of the most featuring needs of a modern measuring equipment are high accuracy, short measuring time and computational capabilities to process collected data.

Laser-based equipment, controlled by microcomputers is a valuable tool in fine mechanics, due to the excellent stability of the used length standard, in this case the wavelength of a He-Ne laser and due to the microcomputer which gives flexibility to the whole equipment.

Two laser-based instruments controlled by microcomputer will be further described, together with a recent advance in length measuring accuracy.

DIAL INDICATORS VERIFICATION EQUIPMENT

Dial indicators are mechanical measuring instruments used very often in accurate machining. Manufacturers of these instruments are interested to get a quick and complete characterization of their products: accuracy error, the hysterezis error and the fidelity error. The user needs an error diagram of the dial indicator over the whole measuring range, to know the most accurate interval in the measuring range.

Starting from these goals, a method and an equipment for dial indicators verification were developed. [1,2] According to the method, the linear displacement of the dial indicator rod is accurately measured in certain verification points over the measuring range, in forward and backward movement. The difference between the measured displacement and the expected displacement is the accuracy error in every verification point. For each type of dial indicator is a definite number of verification points, and an accuracy error for every point. From these accuracy errors in every point, a microcomputer computes the accuracy error for the whole range, the hysteresis error and the fidelity error.

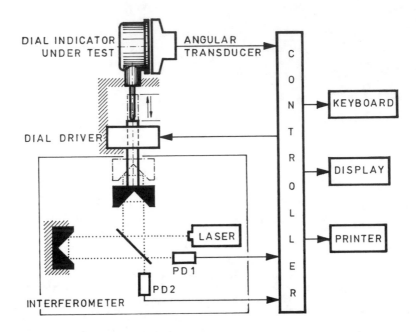

Fig. 1   The block-diagram of the equipment for dial indicators
         implication

The block diagram of the equipment for dial indicators verification
is shown in figure 1. The dial indicator under test is kept in a holder.
Its rod is in mechanical contact with the driving spindle of a driving
mechanism. A special angular transducer is put over the dial indicator
face, for remote sensing the position of the dial. Signals given by this
angular transducer, when the dial indicator rod is moved forward and backward, correspond to the verification points. The linear displacement of
the dial indicator is measured by a Michelson interferometer, with a
He-Ne laser.

The operation of the whole equipment is governed by a controller (a
microcomputer). It sends a command to the dial driver, to move forward and
backward the dial indicator rod, and receives data: signals from the angular transducer considered as the verification points, and from the
interferometer photodetectors PD1, PD2, considered as the linear displacement signals. From these data are depicted the accuracy error, the hysterezis error and the fidelity error, displayed at the end of the verification. The results are also available on a printer. The controller receives commands from the operator via a keyboard console.

Main technical data:
     - verification accuracy less than 1/3 graduation of the dial indica-
tor under test, but not better than 0.0004 mm;
         - displacement speed of the dial driver spindle is about 1 mm/sec;
         - up to five verification cycles for each dial indicator under test.

480

GRADUATED RULES VERIFICATION EQUIPMENT

Graduated rules are the key-element giving the measuring accuracy of the optical readers, used mainly in machine building.

The verification of the graduated rules by an operator is a time-consuming task, and is affected by operator's errors. The automated verification avoids all the human errors and speeds-up the measurement. The goals of an equipment for graduated rules verification are: checking the graduation error over an operator-defined legnth, plotting the error diagram either over the whole or over a limited length, and taking a decision of accepted or rejected rule, according to defined criteria.

A laser-based quipment for graduated rules verification was developed, according to Abbe's principle (figure 2).

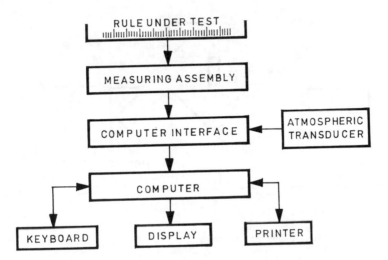

Fig. 2   Block-diagram of the graduated rules verification equipment

The spacing between any two graduations of the rule is measured with a Michelson interferometer, having the beam in the moving arm, strictly parallel with the rule under test. The measuring assembly has a remote-sensor for rule graduations, held on a carriage that moves over the verification length of the rule, under computer control. Due to the long measured length and to the high accuracy needed, the measurements made with the laser interferometer must be corrected with the atmospheric parameters: temperature, pressure and humidity. These corrections are performed automatically using the signals given by temperature, pressure and humidity transducers. The temperature of the rule under test is also taken into account, to express the length at the reference temperature. The measurement results are displayed on computer cathode-ray tube in alphanumeric and graphic form and are available also to a graphic printer.

Main technical data:

- verification range: 1 - 300 mm
- verification accuracy: 0.002 mm/m
- operating conditions:
    temperature: $15^{\circ}C - 25^{\circ}C$,
    air pressure: 860 - 1060 mbar
    relative air humidity: 20% - 80%.

ELECTROOPTICAL PHASE SHIFT CONTROL

Interferometric measurements of linear displacements have an accuracy of $\lambda/2$ or $\lambda/4$. Very few industrial measuring systems have an accuracy of $\lambda/32$ [3] got by electronic means (PLL technique). A true fringe interpolation having a much better accuracy was obtained by an electrooptical phase shift control [4], into a Michelson interferometer (figure 3).

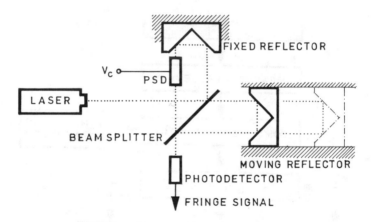

Fig. 3   The optical setup for interferometric measurements using electrooptical phase shift control

We made an electrooptical phase shifting device (PSD), located in the unmoved arm of the interferometer which introduces an optical phase shift in $0 - 2\pi$ range, when a voltage V_c is applied to its electrodes. The optical phase shift is linear dependent of the control voltage V_c.

From the preliminary experiments made with the electrooptical phase shift control into an optical setup as in fig. 3, can be depicted two main features:
- improved accuracy $\lambda/100$ in true interpolation without averaging;
- the use of a single frequency stabilized laser as a coherent light source.

REFERENCES

1. N.Miron, V. Vlad, D. Sporea, and M. Petre, Romanian Patent No. 71070.
2. N. Miron, V. Vlad, D. Sporea and M. Petre, U.S. Patent No. 4378160
3. R. C. Quenelle, B. G. Gordon and A. F. Rude, Hewlett Packard Journal, 34: 4, 3 (1983)
4. N. Miron, Signal Processing Methods in Laser Interferometry and Anemometry, Ph.D.Thesis, Central Inst.of Physics, Buch.,1983

LOCAL HEAT TRANSFER SENSOR FOR

MACHINE TOOLS

Jerzy Jędrzejewski, Kazimierz Buchman,
Bogusław Reifur

Wrocław Technical University,
Institute of Production Engineering
50-370 Wrocław, Poland

INTRODUCTION

Machine tools and other precise machines impose strict
conditions on heat transfer processes within their structure.
Heat transfer   is mainly due to convection forced by
rotating parts and is still far from being adequately acco-
unted for at the design stage[1]. Local values of the heat
transfer coefficient /h.t.c./ need to be controlled according
to operating conditions of a machine tool. It is also desirable to have them given in an analytic form so that they
could be used in computer evaluation of temperature distribution and thermal displacements done by a machine tool designer. Accuracy of the numerical values of the h.t.c. introduced into computer programs has a considerable effect on the
calculated temperatures and displacements. There is a need
for a new, simple kind of h.t.c. sensor since classical
alphacalorimeters mounted in the body walls of a machine tool
are difficult to operate and destructive to the tested object.

THEORETICAL PRINCIPLES OF THE SENSOR CONSTRUCTION

The sensor /also called alphacalorimeter/ has the form
of a metal disc placed in a seating made of thermal insulator,
the two elements being mounted on a level with the so-called
substitute wall /Fig.1/. As can be seen, the method requires
no holes in the tested wall. The substitute wall must be
shaped in such a way so as not to disturb the actual air flow
and in addition to this its temperature must be approximately
the same as that of the tested wall, the latter condition
being satisfied by using an electric heater 1. The front face
of the alphacalorimeter forms a common plane with the substitute wall and its heat transfer characteristics are the same
as those of the substitute wall and therefore the same as
those of the tested wall. Measuring the α coefficient for the
sensor face is therefore meant to be eqivalent to measuring
the real value of α .

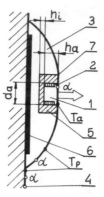

Fig.1 Alphacalorimeter placed in the substitute wall, where
1-alphacalorimeter, 2-measurement surface, 3-substitute wall surface, 4-tested surface, 5-insulator,
6-heater 1, 7-heater 2

From the theory of a steady thermal state the relationship can be derived

$$\alpha = \frac{C}{A \cdot \gamma} \cdot m \qquad (1)$$

where the following characteristics of an alphacalorimeter
are used:

    C - heat capacity
    A - measuring surface area
    γ - coefficient of nonuniformity of temperature distribution
    m - rate of cooling

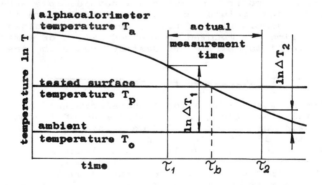

Fig.2 Temperature changes involved in the α coefficient
measurement

In order to determine the value of α, the rate of cooling
must be found first

$$m = \frac{\ln \Delta T_1 - \ln \Delta T_2}{\tau_2 - \tau_1} \qquad (2)$$

where $\Delta T_1, \Delta T_2$ are the increases in alphacalorimeter tempera-

484

ture in excess of temperature of the surroundings measured
during the time $\tau_2 - \tau_1$ /see Fig.2/. The time $\tau_2 - \tau_1$ should
be so chosen as to make sure that the function $\ln\triangle T$ is a
straight line. This means that a sensor is in a steady thermal state.

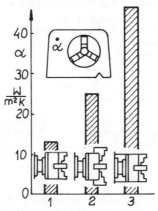

Fig.3 Effect of the spindle configuration on   coefficient
     value

The coefficient γ in Eq.$\left(1\right)$ is given by

$$\gamma = \frac{d/\triangle T_A/}{d/\triangle T_v/} = \frac{v}{A}\frac{\iint_A \triangle T_A dA}{\iiint_v \triangle T_v dV} \qquad \left(3\right)$$

and is the ratio of the mean temperature increase $\triangle T_A$ of the
object surface A to the mean temperature increase $\triangle T_v$ occurring within the volume v. It is a measure of inhomogeneity in
temperature distribution over the investigated object, and in
the Biot number.
The assumption $\gamma = 1$ applied to the alphacalorimeter under
consideration can lead to considerable errors in determining
the coefficient   value. This is due to the fact that the
sensor is not located within a homogeneous environment - its
reference face, giving out heat to the surroundings, is in
contact with the air, while the other parts are in contact
with heat insulation 5 /Fig.1/.
Determination of the coefficient γ value for a given alphacalorimeter is a time - consuming task. An assessment of the
influence of design factors /such as the alphacalorimeter
dimensions and materials and the thickness and material of the
insulation/ as well as of operational factors and actual
measurement time was made on the basis of numerical analysis
of temperature distribution over the machine body in which
the alphacalorimeter was mounted.
A γ value close to unity can be attained if the alphacalorimeter is built according to the requirements of Table I.
Then the error in determining the γ value will not be greater
than 10%. A more precise estimate can be obtained by using
relationship $\left(2\right)$. The maximum error to be expected is then
less than 4%.

Table 1

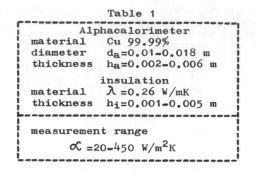

```
Alphacalorimeter
material     Cu 99.99%
diameter     da=0.01-0.018 m
thickness    ha=0.002-0.006 m

insulation
material     λ =0.26 W/mK
thickness    hi=0.001-0.005 m

measurement range
    α =20-450 W/m²K
```

Prior to measurement, the sensor temperature must be raised by heater 2 /see Fig.1/ above the temperature of the tested wall, the excess of temperature being such that on cooling during the time $\tau_2 - \tau_1$ the sensor must maintain its steady state. The measurement should be carried out at the time τ_b, when temperatures of the sensor and of the tested wall are close.

The temperature of the substitute wall can be made to match that of the actual headstock wall through the use of heater 1. Temperature rises ΔT can be measured by thermocouples connected with a digital meter and a printer. Further improvement can be achieved by coupling the two devices with a minicomputer that chooses by itself an optimum value of the measurement time $\tau_2 - \tau_1$.

The coefficient γ/Eq.1/ for the present sensor is slightly different from unity. A maximum error due to assuming that $\gamma = 1$ is approximately equal to 10%. The computer aided experiment can account for a real value of the γ coefficient and then the error can be reduced to a few percent. If the sensor is built according to specifications of Table 1 the maximum error introduced by the remaining quantities in Eq. (1) will not exceed 4%.

An unquestionable advantage of the presented method is a short measurement time, not exceeding a few tens of seconds[2], as was shown in our earlier experiments.

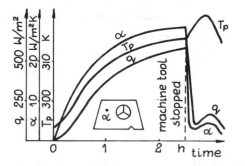

Fig.4 Change in α value, in heat flow q and body temperature T_p for an operating machine tool

The alphacalorimeter is intended to determine the rate of cooling m, as obtained from measurements of the temperatures

of the surface and of the surroundings, the heat capacity of
the alphacalorimeter and the heat transfer coefficient α
value being given. The alphacalorimeter can therefore be
used to determine:
- the rate of cooling m
- the heat transfer coefficient α and
- the heat flux $q = \alpha/T_o-T_A/$ from the investigated surface
  to the surroundings.

APPLICATION EXAMPLES

The present sensor may be used to determine local values
of the h.t.c. for heat flow given out by the machine tool body
to the surroundings and within the body. A designer is thus
given a quantitative picture of thermal processes occuring
within the machine tool structure and is able to control them
in order to achieve the desired dimensional stability.

Fig.5 The α coefficient value distribution

Fig.3 shows the effect the protrusion of the lathe chuck jaws
has on the α coefficient and temperature at a chosen point
of the headstock /bar 1,2/ and the effect of additional elements causing air flow /bar 3/.

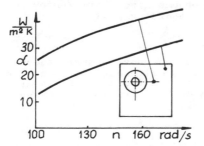

Fig.6 Effect of the spindle rotating speed on α value

Fig.4 shows changes in value of α, in temperature and stream
characteristic at a fixed point within time, during which a
machine tool was operated at a constant rotational speed of

the spindle. Such relationship permits local conditions of
unsteady heat transfer to be described exactly. The distribution of α value over the body walls is usually very complex
/see Fig.5/ and the exact description of the heat transfer
distribution is difficult to achieve. A general law governing
the heat transfer would also have to allow for the effect of
rotational speed that, as judged from Fig.6, can be very
important.

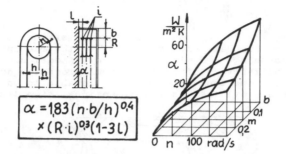

Fig.7 Heat transfer forced by an operating belt transmission

A sufficient amount of experimental data can, however, yield
working mathematical relationships to be used in determining
temperatures and thermal deformations. An example of such a
relationship found for a belt transmission is presented in
Fig.7.
Basically, the alphacalorimeter can be used to measure the α
coeficient over the range from 20 to 450 W/m^{2}K. Measurements
below 20 W/m^{2}K are subject to unacceptably large errors due
to the inherent indeterminability of temperature evaluation.
Temperature changes within the sensor associated with low
values of α are so subtle that even minor inaccuracies give
rise to large errors in determining the rate of cooling.
This difficulty can be overcome to some extent by using more
sensitive equipment.
The upper limit value of 450 W/m^{2}K is dictated by the measurement time which becomes too short /under 20 seconds/ and
thus requires fast recording devices.

CONCLUSION

     The meter presented above assures adequate accuracy and
the measurement time short enough to monitor nonstationary
thermal processes. It can therefore serve as a useful device
for quick analysis of heat transfer processes within the
machine tool structure.

REFERENCES

1. Jędrzejewski J., Buchman K.: Heat transfer in lathe headstock /in German/ Industrie Anzeiger 1977, 77, pp 1436-1439
2. Buchman K., Reifur B.: Alphacalorymetric measuring method
   of coefficient of heat transfer from machine tool body to
   environment. Scientific Papers of the Institute of Machine
   Building Technology of Wrocław Technical University 1985,
   No 29, Monographs No 5

DYNAMICAL BEHAVIOUR OF **SHEATHED** THERMOCOUPLES UNDER FIELD CONDITIONS

Bernhard Sarnes

Lehrstuhl für Elektrische Meßtechnik
Technische Universität
München, **Federal Republic of Germany**

1. INTRODUCTION

In some applications a deconvolution circuit is used to speed up the dynamics of thermocouples. The speed up circuit needs exact thermocouple time-constants for correct operation. Wrong time constants produce over- or undercompensation of thermocouple inertia. Pre-determination of time constants in laboratory is insufficient, as the found time constants are valid only for one well defined operating situation (surrounding medium, temperature, speed, turbulence). Under field conditions these parameters vary considerably and cause a change of time constants. The following table shows the variation for an 8 mm thermocouple (iron constantan)[2].

Table 1   Time constants for different operatingsituations

operating situation		90 % time-constant
water speed	0,2 m/s (30 °C)	1,68 s
	0,35 m/s (30 °C)	1,15 s
water temperature	20 °C (0,2 m/s)	1,97 s
	40 °C (0,2 m/s)	1,63 s

Therefore, to reach acceptable deconvolution results, on-line identification of thermocouple dynamic behaviour is unavoidable. One possible method is to heat up the thermocouple internally and measure time-constants. Identification also provides the possiblity to observe the conditions of the surrounding medium with the same sensor.

2. DYNAMIC BEHAVIOUR OF THERMOCOUPLES

The Fourier differential equation and the heat-transfer coefficient between thermocouple and medium describe the thermocouple dynamic behaviour. Because of material constants not exactly known and the great computation effort a modell with concentrated parameters is used to approximate the thermocouple. This model results in a linear system:

$$H(p) = \frac{1}{1+pT_1} \cdot \frac{1}{1+pT_2} \cdot \dots \cdot \frac{1}{1+pT_n}$$

$H(p)$ = transfer function
T_i = time constants

In many applications a second order transfer function

$$H(p) = \frac{1}{1+pT_1} \cdot \frac{1}{1+pT_2}$$

is an adequate model of the thermocouple. Figure 1 shows the components for forward operation.

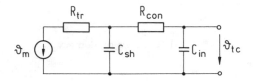

Fig. 1   Forward operation of a thermocouple
R_{tr} = thermal resistance between thermocouple and medium
R_{con} = thermal resistance for conduction in thermocouple
C_{sh} = thermal capacitance of sheath
C_{in} = thermal capacitance of insulation and wires
ϑ_m = temperature of medium
ϑ_{tc} = temperature at thermo-junction

$$H_f(p) = \frac{\vartheta_{tc}(p)}{\vartheta_m(p)} = \frac{1}{p^2 \cdot R_{tr}R_{con}C_{in}C_{sh} + p(R_{tr}C_{sh} + R_{tr}C_{in} + R_{con}C_{in}) + 1}$$

Operating in inverse mode a electric current heats up the thermocouple internally. The current produces a constant heat transfer rate q.

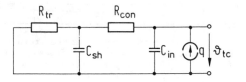

Fig. 2   Inverse operation of a thermocouple

The transfer function for inverse operation is

$$H_i(p) = \frac{\vartheta_{tc}(p)}{q(p)} = \frac{R_{tr} + R_{con} + p \cdot R_{tr} \cdot R_{con} \cdot C_{sh}}{p^2 R_{tr}R_{con}C_{in} \cdot C_{sh} + p(R_{tr}C_{sh} + R_{tr}C_{in} + R_{con}C_{in}) + 1}$$

490

The denominators of both transfer functions are identical. Therefore internal heating is a way to find the thermocouple time constants. The numerators are different. This produces different thermocouple step responses in the time domain. H_f corresponds to a difference of 2 exponential functions (delay time), H_i to a sum of 2 exponential functions (no delay time).

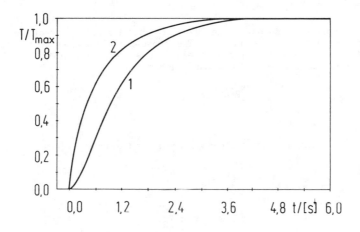

Fig. 3 Thermocouple step response for forward (1) and inverse (2) operation

3. CIRCUIT CONCEPT

The equipment consists of a power amplifier for heating up the thermocouple, the measuring amplifier, an analog digital converter and the microcomputer with display and keyboard. The power amplifier produces a pseudo random binary signal. Alternating current with 1 kHz frequency switched on and off causes Joule heating, direct current with two polarities causes Peltier heating. For the very short measuring time the power amplifier is in high impedance state and the measuring amplifier is connected with the thermocouple. The power amplifier produces temperature steps of 1 K in water and 10 K in air.

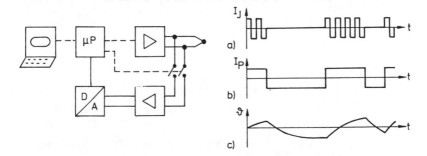

Fig. 4 Equipment for internal heating technique and signals:
a) Joule-heating   b) Peltier-heating   c) junction-temperature

4. SIGNAL ANALYSIS

The thermocouple excitation signal and the response signals are sampled and stored. The parameters (a_1, a_2, b_1, b_2) of the z-transformation

$$H(z) = \frac{b_1 z^{-1} + b_2 z^{-2}}{1 + a_1 z^{-1} + a_2 z^{-2}}$$

are computed by means of the COR-method[3]. This method involves correlation and least-squares-fitting. The cross-correlation between excitation and response signal suppresses extranoues noise and drift. In the next step H(z) is divided into partial fractions

$$H(z) = \frac{K_1 \cdot z^{-1}}{1 + C_1 z^{-1}} + \frac{K_2 \cdot z^{-1}}{1 + C_2 z^{-1}}$$

The correspondence between z-domain and p-domain

$$\frac{1}{1 - e^{-\Delta t / T_i} \cdot z^{-1}} \quad \leftrightarrow \quad \frac{K}{1 + p T_i}$$

turns out the time constants T_1 and T_2.

5. MEASURING RESULTS

The measuring time is dependent on the thermocouple diameter and the medium. Typical measuring times for a 3 mm Chromel/Alumel thermocouple are 2 min in water and 20 min in air. The random measurement error amounts to ±5 % of the time constant ±0,05 s. For Chromel/Alumel thermocouples (not grounded) the following results were obtained:

Table 2.   Time constants of Chromel/Alumel thermocouples

Diameter	Medium (23 °C)	T_1 [s]	T_2 [s]
3 mm	water 0,2 m/s water natural convection air turbulent convection	0,11 0,14 4,1	0,73 1,90 24,0
1 mm	water 0,2 m/s water natural convection air turbulent convection	0,02 0,04 0,69	0,15 1,30 3,91

6. REFERENCES

1.   D. Hofmann, Dynamische Temperaturmessung, VEB Verlag Technik, Berlin (1976).
2.   D. Huhnke, Über Methoden zur Messung der dynamischen Eigenschaften von elektrischen Berührungsthermometern, Doctoral Thesis (1971).
3.   R. Isermann, Prozeßidentifikation, Springer, Berlin-Heidelberg-New York (1974).
4.   E. Schrüfer, Elektrische Meßtechnik, Hanser Verlag, München (1983).

492

A NEW SYSTEM FOR MEASURING AND CONTROLLING THE FABRIC TEMPERATURE PROFILE

DURING DRYING AND FINISHING OF FABRIC IN TENTER RANGES

Dieter Rall

Trans-Met Engineering, Inc.
Anaheim, California, U.S.A.

INTRODUCTION

In the processing of fabric there are basically two operations which are performed in large ovens called tenter ranges. The first operation consists of drying the fabric after completing operations such as bleaching, dyeing or the application of finishing resins. The second operation consists of curing the finish on the fabric. As the fabric passes through the oven, heated air is blown against it to drive off the moisture. The ovens are generally zoned and air temperature is controlled in each zone. During drying the fabric is actually cooled by the evaporation process. The distance from the entrances, at which the free moisture removal is completed and the fabric begins to approach the air temperature, will vary with such factors as: initial moisture content of the fabric, weight of the fabric, temperature and relative humidity of the heated air, velocity and manner of air impingement on the fabric, and speed of the fabric through the range.

To make the ovens as energy-efficient as possible, it is necessary to know the point where the fabric is dry to prevent over- or under-drying of the fabric. Over-drying wastes energy and may damage the product while under-drying wastes energy because the product will need to be rerun. The point of dryness can be identified by a sudden rise in fabric temperature above the wet bulb drying temperature.[1] The following discussion describes a system which employs a unique method of surface temperature measurement to determine this point.[2] A graphical display presents a real-time fabric temperature profile along the fabric path to assist the operator in maintaining proper range operation. A cascade system for automatic control of the fabric temperature and hence, point of dryness, is also described.

SYSTEM DESCRIPTION

The system consists of a central graphical display together with a series of sensing heads located along the path of the fabric through the tenter range (see figure 1). The necessary signal-conditioning electronics for each sensing head is an integral part of the graphical display system (see figure 2). A typical sensing head installation is shown in figure 3.

The sensing head consists of an aluminum sensor body and associated housing in which is embedded a platinum RTD (Resistance Temperature Detector) to accurately measure its temperature. A heat flow sensor is embedded in the face of the body. This sensor is a fast-response differential thermopile which generates a millivolt signal directly proportional to the rate and direction of heat flow into or out of the face of the sensor body. When the sensor is placed in close proximity to the fabric, this signal is directly proportional to the temperature difference between the body and the surface of the fabric. The position of the sensor, relative to the fabric, is fixed by a Teflon-coated guide dome which rides on the fabric. The fabric surface temperature is obtained by the algebraic sum of the temperature difference and the sensor body temperature. Each measured temperature point is presented on the bar graph display to allow rapid and convenient interpretation of the fabric temperature profile. Each point can also be read digitally by means of a pushbutton selector switch.

THE SENSING HEAD

Figure 4 is a detailed crossectional diagram of the sensing head. No attempt is made to actively control the body temperature. Rather the sensing head body is allowed to passively reach an equilibrium temperature. This sensing head temperature results from the guide dome contact with the fabric and the hot air from the oven. The insulation between the body and housing assures that the body will run closer to the fabric temperature than the air temperature. Under normal operating conditions the sensing body runs from 5 to $30^{\circ}F$ warmer than the fabric.

PRINCIPAL OF OPERATION

The heat flow sensor is used to compare the unknown fabric temperature to the known sensor body temperature by measuring the convective heat flow across a narrow air gap.[3] Referring to figure 4, the rate of heat transfer across the gap is expressed as:

$$\dot{Q}=(\frac{k}{d}+h)\ T,$$

Where: $\dot{Q}$=heat transfer across the gap, k=thermal conductance of the gap, d=distance across the gap, h=effective heat transfer coefficient, and $\Delta T=T_p-T_H=$ (the temperature difference between product surface and sensing head). The thermal conductance of the gap (k) can be considered constant for sensing head temperatures within $+50^{\circ}F$ of the product temperature. If the air gap distance (d) is small and constant, the heat transfer coefficient (h) turns out to be invariant and essentially zero over a broad range of product velocity. Consequently, the heat transfer across the air gap is directly proportional to the temperature difference:

$$\dot{Q}\propto\Delta T.$$

The rate of heat transfer across the air gap is measured by the heat flow sensor in the face of the sensor body. This signal is pre-calibrated in the laboratory. The resulting algebraic sum of the calibrated heat flow signal and the head temperature gives the fabric surface temperature:

$$\Delta T + T_H = (T_P - T_H)+ T_H = T_P$$

The accuracy of the product temperature measurement will be within $\pm 1/2\%$ of measuring range, $(0-500^{\circ}F)$, as long as the sensing head temperature is within $\pm 50^{\circ}F$ of the product temperature. This condition is assured by the sensing head design.

494

The response time of the system is determined by the time constant of the heat flow sensor. Sensors are designed for a time constant of one second. Time constants as fast as 0.250 sec. have been achieved. (Time constant = 63.2% of a step in surface temperature).

The heat flow sensor signal is calibrated against a standard reference temperature body to establish a known ΔT. The temperature sensor in this reference body is periodically calibrated against standards traceable to the National Bureau of Standards.

AUTOMATIC CONTROL

Tenter ranges are generally divided into several zones in which the air temperature is individually controlled. it is not practical nor safe to attempt fabric temperature control by direct feedback to the burner control. Instead, a cascade scheme of control is employed. This consists of using two controllers. One controller sets the basic air temperature in the zone while the second generates a control signal, based on desired fabric temperature. The output of the second controller resets the set-point of the first air temperature control. Limits on the range of the cascade control signal prevent creating unacceptably large swings in air temperature. Figure 5 shows the cascade control in diagramic form. Full 3-mode (PID) control is available with bumpless transfer when changing from manual to automatic control.

TEST RESULTS

Typical results for processing fabrics of the same weight at different speeds are shown in figure 6. In these particular cases the range was operated to first dry the fabric and then heat it to the cure temperature of the finish. The finish had been applied by two different methods, pad application and kiss roll application, resulting in different amounts of initial moisture in the fabric at entry. It is apparent that for the pad application, and a speed of 65 ypm, the fabric was dry within the first 10 feet from the entrance. Because the air temperature in the following zone #2 was too high, significant over-heating occurred during this phase. In addition, this high curing temperature was held for an excessive period of time. These operating conditions were typical for the ranges tested before benefit of fabric temperature information. With the reliable temperature profile information, the operator was able to increase speed form 65 ypm to 100 ypm (an increase of 54% in speed) and reduce the air temperature in the second zone to achieve dryness at about 30 feet from the entrance, prevent over-heating and allow enough time for the fabric to rise to the desired sure temperature for the required dwell time. Increasing the speed to 120 ypm forced dryness to be completed at about 45 feet from the entrance but did not allow sufficient dwell time for the fabric to reach the desired cure temperature prior to exiting the frame. If only drying were to be achieved the temperature profile would look like the bottom cure in figure 6.

SUMMARY

The key results can be summarized as follows:
1. Fabric temperature can be accurately and precisely controlled to eliminate over-heating or under-drying, thereby assuring consistent quality and eliminating reruns.

2. By monitoring and controlling fabric temperature, tenter ranges can be operated more efficiently to increase production, increase profit margins and minimize energy consumption. Average operating speeds have commonly been increased by 25% with this system of control.

REFERENCES

1. Steele, B., Controlling Fabric Moisture Content, American Duestuff Reporter, October 1977, Page 35, 36, 75.

2. Hornbaker, D.R., Rall, D.L., The Convective Null-Heat Balance Concept for Non-Contact Temperature Measurement of Sheets, Rolls, Fiber and Wire, "Temperature: It´s Measurement and Control in Science and Industry," Vol. 4, Page 737-747, Instrument Society of America, Pittsburgh, 1972.

3. United States and Foreign Patents Issued.

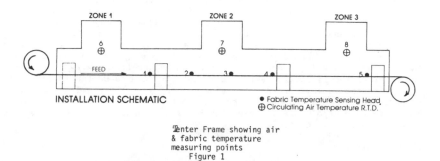

Tenter Frame showing air
& fabric temperature
measuring points
Figure 1

Figure 2

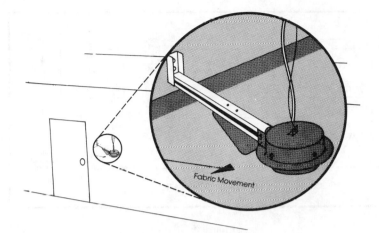

Figure 3

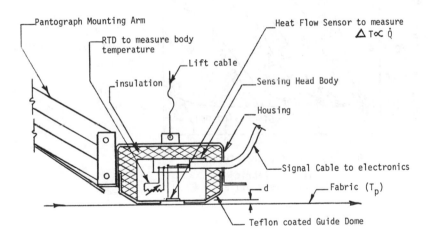

Pantograph Mounting Arm

RTD to measure body temperature

insulation

Lift cable

Heat Flow Sensor to measure $\Delta T \propto \dot{Q}$

Sensing Head Body

Housing

Signal Cable to electronics

d

Fabric (T_p)

Teflon coated Guide Dome

Crossection of sensing head
Figure 4

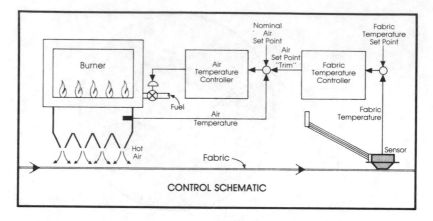

Figure 5

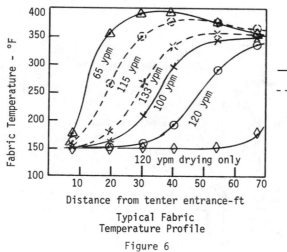

Pad application

--- Kiss Roll application

Typical Fabric
Temperature Profile

Figure 6

CAPTURE RANGE MEASUREMENT IN ANALOG IC TEST SYSTEM

Andrzej Krzysztof Wach

Industrial Institute of Electronics
ul. Długa 44/50
00-241 Warszawa, Poland

INTRODUCTION

The world market offers a large number of analog
IC types which are phase-locked loop /PLL/ circuits on their
own or include a PLL circuit in their structure. Capture range
is one of the most important PLL parameters to   be measured
in the production process of these ICs.

MEASUREMENT METHOD

The capture range of a PLL is determined by measuring the
upper and lower limit frequencies of this range. The use of
capture range measurement in ICs production process has imposed
a necessity of developing such a measurement method that would
permit the measurement time minimization and be suited for
automatic performance.
Consequently, a stepwise tuning of test signal frequency has
been accepted in the method presented.
A test signal having a definite frequency is generated for a
time period longer than the PLL pull-in time. Then, from the
response of the tested loop, a new frequency is generated
again. Before each step /new value of test signal frequency/,
some initial condition are set; the term "initial conditions"
should be understood as the VCO tuning-in to the center fre -
quency and, in case the lock-in condition has been acquired in
previous step, the throwing of the tested circuit out of lock.
The number of steps performed depends on the search method
used and the measurement accuracy required. From the single-
-parameter search method available, binary search method has
been chosen as being one of the most effective ones and very
convenient for programmable realization.

MEASURING CIRCUIT

The measuring circuit /Fig. 1/ consist of a programmable-
-frequency generator and microcomputer circuit. The generator
is arranged using a PLL loop and a three-decade programmable,

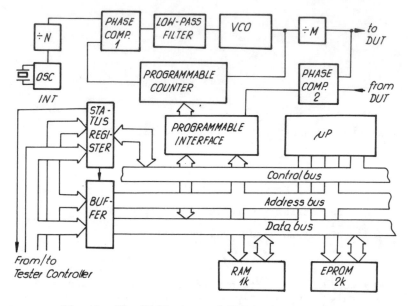

Fig. 1 Block diagram of PLL capture range
measurement circuit

reversible counter. A crystal-controlled oscillator is used
as frequency standard.

Having written zero to this counter, the generator forces, at
the DUT input, a signal having its frequency equal to the DUT
center frequency, irrespective of the counter counting direction. The programmable state of the counter is thereby a direct
percentage value of frequency deviation from the center frequency.

The microcomputer is built around a microproccesor, RAM memory
used as a stack, programmable parallel interface which stores
the counter settings and the DUT response, and an EPROM which
stores the program to control the operation of entire system.
The operating principle of the measuring circuit consist in
the generation of a test signal of a definite frequency, applied to DUT input, and monitoring, in a defined time, the state of
a phase comparator /2/ which flags the DUT VCO lock-in condition with test signal frequency. The circuit permits the determination of both capture range limits by programming the
change in counting direction and repeating the execution of
the system control program two times.

MEASUREMENT ALGORITHM /FIG. 2/

The measurement of one limit is made in form of n steps
where n depends on measuring range span and measurement error
accepted. Initial conditions are set before each step, and the
center frequency of DUT is automatically set by the circuit
included in the measuring system head. Then, in the first
step, a test signal is generated having its frequency deviation
from the center frequency equal to half the measurement range.
If the circuit does not lock within a time than pull-in time,

500

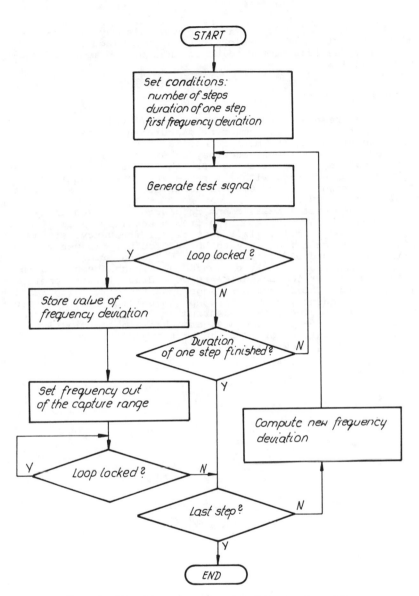

Fig. 2 Flowdiagram of measurement algorithm
for one limit of capture range

initial conditions are re-set, and the next step provides the
generation of a test signal with a frequency deviation equal
to half of the former setting. If the loop has locked, the
current setting is stored in memory, and in the next step, a
test signal is generated with a new setting equal to the sum
of former setting.

The number of n steps is performed this way and all the values of deviations for which locking took place are stored in memory. The last stored value is that of the measured capture range limit.

MEASUREMENT ERROR

It has been assumed, at the stage of development of measuring method that the number of measurement steps will chosen according to the measurement range span and accuracy required. The measurement accuracy is directly improved by increasing the number of steps with the measurement range span unchange. For example, with 10% range and seven measurement steps, the resolution of 0,1% is achieved; with ten steps, the resolution is 0,01%.

The theoretical measurement resolution is essentially affected by the errors resulting from a specific realization of measurement circuit, viz.: test signal frequency setting error, test signal frequency to DUT VCO output frequency comparison error and DUT center frequency tuning error.

The measuring circuit presented herein has been used in a test system for the measurements of stereo circuits emplo ing PLL loop. The measuremet range amounts to 10% with respect to center frequency, number of steps amounts to ten, and resolution is 0,01%. The total measuring error is 0,005% with reference to the center frequency, and the main contributive components to this error are: the test signal frequency setting and frequency comparis on errors.

COMPUTER AIDED POLYMERIZATION BASED ON ULTRASONIC SENSORS

Peter Hauptmann

Technische Hochschule Magdeburg
Sektion Technische Kybernetik und Elektrotechnik,   3010 Magdeburg, GDR

INTRODUCTION

The various applications of ultrasonic methods in science,
technology, medicine, and other fields have been known for
a long time. The fact that considerable advances have been
made in the last ten years in the use of ultrasound, demonstrate
its capability and usefulness. On the other hand, detailed
ultrasonic studies for monitoring industrial processes are
scarcely known. The aim of this paper is to illustrate an
ultrasonic velocity method for the process control during
polymerization reactions. Many polymers are produced in great
quantities in different polymerization processes. But the
knowledges on the process condition are rather small. The
main reason for this unsatisfactory situation is the deficit
of suitable process parameters or the impossibility of their
measuring during the process. The ultrasound presents one
possibility to get an additional information about the momentary
polymerization state. A real time computer based system can
be realized.

METHOD

Ultrasonic methods can work in the cw- or aw-regime.
For the process control the pulse methods are applied most.
In this technique the transducer converts the electrical pulse
into an acoustical pulse, that is transmitted into the medium
and is received by a second transducer (pulse travelling method).
Piezoceramic transducers of the resonance frequency 1 MHz are
the applied sensors. The ultrasonic velocity or the absorption
are the measuring parameters. Taking as a basis the alternating
sonic pressure in the investigated medium it holds

$$P = P_o e^{j\omega (t-x/c)} e^{-\alpha x} \tag{1}$$

c is the ultrasonic velocity and α is the absorption coefficient.
t is the time, ω is the angular velocity and x is the way passed
from the sonic wave, P_o is the alternating sonic pressure
at x=0.

From equation (1) we get

$$c = x/t \qquad\qquad\qquad \text{and} \quad (2a)$$

$$\alpha = 1/x \ \ln P_0/P \qquad\qquad (2b)$$

The velocity measurement is reduced to a pure time measurement
if the fixed distance x of the transducers are known. Such a
time measurement was realized with high accuracy (up to 10^{-5}).
A pulse travelling technique was developed for the c measurement[1]. c can be measured with a precision of 1 part in 10^4.
The precision of α -measurements is only in the percent range.
That is the reason why the velocity measurement was used.

EXPERIMENTAL RESULTS

The prediction that the acoustic parameters in a poly-
merizing system during the reaction will be changed could be
confirmed by investigations of different systems. The magnitude
and the kind of change depend on the technology, the components
of the system, the acoustic properties of these components,
and the state of the system (temperature, pressure). In general
the magnitude of changes is a few orders of magnitude higher
then the accuracy of the measuring principle. So a precise
observation of the reaction is possible.

As an example the results of different kinds of Vinylace-
tate-polymerizations are shown in Figure 1. Every state of
the polymerization is characterized by a definite c-value.
In the case of the semicontinuous polymerization, the ultrasonic
velocity generally decreases with the polymerization time and,
in the case of discontinuous polymerization, the velocity increases. This different behaviour is determined by the influence of the components responsible for the polymerization.
Such systems are complicate, heterogeneous systems with at least
three different components. The ultrasonic velocity gives
overall information about the state of such systems. Although
the velocity reacts very sensitive to material or phase changes
this information is unspecific. If one wants to get a quantitative statement on an interesting value, for instance the
monomer conversion, the components of the system have to be
investigated in detail in order to understand their influence
or to find empirical relationships or algorithms. The experiments showed that it might be possible to make quantitative
predictions if the influence of the major components are known.
For a semicontinuous polymerization as shown in Fig. 1 the following relation between c and the components of the polymeri-
zing system is valid[2]

$$c = c_0 + \left(\frac{\Delta c}{\Delta k}\right)_{M<L} k_M^L + \left(k_M^{all} - k_M^L\right)\left(\frac{\Delta c}{\Delta k}\right)_{M>L} + k_P\left(\frac{\Delta c}{\Delta k}\right)_P +$$
$$+ k_I\left(\frac{\Delta c}{\Delta k}\right)_I + k_S\left(\frac{\Delta c}{\Delta k}\right)_S + k_E\left(\frac{\Delta c}{\Delta k}\right)_E \qquad (3)$$

Where c_0 is the velocity of the disperison medium, k_M^L is the
limit concentration of the solvated monomer, k_M^{all} is the total
used monomer concentration, k_P is the concentration of polymer,

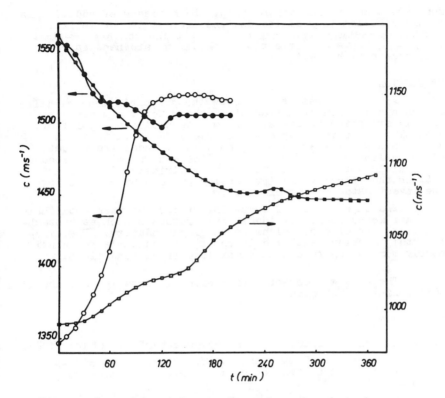

Fig. 1   Dependence of c on the polymerization time
         (● ,■ semicontinuous ; o discontinuous and □ solution
         polymerization)

k_I is the initiator concentration, k_S is the stabilizer concentration and k_E is the concentration of added electrolytes.
($\triangle c/\triangle k$) are the corresponding concentration coefficients.
When the concentration coefficients, especially the temperature dependences, of the different components are known, k_M
or k_P can be determined immediately in the case of computer
application. For other kinds of VAC-polymerizations or other
systems, such as VC, SB modified algorithms or empirical
relationships were got [3].

        For the experiments the 8bit computer MPT 179 (ZfK
Rossendorf) was applied. In the reactor the transmitter sensor
releases the time measurements which are stopped with the
receiving signal. The time τ , proportional to c, is given
to the input of the computer. Simultaneous the computer
receives the temperature T, which is very important for the
correction of the velocity values and/or their concentration
gradients and the temperature control of the process. So the
temperature inside the reactor is controlled by the computer
on the base of a given programme. On the other hand the computer produces determined programmable time differences and

monitors a controlled volume pump. So a dosage of additionals
to the reaction is realized and the process is controlled.
Different software was developed to fit the conversion-time
behaviour. Several correction programmes were necessary to
guarantee a high precision.

CONCLUSIONS

An attempt was made to show the principle possibilities
of the application of ultrasonic methods for the monitoring
of polymerization processes. From a technological point of
view ultrasonic velocity methods have a few advantages in
comparison with other methods, especially under the aspect
of an on-line measurement. An on-line control of the process
is possible when an algorithm or an empirical relationship
between the velocity and interesting process parameters for
the investigated system exists.

The experiences of these investigations were applicable
to other measurements of industrial processes such as biotech-
nological, other chemical or in the food industry. The experimental technique employed is robust and suitable for continously monitoring reactions under industrial conditions.

The work was done together with S. Wartewig (Merseburg)
and F. Dinger (Schkopau).

REFERENCES

1. R. Säuberlich, documentation "velocimeter" Merseburg (1985)
2. P. Hauptmann, F. Dinger and R. Säuberlich, A sensitive
   method for polymerization control based on ultrasound,
   polymer 26: 1541 (1985)
3. P. Hauptmann, F. Dinger and S. Wartewig, polymer synthesis
   as studied by ultrasonic methods, in "polymer yearbook",
   harwood publishers, in press

SENSORY EVALUATION OF FOOD INTELLIGENT MEASUREMENT

Anita Kochan

TU Dresden

Mommsenstr.13
8027 Dresen, GDR

INTRODUCTION

Sensory evaluation is the decisive basis for quality assurance in food industry of the GDR /Fig.1./ an quality assurance requires both the equipment and means to measure such things as colour, flavour, odour and texture.

But too small part of the equipment and simple procedures for these measurements has reached a point where routine objective tests can be made in production situation by computer aided quality control. Therefore it will be necessary to improve sensory evaluation to a level of intelligent measurement. A team of exports and scientists of the collective combines of NAGEMA and SWEETS as well as of the Dresden University of Technology worked on the following way in a research cooperation /Fig.2./. Suitable sensory methods are selected for reproducible measurement and evaluation of the quality expected by the consumer. According to Szczesniak[1] user profile analysis is used for this purpose.

After the selection of essential properties determining quality of food by consumers and expoerts the next step will follow. Special attention is paid to the problems of qualification test persons. The goal is the determination of mathematical functions between concentration of responses. The technicall usable graduation necessery for this purpose requires interval level. This means the intervals on the scale are equal in terms of the characteristic to be scaled.

This won't cause any problems if the characteristics are directly measurable.

The sensorial paramters, however, are subjective values detectable and estimable using indirect graduation methods according to Fechner[2] - theoretically founded by Thurstone[3].

Figure 1   Proportion - Qualitycoefficient KQ_i

		sharing [%]
KQ_1	Detectability value	60
KQ_2	Instrumental analytical measurements (chemical, physical)	40
KQ_3	State of packing	10
KQ_4	Microbiological evaluation	20
KQ_5	Durability	10
KQ_6	Nutritional value	10

Figure 2   Algorithm "Sensory Evaluation"

Carrying out a consumer analysis for chocolate

↓

User profile analysis and experts inquiries

↓

Product-specific taster qualification

↓

Exact term determination for all quality-relevant characteristics and partial characteristics

↓

Modification of the intensity scale according to MOHR for all characteristics and partial characteristics

↓

Detection and determination of optimum intensities and weighing factors

↓

Selection and test of suitable mathematical-statistical methods and formation of result evaluation and representation

Figure 3   Modification of the Intensity Scale for the Parameter "Textur"

Intensity	Description of the sensation			
	biting resistivity	melting	smooth-fine	sticky
5	very solid	very fast	very fine polish	very keen sticky
4	solid	fast	fine polish	keen sticky
3	middle solidity	medium	middle-fine polish	medium
2	small solidity	slow	sandy	faint
1	soft (crumbled)	very slow	keen sandy	barely perceptible
0	untypical	melts not	very keen sandy	not sticky

Figure 4   Discriminatory Scheme   Milk Chocolate, sensorical Characteristics

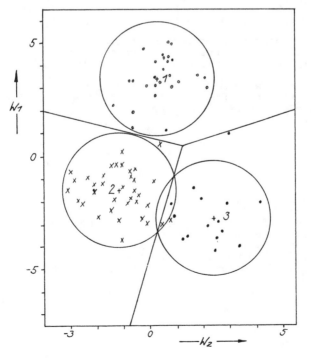

509

On this basis Hoppe presented a differentiated procedure taking into account very slight sensitive differences. The tasters are given two stimuli they will compare with the stronger sample in terms of the test characteristic being marked. According to this procedure desoribed in /4/ a sweet intensity scale with differential threshold values as a unit of measurements can be established and according to

$$R = R_m \quad 1 - \exp. \; (\; - B.S.) \quad , \text{ where}$$

R   is the sensitive intensity in differential threshold
     units
R_m the maximum sensitive intensity
B an exponential coefficient in a reciprocal concentra-
     tion unit   and
S   the concentration

concentration and sensitive intensity are determined.

Practical application of this measuring theory is the sensory profile method for the evaluation of process and material parameter variants as well as quality evaluation of foods.

Figure 3 shows the required modification of the interval scale as a further step towards the solution. The application of this profile method of quality-securing process design of chocolate refined by KONTICONCHE 420 for the first time confirms the fact, that this sensory testing is an equal method for analyzing food-stuff with the aduantage of integrating a variety of food ingredients as intervalscaled sensitive values possible only for human sensors.

Process control principally possible concerning measuring theory requires the correlation between sensory and measuring-technically definable parameters to be modelled.

This is done by modelling through classification by means of cluster analysis (Fig. 4), thus answering the question of the title of our paper with "yes".

REFERENCES

1. Szczesniak, A.S.: Food Texture and Rheology. Edited by
     Shermann, P. Academic Press, London 1979, S. 1-20
2. Fechner, G.T.: Elemente der Psychologie. Breitkopf und
     Härtel, Leipzig, 1960
3. Thurstone, L.L.: Psychol. Rev. 34 (1927), S. 273
4. Hoppe, K.: Lebensmittelindustrie 32 (1985), S. 227-231
5. Böhlmann, S.: Lebensmittelindustrie 32 (1985), S. 62-65

PHYSICAL PROPERTIES AND THEIR RELEVANCE TO NONDESTRUCTIVE

TESTING

Winfried Morgner, Gerhard Mook
and Helmut Stöckigt

Otto von Guericke Technical University
DDR-3010 Magdeburg, German Democratic Republic

Nondestructive testing of materials (NDT) can seldom resort to sensors common in technology for temperature, pressure or length gauging. Instead, direct use is made of the physical properties of materials. In the present paper, only the classical macroscopic physical properties are considered.

For technical purposes of NDT, resorting to the previously common classification of physical properties into structure-sensitive and structure-insensitive properties has appeared to be a valuable aid (Morgner, 1975). For the accuracy of measurement required, all of electrical conductivity, absolute differential thermoelectromotive force, modulus of elasticity, and saturation magnetization are structure-insensitive since they hardly vary under the influence of elastic or plastic deformation and impurities. Still, they are not at all unattractive. In fact, they are excellently suited to detect variations in chemical and phase composition without being influenced by effects such as stresses, plastic deformation and impurities.

In contrast, it is internal friction, permeability, coercive field strength, hysteresis losses and remanence which are structure-sensitive in this light. Based on this philosophy, Table 1 reflects some of the fundamental principles of NDT.

Once in a while, the reverse is desirable, viz. to suppress the detrimental influence of physical properties on the measuring result. For example, this is true in crack testing by means of an eddy current coil where marked variations in permeability, electrical conductivity or surface geometry are liable to occur from specimen to specimen and within a component alike.

These factors as well as unintentional probe lift-off have a bearing on the measuring result in surface crack inspection by means of eddy current and must, therefore, be allowed for by an intelligent crack tester.

TABLE 1   Physical properties as a basis for nondestructive
testing

PROPERTY	SS	PRINCIPLE OF NDT	APPLICATION
Mechanical			
– elastic	NO	US velocity, US goniometer	D,T,R,P
– inelastic	YES	US attenuation, impedance	D,M,S
Electrical	NO	Eddy current, four-point pole method	D,T,M,S,R,P
Thermal	NO	Infrared thermography, photoacoustics	D,P
Thermoelectric	NO	Thermoelectric con-tact voltage (Morgner, 1973)	D,T,M,S
Magnetic			
– Permeability	YES	Adhesion, field distortion, screening power (Kral and Morgner, 1980); impedance, electronic fluxmeter (Morgner and Michel, 1976)	D,T,M,S,R,P
– Coercive field strength	YES	Residual induction (Morgner and Michel, 1976); residual field, residual point pole (Morgner et al., 1982); yoke	M,S,R,P
– Remanence	YES	Residual field, re-sidual point pole (Morgner et al., 1982); yoke	M,S
– Saturation magnetization	NO	Yoke	M
– Hysteresis loss	YES	Electronic integra-tion of hysteresis loop	M,S,R,P

NOTE: D – Defectoscopy; T – Thickness; M – Microstructure;
S – Sorting; R – Residual stress measurement;
P – Property measurement; SS – Structure-sensitive

Part solutions already exist in frequency and coil
optimization, as well as multi-frequency (Mook and Morgner,
1983) and multi-sensor (Morgner, 1982) methods.

An alternative approach adopts the behaviour of a coil
as an integral part of a resonant circuit. When the coil is
lifted off, brought close to a crack, inner or outer edge,
or to a spot having a different permeability, then both amplitude and frequency undergo characteristic changes. A
tester the block diagram of which is shown in Fig. 1 detects

512

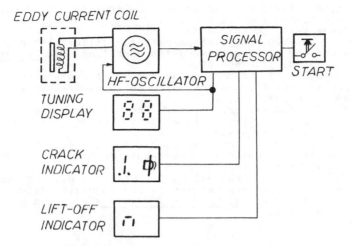

Fig. 1 Block diagram of intelligent crack
        tester

cracks only. Lift-off effect, influence by permeability and
conductivity are eliminated. The detector automatically tunes
to conductivity and permeability of the specimen tested which
may consist of any conductive metal (Pokrovskij et al., (1984).

When the signal processor is started, the tune-in voltage
is changing as long as the most convenient operating point of
the HF-oscillator for the test procedure is reached. Thus,
the measurement range is automatically tuned to the electromagnetic properties of the test material. The corresponding
tune-in voltage can additionally be applied for testing and
evaluating test specimen properties (crack depth, surface
layer thickness, material identity).

Using the test procedure mentioned, the special kind of
signal processing permits cracks to be distinguished from
changes in permeability.

Adopting this circuit, a crack detector was designed to
be housed in the enclosure of a pocket calculator. Thus, the
unit is supposed to be the smallest crack tester in the world.
Further miniaturization by adopting the single-chip microcomputer engineering approach can turn it even into an intelligent sensor.

REFERENCES

Kral, B., and Morgner, W., 1980, Gefügeprüfung von Wälzlager-
     ringen unter Ausnutzung ihrer magnetischen Abschirmwir-
     kung, Wiss. Z. TH Otto von Guericke Magdeburg, 24, No. 1.
Mook, G., and Morgner, W., 1983, Einsatz eines Mikroprozessor-
     systems für Steuerungs- und Verarbeitungsaufgaben bei der
     Mehrfrequenzwirbelstromprüfung, Wiss. Z. TH Otto von
     Guericke Magdeburg, 27, No. 1/2, p. 161.

Morgner, W., 1973, Principles and Applications of Thermoelec-
    tric Nondestructive Testing, <u>Proc. 7th WCNDT</u>, Warsaw,
    Paper E-15, pp. 249-253.
Morgner, W., 1975, Qualitätsparameter metallurgischer Erzeug-
    nisse und ihre zerstörungsfreie Prüfung, <u>Die Technik</u>,
    30, No. 4, pp. 255-259.
Morgner, W., 1982, Prüfanordnung zur Unterdrückung des Perme-
    abilitätseinflusses beim Wirbelstromeinfrequenzverfahren,
    WPG 01 N/243 2836, 19 Nov.
Morgner, W., and Michel, F., 1976, Zerstörungsfreie Gefüge-
    prüfung durch Gleichfeldverfahren, <u>Feingerätetechnik</u>,
    25, No. 1, pp. 41-42.
Morgner, W., Rez, J., and Weiss, J., 1982, Strength and Hard-
    ness Testing Using the Point-pole Method, <u>Proc. Xth
    WCNDT</u>, Moscow, Vol. 5, p. 5.
Pokrovskij, A. D., Chrostov, A. I., Morgner, W., and Mook, G.,
    Entwicklung neuartiger Geräte zur magnetinduktiven Ober-
    flächenrißprüfung, <u>Wiss. Z. TH Otto von Guericke</u>, 28,
    No. 4, p. 5.

SIZE CONTROL OF MACHINE TOOL

Vladimir Chudov

Institute of Machinery Research
Academy of Sciences USSR
Moscow

AFTER CYCLE CORRECTIONS

The machine-tool size set is offsetting from many causes (heating of machine, wearing of instrument and so on). This set must be regulated for keeping the size scattering of machined parts within the bounds of tolerance.

The after-cycle correction is the uneven altering of the machine-tool size set according to the measurement, perfomed after current cycle termination. This type of correction is used in the case of impossibility of continuous measuring of smoothly altering size while issuing the stop command upon reaching the specified value. Last technic is useful only when the direction of instrument's movement coincides with the direction of size measurement, but not any case. Usualy after-cycle correction is performed in centerless grinding, however it may be also used for setting the signal level of size guided devices. What is it's essense?

The executive limit is fixed below the tolerance one and used to init the correction in one of the following cases: when any workpiece violates this limit, or the specified number of consecutive ones do that or the mean value (or median) of the set exceeds it and so on. The second executive limit is fixed when the both-side trend is expected (fig. 1).

Much more simple zero-indifference corrections (ZIC) with both ehecutive limits coinciding (fig. 2) in the middle of tolerance zone proved to be more effective[1]. One of the kinds of such correction algorithms is the sign-sensitive one, where only the sign of offset from the executive level is fed back. It is possible to start the correction procedure after each, each two, three and more consecutive workpiece processings.

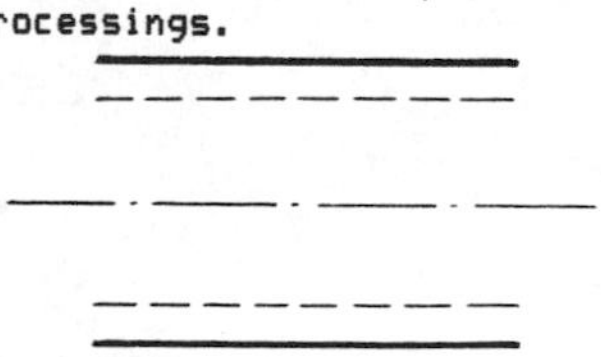

Fig.1 Both-side correction

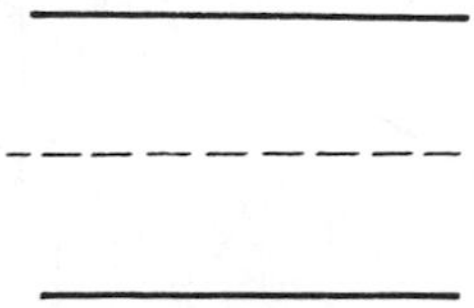

Fig.2 Zero-indifference correction

The following are two another examples of the sign-sensitive ZIC.In the "sizematic" cycle for internal-grinding machine the instrument support feed is carried out up to the hard rest, and then the machining is continued at the definite time. The constant step correction (positive or negative) can be performed after measuring the finished workpiece outside the machine.

We used the modified cycle[2] at the BDA-40 machine (Czechoslovakia) when the revolving abrasive stone received longitudinal oscillations. The across feed was performed up to the rest, and then the machining was continued up to the specified size under the control of the measuring device. The rest position was updated after each cycle (by constant value) to maintain the desired accuracy to productivity relation. The correction sign depended upon was the cycle finished till the deadline of the timer (started by support touching the rest) or not.

The proportional ZIC[3] is even more effective although more complex. The required correction is the offset measured by the factor of -0.3 to -0.7 this case. It puts out not only the systematical trend, but alsow the correlating part of a random error.

COMPUTER-AIDED IMITATION

Optimal ZIC algorithm parameters depend on the current characteristics of the process, on the equipment, instrument, material and so on. The sign-sensitive ZIC algorithm parameters include the offset step value and number of consecutive workpieces with the same offset from the executive limit sign. The only proportional ZIC algorithm parameter is the factor of proportionality. Moreover, it is important to determine the number of radius (diameter) measurements required and the function (minimum, maximum, mean) to be used for correction value obtaining, expecially in common now case of large (with respect to the tolerance zone width) workpiece shape deviations.

The computer aided imitation seems to be the only method of such multy-factor optimization, because it scarcely could be done analitically, while experiments require too much time for every new shape. It was suggested[4] to choose the suitable algorithms using the representative set of measurements performed on the workpieces processed

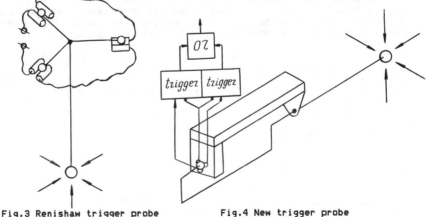

Fig.3 Renishaw trigger probe   Fig.4 New trigger probe
                                                "COROMYSLO"

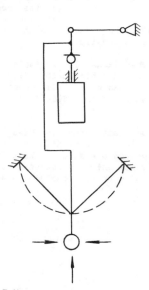

Fig.5 Measuring head "ELKA"

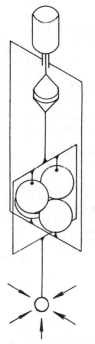

Fig.6 Measuring head "SEDLO"

without any correction. The necessity of only the conformity of all the
workpieces with the tolerance zone, not the best algorithm, simplifies
the task considerably. The only reason for searching for the best
algorithm, ensuring the largest output of fitting workpieces, is the lack
of fully satisfactory variant. In any case the total number of variants
to be tested is't very high due to relatively smooth parameter
dependencies permitting rough discretisation.

As it was shown by imitation, the simpliest (in connection to
mechanical means) sign-sensitive ZIC gives the best result for keeping
the process within the tolerance limits although the proportional ZIC is
more effective for the dispersion minimization. Obviously the
proportional ZIC is more sensitive to occasional throws of set size,
responding with a series of reducing oscillations. If the maximal
correction is restricted to, for instance, half of the tolerance zone
size[5], the process staggering by random factors is eliminated.

APPARATURE

The current growth of interest to the after-cycle correction is due
to measuring probe usage in the digitally controled cutting tools and to
considerable progress in FMS field where it is necesarry to maintain the
specified level of the size set without any manual intervention.

It is common today to supply tool machines with measuring probes
such as Renishaw touch-trigger probes, fig. 3 for example. The following
disadvantages of this probe: measuring force dependency on the measuring

direction (twice) and the existence of the reductional transmission, both leading to accuracy decrease, are eliminated in the probe[6] represented on fig. 4. The possibility of obtaining only single-bit information (touch – no touch) which force some of the tool machines to make two approaches to the object to be measured (at high and then at low speed) is the common disadvantage of the both probes considered. Now we use the measuring probes based on conventional electro-contact or inductive or moire sensors with specially designed transmissions to the tip.

The three-sided band-string measuring probe[7] 'ELKA' is suitable for semi-plane operation (used in lathes) – fig. 5. Another probe (fig. 6) permits the semi-space operation and is more suitable for mill machines. It is based[8] on our original high-order kinematic pair 'SEDLO' (saddle) consisting of two pairs of spheres – fig. 6.

The first output of the two-limit electrocontact sensor is used for speed reduction, the second – for coordinate measurement. The moire sensors supply sufficient information for improved operations. Last case the actual coordinates could be obtained as a sum of the sensor output and the machine coordinates itself when it is stopped completely after moving at a high speed.

INDEX

520